ATOMIC ABSORPTION
SPECTROCHEMICAL ANALYSIS

B.V. L'VOV

Soviet Academy of Sciences, Leningrad

Atomic Absorption Spectrochemical Analysis

With a Preface by
A. WALSH, F.R.S.

*Commonwealth Scientific and Industrial Research Organization,
Melbourne, Australia*

Authorized translation by
J. H. DIXON, F.I.L. (Russ)

AMERICAN ELSEVIER PUBLISHING COMPANY INC.
NEW YORK

© Adam Hilger Ltd, 1970

Translation authorized by the 'Nauka' Press, Moscow

Published in the United States by
AMERICAN ELSEVIER PUBLISHING COMPANY INC.
52 Vanderbilt Avenue, New York, New York 10017

ISBN 0 444 19618 8

Library of Congress Catalog Number 73-141641

Printed in Great Britain

PREFACE

A. WALSH
CSIRO, Clayton, Australia

It is a privilege indeed to be invited to write a preface to the official English translation of Professor L'vov's book. The past fifteen years have witnessed a remarkable renaissance of atomic absorption spectroscopy, a branch of optical spectroscopy which had been largely neglected since the classic work of Bunsen and Kirchhoff one hundred years ago. In large part the renaissance has been due to a surprisingly belated realization of the potentialities of atomic absorption methods of chemical analysis, now extensively used throughout the world. Probably more than 99 per cent of all analyses now carried out by such methods are still based on the use of a flame for the production of an atomic vapour, although the limitations and difficulties associated with flame methods are now well recognized. There have been many attempts to develop alternative methods of atomization but progress has been disappointing. The most important advance has undoubtedly been that made by Professor L'vov in his development of the graphite cuvette, usually referred to as the L'vov furnace. His technique has now been developed to the stage where it constitutes a powerful weapon in the armoury of the analytical chemist. If this book contained nothing more than a detailed account in English of the theory and practice of this important method it would be assured of a warm welcome, since spectroscopists outside Russia have been slow to appreciate the full potentialities of the furnace technique.

But this book is far more than an account of Professor L'vov's own contributions, and is indeed the first detailed presentation of the physical basis of modern methods of atomic absorption spectroscopy. Furthermore, it indicates an enlightened viewpoint in its obvious appreciation of the fact that the subject is not to be regarded, even from a purely analytical point of view, merely as a branch of flame spectroscopy. Professor L'vov also convincingly demonstrates the power of modern methods of atomic absorption measurement in various physical investigations and many of his own contributions, particularly the determination of oscillator strengths, are discussed in detail.

Workers in atomic absorption spectroscopy will appreciate Professor L'vov's views on various aspects of the subject, and will find in this book many references to important Russian publications of which they have not previously been aware. The subject will benefit greatly from the English version of a book which will undoubtedly be a standard work of reference for many years to come.

PREFACE TO THE RUSSIAN EDITION

In recent years atomic absorption spectrochemical analysis has become widely used in spectrographic analysis practice. Several hundred publications have dealt with this subject. Interest in it is no accidental matter. In many respects the atomic absorption method is preferable to the emission method, for it affords definite prospects of solving the principal problems of modern analysis. These problems include the development of methods of determination which do not involve the use of standards, and the analysis of ultra-pure materials.

Until now, however, the results of experimental research have nowhere been thoroughly and completely set forth, except for a few reviews of a bibliographic nature and the very short book *Atomic Absorption Spectroscopy* by Edwell and Gidley, which related to work carried out before 1960.* The situation is even worse as regards the theory behind the method. Most research workers have based their investigations on fragmentary and extremely approximate information given in the earliest published works on the subject. This state of affairs is not only hindering the wider use of this method of analysis and impeding its further development, but also sometimes results in wrong interpretation of results and incorrect attempts at improving the method.

The purpose of my work has been not only to systematize the results obtained, but also, as far as possible, to interpret and generalize them theoretically. Such a method of approach seems, to me, better than baldly describing the results given in individual published works, since spectrographic analysis practice has proved that accurate reproduction (copying) of work done by other research experts is, in many cases, a difficult and quite unjustifiable task.

The book contains a brief exposition of certain matters related to physical optics and physical chemistry; these sections, which make no claim to originality, will make it easier to understand the processes on which the method of analysis is based, or alternatively will serve as a guide. At the same time the book does not deal with certain matters relating to methods, such as techniques for handling flames, since these have been dealt with in detail in books on flame photometry.

This text will be of interest to engineers and scientists engaged on spectrographic analysis, to physical chemists in view of the prospects of using the method for measuring atomic constants (oscillator strengths and resonance line widths), diffusion factors, elasticities of metal vapours, etc. The book will be useful for students working on these special subjects.

The details in the book are largely based on theoretical and experimental research carried out by the author, starting in 1956, in the spectrographic laboratory at the State Institute for Applied Chemistry.

A first attempt at systematizing results is bound to contain errors, and this applies

*Here I am referring to the first edition of the book published in 1961.

even more to the theoretical interpretation of results; I shall be extremely grateful to be informed of such mistakes.

The author wishes to take this opportunity of expressing his deep gratitude to Professor L. V. Lipis, Dr V. I. Mosichev and Dr E. N. Vitol' for the helpful comments they made when reading the manuscript of this book.

LENINGRAD B. V. L'VOV.
December 1965

PREFACE TO ENGLISH EDITION

Although little more than four years has elapsed since this book was first published in the Russian language, and only five years since work ended on it, a considerable proportion of the information in the book has become so out-of-date that I have been obliged to rewrite many of the sections in their entirety. This applies to the whole of Chapter 5, and to parts of Chapters 3 and 6. Many alterations have been made to the remaining sections of the text. For example, it was necessary, in Chapter 1, to clarify my remarks on the various methods of measuring absorption, and additional information on lamps to which pulsed current is supplied has been added to Chapter 2, as well as a few additions regarding light sources with boosted outputs; the short and obviously outdated section on commercial instruments has been omitted. Alterations have been made to Chapter 4: in the section on sensitivity of measurements terminology has been clarified, a section on the integration method of measuring absorption in a flame has been added, and the deduction of an equation which portrays the effects of excess of a substance on the results of measuring in a flame (p. 181) has been made more accurate.

Though additions and corrections to Chapters 2 and 4 are fewer than for the other chapters, this does not mean that other important, or merely useful, results have not been obtained in the fields concerned. A vast amount of factual information has been published since the Russian edition of my book was published, concerning the use of atomic absorption for analysing different substances. This has given us a more accurate and extensive picture of the possibilities of the method, and there is no doubt that it is of great practical value to analysts.

The systematic analysis and generalization of this information would, however, be attended by a certain amount of risk that, whilst work was being done on it, fresh ideas and new and even more interesting results might appear; this might have meant that the work of processing all the information would have to begin again. The author has also taken into consideration the fact that English readers have at their disposal three comparatively recently published monographs written by Robinson (1966), by Elwell and Gidley (1966), and by Ramírez-Muñoz (1968), which to a large extent fill this gap.

Thanks are due to Dr A. C. Menzies for his courteous proposal that my book should be translated into English, to Mr N. Goodman and Mr D. Tomlinson for the trouble they took in having the book translated and printed, and to Dr A. Walsh for kindly writing a preface.

The remarks kindly made by Professor C. T. J. Alkemade (State University of Utrecht) concerning interpreting the effects of foreign substances on the results of measurements made in a flame, also the remarks made by Dr L. M. Ivantsov (USSR Academy of Sciences Geokh I), have greatly assisted me in rewriting this book. I am very grateful to Dr W. Slavin (Perkin–Elmer) for sending me copies of every issue of the periodical *Atomic Absorption Newsletter* (published by Perkin-

Elmer), to Dr B. F. Scribner (National Bureau of Standards) who sent the papers of the First Symposium on Analysis of Trace Elements, which was held in October 1966 at the NBS Institute for Materials Research, and to Professor V. A. Fassel (State University of Iowa) for supplying details of research into atomic absorption spectroscopy carried out in his laboratory.

I am deeply grateful to my colleagues, especially to Dr V. I. Mosichev for the useful remarks he made after reading the amended sections of the book, and to D. A. Katskov, A. D. Khartsyzov, G. G. Lebeder, and Y. V. Plyushch, who have, during recent years, taken part in the different stages of our joint research.

LENINGRAD
June, 1970

B. V. L'vov.

CONTENTS

	Introduction	1
1	Theory of Atomic Absorption Measurement	5
2	Atomic Absorption Spectrophotometers	36
3	Atomization of Samples—General Principles	115
4	The Flame	135
5	The Graphite Cuvette	193
6	Special Fields of Application for the Atomic Absorption Method of Measurement	253
	Conclusion	289
	Appendices	295

    1 Mathematical and Physical Constants
    2 Correspondence between Spectral and Energy Units and Quantities
    3 Physical and Chemical Constants of Elements
    4 The Most Sensitive Resonance Lines of Elements

References	307
Index	315

INTRODUCTION

The discovery of atomic absorption and the history of research into it are integral parts of the entire history of spectroscopy and spectrochemical analysis. In 1802 Wollaston repeated Newton's experiment of analysing the continuous solar spectrum, and revealed that, if a beam of light from the sun passes not through a circular aperture in a shutter but through a slit, the solar spectrum is intersected by several dark lines. This discovery was not, however, considered important. Fifteen years later, working independently from Wollaston, Fraunhofer again found the dark lines in the sun's spectrum; in his honour they were called Fraunhofer lines.

Not till 1859 did Kirchhoff establish the origin of these dark lines. Kirchhoff was the first person to draw the clear conclusion that the chemical composition of a substance could be determined from its spectrum. Working with Bunsen, Kirchhoff gave numerous examples of the use of spectra for determining alkaline metals in a flame. Bunsen and Kirchhoff are thus rightly considered to be the founders of spectrochemical analysis.

In 1861, Kirchhoff published a paper on the spectral analysis of the chemical composition of the sun's atmosphere; in this paper he established the presence of certain elements in the sun from the fact that the emission lines for these elements coincided with the Fraunhofer lines in the solar spectrum. As a result, astrophysics and astrochemistry, which determine the chemical compositions and physical states of heavenly bodies, and the manner in which they move, are the most important field of application for atomic absorption spectroscopy.

The first twenty years of the twentieth century were marked by considerable achievements in the field of atomic absorption theory. During this period the fundamental relationships between absorption and the atomic constants were established (Kravets,[1] Füchtbauer,[2] and Ladenburg[3]), and the theory for the pressure broadening of lines was formulated (Lorentz[4]); a relationship was obtained for the shape of the absorption line under the combined influence of several broadening effects (Voigt,[5] Reiche[6]), and methods of measuring atomic absorption were developed (Ladenburg and Reiche,[7] and Füchtbauer and his colleagues[2,8]).

A great proportion of the theoretical results were obtained on the basis of Lorentz's classic electronic theory. Developments in quantum theory, in particular Einstein's quantum radiation theory, made it possible to establish the physical significance and connection between the different radiation and absorption processes taking place in atomic systems; as a result, certain concepts introduced in the classic theory in an extremely arbitrary manner (for instance, the oscillator strength) could be interpreted.

The theoretical basis found for the processes of absorption made it possible to obtain quantitative data regarding solar and stellar atmospheres: their chemical compositions, temperatures, electron concentrations, etc.[9]

The absorption method was used in interpreting complex spectra, since only lines which begin at low energy levels are observed in absorption. To carry out this research, King designed a special graphite furnace.[10] The spectra of many elements were identified using King's furnace, in particular certain lines in the complex spectra of uranium and plutonium.[48,11]

It is important for the purposes of astrophysics, research into plasma, and determining special features of atomic structure, to know the lifetimes of atomic excited states and the effective cross-sections of atoms when they collide with molecules of foreign gases. These constants are determined by different methods, including absorption spectroscopy methods.

Absorption measurements are also used for investigating the hyperfine structure of atomic lines and the Zeeman effect, i.e. in cases in which very narrow spectral lines are necessary for recording a phenomenon.[12]

The use of atomic absorption in analytical chemistry began in the forties, when it was applied to the determination of mercury vapour in air.[13] In 1954, Bochkova published a paper[50] concerned with the use of atomic absorption for analysing gases. Up to 1955, the use of atomic absorption spectroscopy for analysis purposes was confined to the work described in these references.

The reasons for this backwardness, which was particularly noticeable in comparison with the rapid development of absorption methods for determining molecular structures, were firstly the very rapid development of emission methods of analysis from atomic spectra, which satisfied analysis requirements sufficiently well, and secondly the absence of a method sufficiently simple for use in determining the large number of elements in samples with different compositions. Actually, in spite of the great number of methods by which atomic absorption was measured in experimental physics, none of these methods, as we shall see later, satisfied the practical requirements for analysis from the point of view of the range of elements embraced, sensitivity and simplicity. It is even more important that the advantages of absorption methods of analysis by comparison with emission methods were not clarified to such an extent as to stimulate research in the field of atomic absorption spectrochemical analysis.

In 1955, Walsh (in Australia) and Alkemade and Milatz (in Holland) independently published papers indicating the substantial advantages of absorption methods over emission methods for quantitative spectrochemical analysis. Alkemade and Milatz[14] considered only the question of selectivity in atomic absorption methods, but Walsh[15] discussed general problems associated with the development of absolute methods of spectrochemical analysis. Walsh showed that absorption methods have very great advantages over emission methods and proposed a rational method of recording atomic absorption; he advanced the design of an apparatus for conducting analyses. Somewhat later (in 1957), Walsh and his colleagues published the results of the experimental development of his method.[16]

During the following fifteen years, more than two thousand works were published on atomic absorption spectrochemical analysis. These were concerned both with research (studying the possibilities of the method, improving the technique of the analysis, finding new methods of producing atomic vapour, etc.) and the extension of the field for using procedures which had already been perfected for analysing new substances. The method became widely used in many countries. The largest optical equipment firms started mass-producing apparatus. The atomic absorption method is now included as a separate section of spectral analysis in conference programmes and symposia, also in reviews of methods of analysis and in bibliographic manuals.

Work on atomic absorption in the U.S.S.R. covers not only the use of the method for analysing the elementary compositions of substances, but also other lines. Absorption methods of analysing gases are being developed as well as simplified methods of determining the isotopic compositions of elements. The relative and absolute values of oscillator strengths and resonance line widths, and the diffusion coefficient for elements present in free atomic form in inert gases, are being measured.

Chapter 1

THEORY OF ATOMIC ABSORPTION MEASUREMENTS

EMISSION AND ABSORPTION OF LIGHT

The emission and absorption of light are associated with the processes of transition of atoms from one steady state to another. For the case of steady states i and k, with energies of E_i and E_k, when $E_k > E_i$, the $i \rightarrow k$ transition results in the absorption of light, and the $k \rightarrow i$ transition results in the emission of light with a frequency

$$\nu_{ki} = \frac{E_k - E_i}{\mathbf{h}} \qquad (1.1)$$

where $\mathbf{h}$ is Planck's constant.

According to Einstein's quantum theory of radiation, there may be three types of transition between levels i and k:

1. Emission $(k \rightarrow i)$ transitions from the excited state into a lower energy state, taking place spontaneously.

2. Absorption transitions $(i \rightarrow k)$ from a lower into a higher energy state, taking place in response to the action of external radiation with a frequency ν_{ki}.

3. Emission $(k \rightarrow i)$ transitions from an excited state into a lower energy state, stimulated by external radiation of the same frequency ν_{ki}.

The $k \rightarrow i$ emission transitions thus include two types of transition: spontaneous transitions taking place without any external cause and transitions stimulated by external radiation. The $i \rightarrow k$ absorption transitions are always stimulated by external radiation.

The inverse of absorption is not spontaneous emission but stimulated emission, showing as amplification of the beam of light passing through the medium concerned. In early years, this phenomenon was regarded as a theoretical curiosity, but it is now utilized in lasers.

For many years, only one of the three possible processes which may take place in atomic systems, namely spontaneous emission, was used in atomic spectrochemical analysis. Atomic absorption spectrochemical analysis is based on a phenomenon of an entirely different nature, the stimulated process. Atomic emission spectrochemical analysis and atomic absorption spectrochemical analysis are therefore associated not with opposing physical processes, as is often suggested in published works on the analytical aspects of spectroscopy,[17] but with absolutely different processes, which are however interconnected.

The phenomenon of stimulated emission has not yet been used directly for the purpose of analysis (here we have in mind the phenomenon as itself, and not the apparatus developed on this principle and used, for example, for the local atomization of samples by means of powerful light pulses[18]). In this book, we shall not discuss possible methods of using stimulated emission for analytical purposes. Methods of this type are certainly interesting, although it seems to me that they cannot provide anything newer in principle than absorption methods, since we are dealing only with opposite aspects of the same phenomenon.*

THE CONCEPTS OF THE ABSORPTION LINE AND ABSORPTION COEFFICIENT

The absorption of light by the atoms of any elements can be observed by passing a beam of light from a light source with a continuous spectrum through a medium in which there are free atoms of the element concerned. If an instrument with a high resolving power is used, it is easy to locate regions of lower intensity in the continuous spectrum corresponding to the energies of transition of the atoms from the lower to the higher energy state according to equation (1.1).

As opposed to the wide absorption bands observed in molecular absorption spectroscopy, absorption by atoms takes place within very narrow spectral regions, of the order of hundredths of an angstrom. Atomic absorption therefore shows up in the form of individual narrow lines on spectrograms of continuous sources.

Atomic absorption follows an exponential law for the intensity I of transmitted light against absorbing volume length l; this law is similar to Lambert's law in molecular spectroscopy:

$$I = I_0 e^{-k_\nu l} \qquad (1.2)$$

Here I_0 is the intensity of the incident beam of light and k_ν is the absorption coefficient at the frequency ν.

The absorption coefficient characterizes absorption lines, just as intensity characterizes emission lines. The laws for the distribution of absorption coefficients across absorption lines are similar to the laws for the distribution of intensity across emission lines. In emission spectroscopy, lines are considered strong or weak according to their integrated intensities; in absorption spectroscopy, similarly, they are classified according to their integrated absorption coefficients.

For the purpose of making measurements, a useful parameter is the absorbance A, defined as

$$A = \log (I_0/I) \qquad (1.3)$$

Using equation (1.2), we get:

$$A = k_\nu l \log e = 0.4343 \, k_\nu l \qquad (1.4)$$

It follows that the absorbance is directly proportional to the absorption coefficient.

* It is interesting to note that, although the development of atomic absorption spectrochemical analysis began, and is continuing, simultaneously with work in the laser field, to judge from published literature no importance has been ascribed to the similarity between these fields of spectroscopy.

It was pointed out earlier that atomic absorption corresponds to transitions of atoms from lower to higher energy states. Naturally, therefore, the degree of absorption depends on the degree to which the lower level of a particular transition is populated. With thermodynamic equilibrium in a system, the level population is determined by Boltzmann's law:

$$N_i = N_0 \frac{g_i}{g_0} e^{-(E_i/kT)} \qquad (1.5)$$

where N_i is the number of atoms at the level with an energy E_i, N_0 is the number of atoms at the ground (unexcited) state with an energy $E_0 = 0$, and g_i and g_0 are the statistical weights of the ith and ground states respectively.

The excited-level population is small in comparison with the ground state, as is illustrated in Table 1.1 for elements with different distances between the first excited level and the ground level.

TABLE 1.1 Values of N_i/N_0 for different resonance lines[15]

Lines (Å)	Excitation energy (eV)	N_i/N_0			
		$T=2000°K$	$T=3000°K$	$T=4000°K$	$T=5000°K$
Cs 8521	1·46	$4·44 \cdot 10^{-4}$	$7·24 \cdot 10^{-3}$	$2·98 \cdot 10^{-2}$	$6·82 \cdot 10^{-2}$
Na 5890	2·11	$9·86 \cdot 10^{-6}$	$5·88 \cdot 10^{-4}$	$4·44 \cdot 10^{-3}$	$1·51 \cdot 10^{-2}$
Ca 4227	2·93	$1·21 \cdot 10^{-7}$	$3·69 \cdot 10^{-5}$	$6·03 \cdot 10^{-4}$	$3·33 \cdot 10^{-3}$
Zn 2139	5·80	$7·29 \cdot 10^{-15}$	$5·58 \cdot 10^{-10}$	$1·48 \cdot 10^{-7}$	$4·32 \cdot 10^{-6}$

Absorption is therefore greatest in lines resulting from transitions from the ground state. These lines are called resonance lines in atomic absorption analysis. It must be noted that the term resonance lines is here used in a very loose sense in comparison with the concept introduced, when Wood first observed resonance radiation, to denote the one or more lines corresponding to emission transitions from the lowest excited levels to the ground level. From now on we shall use the term resonance lines in its wider sense—for any absorption lines with an unexcited lower level, whatever the position of the upper level.

ABSORPTION LINE PROFILE

Like emission lines, absorption lines are not monochromatic and infinitely fine lines, but have a certain finite width. By the half-width Δv of an absorption line, we mean the width of the profile at the point at which the absorption coefficient k_v is halved (Fig. 1.1).

It is convenient to express the half-width of a line in terms of frequency (cm^{-1} or sec^{-1}) rather than wavelength, since the half-width in terms of frequency expresses the properties of an absorption line, whatever its wavelength.

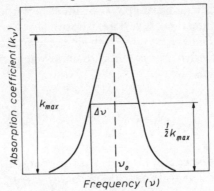

FIG. 1.1. Profile and half-width of an absorption line.

It is easy to establish a relationship between the half-width expressed in wavelength units and the half-width expressed as a frequency. From known relationships,

$$\left.\begin{array}{r}\nu(\sec^{-1}) = \dfrac{c}{\lambda} \\[6pt] \nu(\text{cm}^{-1}) = \dfrac{1}{\lambda}\end{array}\right\} \quad (1.6)$$

where **c** is the velocity of light. After differentiation, we get the following equations for finite intervals:

$$\left.\begin{array}{r}|\varDelta\nu(\sec^{-1})| = \dfrac{c}{\lambda^2}\varDelta\lambda \\[6pt] |\varDelta\nu(\text{cm}^{-1})| = \dfrac{1}{\lambda^2}\varDelta\lambda\end{array}\right\} \quad (1.7)$$

A line profile is governed almost entirely by the combined effect of the following factors:
 1. Natural broadening
 2. Doppler broadening
 3. Lorentz broadening

Other possible causes for the broadening of spectral lines are, for instance, the interactions of atoms with electrically charged particles or with one another, but these are not important in the methods most used for obtaining atomic vapour. We shall therefore confine ourselves to the three types of broadening listed above.

The natural broadening of lines

The natural broadening of lines is associated, from the point of view of quantum electrodynamics, with the degree to which energy levels are broadened. The broadening is a result of the finite lifetime (τ) of the levels between which a transition takes place. The ground level is stable ($\tau = \infty$), and the width of the upper level is therefore the only width of significance for resonance transitions. Thus, the natural width of a resonance line can be defined as

$$\Delta\nu_N = \frac{1}{2\pi\tau} \qquad (1.8)$$

A line profile governed by natural broadening is of the dispersed form described by the equation:

$$k_\nu = k_0 \frac{(\Delta\nu_N)^2}{(\Delta\nu_N)^2 + 4(\nu-\nu_0)^2} \qquad (1.9)$$

where k_0 is the absorption coefficient at the centre of the line.

Different spectral lines have different natural widths. For instance, the natural widths for the Hg 2537 Å, Na 5896 Å and Cd 2288 Å lines are, according to equation (1.8), $0.53 \cdot 10^{-4}$ cm^{-1}, $3 \cdot 10^{-4}$ cm^{-1}, and $2.7 \cdot 10^{-3}$ cm^{-1}. In most cases the natural width $\Delta\nu_N$ does not exceed 10^{-3} cm^{-1}; in comparison with the other causes of broadening, the natural width can therefore be ignored.

Doppler broadening

Doppler broadening is associated with the random thermal motion of atoms relative to the observer. As a result of the motion of an atom at a velocity in the line of sight equal to v_x, the observed frequency of absorption by the atom is displaced by

$$\Delta\nu = \frac{v_x}{c}\nu_0 \qquad (1.10)$$

If the motion of atoms in an atomic vapour is subject to the Maxwell distribution, as is always valid for systems in thermodynamic equilibrium, the distribution of the absorption coefficient k_ν is governed by the equation

$$k_\nu = k_0^{(D)} \exp\left\{-\frac{Ac^2}{2RT}\left(\frac{\nu-\nu_0}{\nu_0}\right)^2\right\} \qquad (1.11)$$

where A is the atomic weight, $\mathbf{R}$ the gas constant, T the temperature, and $k_0^{(D)}$ the absorption coefficient at the centre of the line. The magnitude of $k_0^{(D)}$ is determined by the equation

$$k_0^{(D)} = \frac{2\sqrt{(\pi \ln 2)}e^2}{\mathbf{m}c} \frac{Nf}{\Delta\nu_D} \qquad (1.12)$$

where f is the oscillator strength, $\mathbf{m}$ is the mass of an electron, and N is the population of atoms.

The Doppler half-width of the line depicted by equation (1.11) is equal to

$$\Delta\nu_D = \frac{2\nu_0}{c}\sqrt{\left(\frac{2\ln 2\, RT}{A}\right)} \qquad (1.13)$$

or, if numerical values are substituted for the constants,

$$\Delta\nu_D = 0.716 \cdot 10^{-6}\nu_0 \sqrt{\frac{T}{A}} \qquad (1.14)$$

Table 1.2 contains values of $\Delta\nu_D$ for some resonance lines at different temperatures. The figures in this table show that, even at low temperatures, the Doppler width is much greater than the natural width.

TABLE 1.2 Doppler half-widths of certain resonance lines

Resonance line (Å)	Atomic weight (A)	$T = 200°K$	$\Delta\nu_D$ (cm^{-1}) $T = 500°K$	$T = 3000°K$
Cs 8521	132·91	$1·0 \cdot 10^{-2}$	$1·6 \cdot 10^{-2}$	$4·0 \cdot 10^{-2}$
Li 6708	6·94	$5·7 \cdot 10^{-2}$	$9·1 \cdot 10^{-2}$	$2·2 \cdot 10^{-1}$
B 2498	10·82	$1·2 \cdot 10^{-1}$	$2·0 \cdot 10^{-1}$	$4·8 \cdot 10^{-1}$

Lorentz broadening

We shall now consider Lorentz broadening caused by interaction between the absorbing atoms and the molecules of a foreign gas. This, together with the Doppler effect, makes the greatest contribution to the shape, width and position of absorption lines.

The principal experimental data obtained with regard to this type of broadening amount to the following.

1. Increase in the pressure of the foreign gas causes broadening of the line ($\Delta\nu_L$), displacement of the maximum of the lines relative to its initial position ($\Delta\nu_s$), and asymmetry of the line profile. Fig. 1.2 is a diagrammatic representation of changes in the line profile as the pressure is increased. Line asymmetry can be gauged quantitatively from the ratio of widths for the red and violet halves of the line profile.

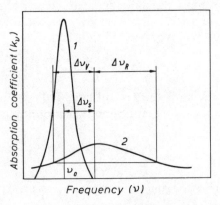

FIG. 1.2. Effects of the pressure on line profiles. 1. Unaltered profile. 2. Broadened and shifted profile.

2. The changes in the half-width and displacement of a line are proportional to the change in the pressure of the foreign gas.

3. Different gases have different effects on the broadening and shifting of lines. Table 1.3 contains the results of experimental measurements of line half-widths for the second doublet of the resonance series of rubidium with different gases.[19]

In heavy gases (argon and nitrogen are among these), the shift is towards the red end of the spectrum, while in light gases it is towards the violet end.

TABLE 1.3 Shift and half-width of the Rb 4216 Å line, calculated per unit of relative gas density (0°C, 1 atm)

Gas	Shift $\Delta\nu_s$ (cm^{-1})	Half-width $\Delta\nu_L$ (cm^{-1})	$\Delta\nu_L/\Delta\nu_s$
Ar	−1·2*	2·21	1·8
N_2	−0·52	1·51	2·9
Ne	+0·22	1·30	5·9
H_2	+0·45	1·87	4·1
He	+0·93	2·77	3·0

* The + and − signs indicate that the lines are shifted towards the violet or red ends of the spectrum respectively.

4. The ratio of the half-width to the line shift for the different lines for heavy gases is a constant between 2 and 3. This fact is confirmed by the results given in Table 1.4.

TABLE 1.4 The $\Delta\nu_L/\Delta\nu_s$ ratio, in argon and nitrogen, for the resonance lines for certain elements

Line (Å)	Gas	$\Delta\nu_L/\Delta\nu_s$	Reference
Hg 2537	N_2	2·2	20
Cs 4555	N_2	2·3	20
Na 5890–96	N_2	2·25	21
K 7665	N_2	2·7	22
Rb 4216	N_2	2·9	23
Na 5896	Ar	2·9	21
K 7699	Ar	2·4	22
Mn 4031	Ar	2·8	24
In 4102	Ar	3·0	25
Ag 3281	Ar	2·3	26
Hg 2537	Ar	2·6	8

Mean 2·6

Two approaches are at present made towards interpreting the broadening processes; one of these is based on considering instantaneous changes taking place during collisions between particles (the collision theory), and the other is based on the statistical consideration of the effect of the collection of molecules on a particular atom under quasi-steady conditions (the statistical theory). It has been proved that the collision theory is valid for the central parts of a line, the statistical theory for the wings. Since from now on we shall be considering only the central parts of lines, we shall now dwell in greater detail on certain conclusions reached on the basis of the collision theory.

The collision mechanism for broadening was first considered by Lorentz in the year 1905. According to Lorentz, atomic emission resulted from the harmonic vibration of an electron within an atom. At the moment when an atom collided with a foreign particle, the vibration was interrupted, while after the collision it was renewed at the same frequency. The vibrations were thus considered as segments of a sine-wave, with the phases distributed at random in the separate segments.

The Lorentz effect gives lines the same form of profile as natural broadening:

$$k_\nu = k_0^{(L)} \frac{\Delta \nu_L^2}{\Delta \nu_L^2 + 4(\nu - \nu_0)^2} \tag{1.15}$$

Here $k_0^{(L)}$ is the absorption coefficient at the line centre,

$$k_0^{(L)} = \frac{2e^2}{mc} \frac{Nf}{\Delta \nu_L} \tag{1.16}$$

$\Delta \nu_L$ is the Lorentz half-width, which in turn depends on the experimental conditions in accordance with the equation

$$\Delta \nu_L = 2 \cdot 6 \cdot 02 \cdot 10^{23} \sigma^2 P \sqrt{\left[\frac{2}{\pi RT}\left(\frac{1}{A} + \frac{1}{M}\right)\right]} \tag{1.17}$$

where P is the pressure of the gas, A the atomic weight of the atoms, M the molecular weight of the gas, and $\pi \sigma^2$ the effective cross-section for collision between an atom and a molecule.

The Lorentz theory explained the linear relationship, recorded in experiments, between line half-width and the pressure of the foreign gas; it did not, however, provide an explanation for the line shift which accompanied the broadening.

Later Lenz (1924),[27] Weisskopf (1933),[28] and Lindholm (1942),[29] continued the development of the collision theory. According to modern ideas, interaction between particles does not cause a break in vibrations, but merely a change in their phase. All the phase changes caused both by close and distant approaches by the interacting particles are taken into account in the calculations and explain why the lines are shifted by $\Delta \nu_s$ relative to the initial frequency ν_0.

According to Lindholm's theory, the line half-width and shift are given by the equations:

$$\Delta \nu_L = 1 \cdot 30 \, \mathbf{C}_6^{2/5} v^{3/5} N \tag{1.18}$$

$$\Delta \nu_s = 0 \cdot 47 \, \mathbf{C}_6^{2/5} v^{3/5} N \tag{1.19}$$

where $\mathbf{C}_6$ is van der Waals' constant for the interaction between particles, and v and N are the relative velocity of the particles and the number of these particles in unit volume. The Lindholm theory enables us to establish a relationship between $\Delta \nu_L$ and $\Delta \nu_s$. In fact, if we compare equations (1.18) and (1.19), we get

$$\frac{\Delta \nu_L}{\Delta \nu_s} = 2 \cdot 77 \tag{1.20}$$

This agrees well with the mean experimental value of $\Delta \nu_L / \Delta \nu_s$ for certain resonance lines in argon and nitrogen (Table 1.4).

It is interesting to note that the Lorentz and Weisskopf–Lindholm theories provide different relationships of broadening to temperature. With a constant relative gas density (0°C, 1 atm), according to Lorentz

$$\Delta \nu_L \propto T^{1/2} \tag{1.21}$$

and according to Lindholm,

$$\Delta \nu_L \propto T^{3/10} \tag{1.22}$$

Unfortunately, experimental measurements of the relationship of $\Delta \nu_L$ to temperature are too few and too inaccurate to decide whether either theory is effectively confirmed.

Research by L'vov and Plyushch[30] along these lines investigated the second members of the main series of alkali metals. We know the displacement in argon for these lines, measured from absorption spectra at 400–473°K.[31-33] The displacements were measured at high temperatures from emission spectra of d.c. carbon arcs. The spectra were excited in a pressure chamber with argon pressures of 19·2 and 1 atm, and were photographed in a spectrograph with a dispersion of 2 Å/mm (Plate 1). When the mean for 15 to 20 independent determinations was found, the error in measuring displacements did not exceed 2 to 3 per cent. Displacements measured at 400–473°K, and measured in a d.c. arc and related to 5000°K, are compared in Table 1.5.

TABLE 1.5 Line displacements at different temperatures calculated per unit argon relative density (r.d.) (0°C, 1 atm).[30]

Lines (Å)	$\Delta \nu_s$ (cm^{-1})/r.d., absorption			$\Delta \nu_s$ (cm^{-1})/r.d., arc		
	Not allowing for temperature	Allowing for temperature according to Lindholm	Reference	Not allowing for temperature	Allowing for temperature according to Lindholm	Allowing for temperature according to Lorentz
K 4047	0·92	0·78	31	1·96	0·82	0·46
K 4044	0·82	0·69	31	1·77	0·74	0·41
Rb 4216	1·18	1·05	32	2·18	0·91	0·51
Rb 4202	0·90	0·80	32	2·04	0·85	0·48
Cs 4593	2·7	2·3	33	2·88	1·21	0·68
Cs 4555	2·5	2·1	33	2·67	1·12	0·63

When these results are considered, the following conclusions can be drawn:

1. The results of high-temperature measurements for K and Rb agree well with those of low-temperature measurements when the relationship of displacement to temperature given by Lindholm is taken into account.

2. When this relationship ($T^{0.3}$) is taken into account, a considerable discrepancy is found between the results of low- and high-temperature measurements of the displacements for Cs. This discrepancy is due to errors in the low-temperature measurements. The abrupt jump in the value of $\Delta \nu_s$ for Cs in the K–Rb–Cs series confirms this.

3. When the Lorentz relationship of $\Delta \nu_s$ to temperature is considered, there is a discrepancy of 70–100 per cent between the results of the low- and high-temperature measurements for K and Rb, and a discrepancy of 3·3:1 for Cs. To compare

the high- and low-temperature measurements for K and Rb by the Lorentz relationship, it would have been necessary to assume that the temperature of the arc was 15 000°K, and this is unlikely under the conditions indicated.

The experiments which have been described thus confirm that the Lindholm relationship of Δv_s to temperature is true. Before, however, a categorical pronouncement can be made, further research must be conducted on a large number of lines.

The profiles of the absorption lines for temperatures of 1000–3000°K and foreign gas pressures of about 1 atm are thus governed by the Doppler and Lorentz effects. The Doppler and Lorentz profiles differ greatly in form (Fig. 1.3). In the case of the Doppler profile, the absorption coefficient is related exponentially to the frequency. Near the centre of the line, therefore, the absorption coefficient alters slowly, but towards the wings it alters more rapidly. In the case of a Lorentz profile, the reverse is true; the centre of the line is more acute and the wings are gently sloping and wide.

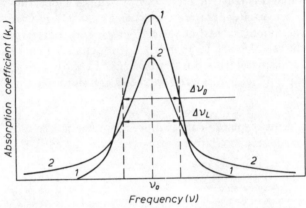

FIG. 1.3. Doppler (1) and Lorentz (2) profiles for the same half-width and integrated absorption.

If both effects operate simultaneously, the central part of the line is governed primarily by the Doppler effect, the wings by the Lorentz effect. The profile of a line, with both factors acting together (natural broadening can be ignored, since $\Delta v_N \ll \Delta v_L$), is expressed by the Voigt equation:[34]

$$k_v = k_0^{(D)} \frac{a}{\pi} \int_{-\infty}^{\infty} \frac{e^{-y^2} dy}{a^2 + (\omega - y)^2} \tag{1.23}$$

where

$$a = \frac{\Delta v_L}{\Delta v_D} \sqrt{(\ln 2)} \tag{1.24}$$

$$\omega = \frac{2(v - v_0)}{\Delta v_D} \sqrt{(\ln 2)} \tag{1.25}$$

$$y = \frac{2\delta}{\Delta v_D} \sqrt{(\ln 2)} \tag{1.26}$$

where δ is frequency displacement from $v - v_0$.

The integral in equation (1.23) is not in definite form and is therefore approximated by, for instance, Harris's method.[35]

HYPERFINE STRUCTURE OF ABSORPTION LINES

The information from the theory of line profiles given above relates to simple atomic lines. Frequently, however, lines consist of several components with different intensities which can be resolved only by using instruments of high resolving power in special experimental conditions, in which the Doppler and Lorentz broadenings do not exceed the distances between components.

For a single-isotope element, the hyperfine line structure is governed by the interaction between P_I, the spin moment of the nucleus, and P_J, the resultant moment of the electron shell. According to the orientation of the moments P_I and P_J, additional magnetic interaction energy is set up between the nucleus and the electron shell, and this causes the energy levels, and consequently also the spectral lines, to split up.

The magnitude of the nuclear moment is associated with the structure of the nucleus. A nucleus with an even number of protons and an even number of neutrons has a moment equal to zero. In the case of these elements there is no hyperfine splitting of the type indicated above. The hyperfine splitting of the resonance lines of elements with a nuclear moment other than zero is of a magnitude between a few thousandths of cm^{-1} and a few cm^{-1} (Bi 3068 Å).

In their natural state, most elements are mixtures of several isotopes with different nuclear moments. The picture of hyperfine splitting therefore represents the hyperfine structures of the individual isotopes superimposed on one another. As an instance, mercury has seven isotopes with the following atomic weights: 196, 198, 199, 200, 201, 202 and 204. The even-numbered isotopes have $I=0$, the isotope Hg^{199} has $I=1/2$, and the isotope Hg^{201} has $I=3/2$. Accordingly the even-numbered isotopes have one component each, the isotope Hg^{199} has two components, and the isotope Hg^{201} has three components.

As a result the 2537 Å line for mercury consists of ten components, some of which however coincide with one another, so that the line actually consists of only five separated components.

Additional complication of the hyperfine structure may finally be caused by isotopic shift of a line. According to modern ideas, the isotopic displacement of levels is associated with difference in the masses of isotope nuclei, or with difference in their volumes. The mass is most important with light elements (H, He, Li, B) and the volume with heavy elements (Hg, Pb, U, etc.). For elements towards the middle of the periodic table, the isotopic effect is small (excluding anomalous shifts in the spectra of rare-earth elements).

The overall profiles of lines with hyperfine structure are thus governed by the form of the structure, i.e. by the splitting and relative intensities of the components, as well as by the Doppler and Lorentz effects. Three main cases should be differentiated with relation to the ratio between the hyperfine splitting (Δv_{hfs}) and the width of the individual components due to the Doppler and Lorentz effects ($\Delta\nu_D$ and $\Delta\nu_L$ respectively):

1. $\Delta\nu_{\text{hfs}} \ll \sqrt{(\Delta\nu_L^2 + \Delta\nu_D^2)}$
2. $\Delta\nu_{\text{hfs}} \simeq \sqrt{(\Delta\nu_L^2 + \Delta\nu_D^2)}$
3. $\Delta\nu_{\text{hfs}} \gg \sqrt{(\Delta\nu_L^2 + \Delta\nu_D^2)}$

In case 1 the quantity $\Delta\nu_{\text{hfs}}$ can be ignored and the line considered to be a simple line, i.e. to consist of one component.

In case 3, all the conclusions regarding profiles which are true of simple lines can be used for each of the hyperfine components of the line; when further calculations or measurements are made, all the operations are performed in the same manner as for several simple and entirely separate lines.

The most complex case is case 2, when the profiles of the separate components are superimposed upon one another and the overall line profiles cannot be expressed analytically. In such instances, the only possible method of portraying an entire profile is to sum the individual profiles graphically.

METHODS OF MEASURING ABSORPTION

It has already been pointed out that methods of measuring absorption are now extensively used in astrophysical research and for the experimental determination of certain atomic constants, for instance transition probabilities. Methods which lend themselves to theoretical analysis, i.e. methods which enable an experimentally measured quantity to be linked theoretically with the required quantity, are used in work of this nature. In order to select a rational method of measurement during spectral analysis, these methods should be studied in greater detail. Methods of measuring absorption are based on determining one of the following quantities:

1. The integrated absorption coefficient of a resonance line.
2. The total energy absorbed from the continuous spectrum by a resonance line.
3. The relative absorption of light from a source with a line spectrum ('line' absorption).

The first method is based on measuring the absorption coefficient for each resonance line as a function of frequency. From the point of view of classical electrodynamics, the integrated absorption coefficient is

$$\int_0^\infty k_\nu \, d\nu = \frac{\pi e^2}{mc} Nf \tag{1.27}$$

Whatever, therefore, the experimental conditions and the reasons for the particular form of the profile of an absorption line, the integrated absorption coefficient is directly connected with the product Nf alone. In this case, when one of these quantities is known, the other quantity can be measured.

The method of making measurements is as follows. A lamp with a known spectral distribution of radiant energy, usually possessing a continuous spectrum, is used as the source of light. The intensity of the light passing through the absorbing medium is measured, and the absorption coefficient $k = \ln(I_0/I)$ is then calculated and represented graphically as a function of frequency. The integral $\int k_\nu d\nu$, corresponding to the area between the curve and the horizontal axis, is then found.

This method involves great technical difficulties, since, with temperatures and foreign gas pressures which are not very high, most of the resonance lines are narrow. To obtain the true distribution of the absorption coefficient by frequencies on a photographic plate, instruments with high resolving powers must be used. If, for instance, the width of a line is 0·1 cm^{-1}, its resolution requires an instrument with a resolving power of about 1 000 000. Resolving powers of this magnitude can be provided only by Fabry–Perot interferometers.

Frequently, this difficulty can be avoided by using lines which have been greatly broadened, for instance by the pressure exerted by a foreign gas. A procedure of this type was used by Füchtbauer and his colleagues[8] when they measured the oscillator strength of the Hg 2537 Å line; the pressure of the foreign gases (Ar, CO_2, H_2, O_2) was varied between 10 and 50 atm. The Hg 2537 Å line proved to be so broadened as a result of the Lorentz effect that a spectrograph could be used for resolving line profiles.

It was found[8] that there was a relationship between the measured quantity f and the pressure of the foreign gas; in order, therefore, to find the true value, the pressure of the foreign gas had to be extrapolated to zero. Following Füchtbauer's experiments, the opinion was widely held that the oscillator strength depended on the pressure of the foreign gas.

Strijland and Nanassy,[36] however, proved that the apparent relationship of f to the pressure of the foreign gas which had been found in Füchtbauer's experiments was actually due to the effects of the foreign gas on the mercury vapour pressure. The method of verification was as follows: the compressed gas was saturated with mercury vapour at constant temperature and pressure, and was then expanded isothermally to a pressure of 15 atm. The values of the integrated absorption coefficients measured for the gas in the compressed and expanded states (allowing for the coefficient for expansion) coincided within the limits of experimental error, confirming that the oscillator strength had not varied. Measurement accuracy was sufficiently high, since the photoelectric method of recording was used. The experiments were conducted with argon, nitrogen, hydrogen and helium, at pressures up to 2000 atm, and at temperatures between 25 and 100°C.

The results of the experiments described above confirm the theoretical assertion that the integrated absorption coefficient is entirely independent of the condition under which experiments are conducted.

The second of the methods indicated earlier consists of measuring the total energy absorbed by a resonance line from a continuous spectrum, the energy being related to the intensity of the incident radiation. By definition,

$$A_\nu = \int_0^\infty \frac{I_0 - I_0 e^{-k_\nu l}}{I_0} d\nu = \int_0^\infty (1 - e^{-k_\nu l}) d\nu \tag{1.28}$$

The advantage of this method is that the total absorption does not, within wide limits, depend on the resolving power of the spectroscopic instrument. If an absorption line, shown by the continuous curve in Fig. 1.4, is not fully resolved, the photometric curve follows the path shown by the broken line. In this case, the area

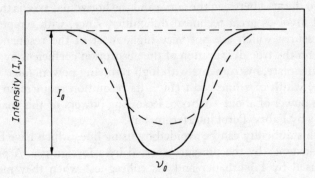

FIG. 1.4. Total absorption with the contour not fully resolved.

above the broken curve is equal to the area above the continuous curve. Standard spectrographs can thus be used for measuring the total absorption.

With low values of $k_\nu l$,

$$1 - e^{-k_\nu l} \simeq k_\nu l$$

whence

$$A_\nu = l \int_0^\infty k_\nu d\nu \qquad (1.29)$$

With low absorbances, therefore, the total absorption method proves to be as good as the integrated absorption method. Taking equation (1.27) into account, we get

$$A_\nu = \frac{\pi e^2}{mc} Nfl \qquad (1.30)$$

The results of measurements using the above methods agree only for very small absorbances. For instance, even with an absorbance of 0·2, the discrepancy between the results is about 10 per cent.

With very great absorbances, the central part of the line proves to be fully absorbed; further absorption, taking place as the concentration of atoms in the atomic vapour increases, occurs in the wings of the line. The distribution of the absorption coefficient in the wings is Lorentzian (1.15), while, with high absorbances, $\Delta v_L^2 \ll 4(\nu - \nu_0)^2$. Therefore

$$k_\nu = \frac{2e^2}{mc} \frac{\Delta v_L}{4(\nu - \nu_0)^2} Nf \qquad (1.31)$$

If equation (1.28) is integrated and equation (1.31) used, we get

$$A_\nu = \left(\frac{2\pi e^2}{mc} Nfl \, \Delta v_L\right)^{1/2} \qquad (1.32)$$

With low absorbances, therefore, the total absorption must be proportion to Nfl, while with great absorbances it will be proportional to $\sqrt{(Nfl)}$. In the transition region, the relationship of A_ν to Nfl cannot be expressed analytically and is determined by numerical integration. Curves linking the total absorption and the concentration of atoms have been plotted by Van der Held[34] for the entire range of absorbances and different values of $a=(\Delta\nu_L/\Delta\nu_D)\sqrt{(\ln 2)}$; these curves are called curves of growth.

The total absorption method is the principal method employed in astrophysical measurements.[9] As an example, this method was extensively used by A. King and R. King and their colleagues for determining the absolute and relative values of oscillator strengths.[37]

The last of the methods of measuring absorption mentioned earlier, the line absorption method, consists of measuring the relative amount of light absorbed from a line source. If we use E_ν to denote the spectral distribution of energy in the incident radiation, and k_ν for the absorption coefficient of an atomic vapour, the line absorption will be expressed by the following equation:

$$A_L = \frac{I_0 - I}{I_0} = \frac{\int E_\nu d\nu - \int E_\nu e^{-k_\nu l} d\nu}{\int E_\nu d\nu} = \frac{\int E_\nu (1 - e^{-k_\nu l}) d\nu}{\int E_\nu d\nu} \tag{1.33}$$

The relationship of the absorption A_L to the product Nfl was calculated for the following two cases.

The identical lamps method

A resonance lamp is used as the source of light, the spectral distribution of which corresponds to

$$E_\nu = c(1 - e^{-k'_\nu l'}) \tag{1.34}$$

where c is a constant independent of frequency, and k'_ν and l' are the absorption coefficient and the discharge length characteristic of the lamp.

If the atomic vapour is in all respects identical with the emitting vapour of the resonance lamp, i.e. if the temperature, pressure, and thickness are identical, then $k'_\nu = k_\nu$, $l' = l$, and equation (1.33) can be re-written as follows:

$$A_L = \frac{\int (1 - e^{-k_\nu l})^2 d\nu}{\int (1 - e^{-k_\nu l}) d\nu} \tag{1.35}$$

If, in addition to this, k_ν is determined solely by the Doppler effect as in equation (1.11), then equation (1.35) can be calculated for different values of $k_0 l$.[34,38] We can thus, by measuring A_L, obtain the value of $k_0 l$, which is in turn connected with Nf by equation (1.12).

This variant of the line absorption method is identical to the self-absorption method; the same source of light is used for the radiating and absorbing medium, and the relative absorption of the beam of light passing through the lamp after being reflected from a mirror is measured. The different variations of the method based on self-absorption in homogeneous media have been reviewed by Frish and Bochkova.[38]

The different lamp method

It is sometimes necessary to interpret the absorption of light from a source for which the vapour pressure and the thickness of the radiating medium are unknown. The following empirical expression can be used for the distribution of intensity in the light source:

$$E_\nu = c e^{-(\omega/\alpha)^2} \tag{1.36}$$

Here, ω relates to the absorption line, not to the emission line, and (as in 1.23) is given by

$$\omega = \frac{2(\nu - \nu_0)}{\Delta\nu_D} \sqrt{(\ln 2)} \tag{1.25}$$

The coefficient α denotes the ratio of the half-width of the emission line to that of the absorption line. A value of unity for α means that the shape and width of the emission line are the same as the shape and width of the absorption line. Values of α greater than unity denote a line broader than the absorption line but of the same shape. A number of authors have solved equation (1.33) for this case and for different values of α and $k_0 l$.[34] To find α we must plot the relationship of $k_0 l$ to Nl graphically. The relationship is bound to be a straight line for a certain value of α, found by trial. This method was, for instance, used in the isotopic analysis of mercury, and in measuring the concentration of neutral atoms in beams of ions, as is described later (pp. 260–278).

The line absorption method can be used for individual simple lines or for lines with well-separated hyperfine components. In the case of a line with overlapping hyperfine components, absorption should be found by graphical integration.

We have discussed here the principal methods of measuring absorption. In addition, the author has proposed and used a combined method of measuring total and line absorption.[39] The total absorption A_ν for the atomic vapour being investigated is recorded photoelectrically by measuring the absorption from a continuous spectrum emitter, using the equation:

$$A_\nu = \frac{I_0 - I}{k I_0} \Delta\nu \tag{1.37}$$

where $\Delta\nu$ is the frequency bandwidth isolated by the monochromator and k is a coefficient for the effect of the wings of the line on the magnitude of A_ν when the instrumental profile is approximately rectangular. The coefficient k is calculated by the method developed by Ostrovskii and Penkin.[40] The final equation for k is as follows:

$$k = 1 - \frac{1}{2\pi p}\left[1 - \frac{1}{24\pi p^2} + \frac{1}{6}r^2 + \frac{1}{15}r^4\right] \tag{1.38}$$

where

$$p = \frac{I_0}{2(I_0 - I)}; \qquad r = 2\frac{s_i}{s_0}$$

In the last equation, s_i and s_0 denote the widths of the entrance and exit slits of the monochromator. Equation (1.38) can be used for calculating the coefficient k with $p > 1.5$ and $r < 1/3$.

The line absorbance A_s (in absorbance units) is also recorded photoelectrically by

Theory of Atomic Absorption Measurements

using a line source of light. It will be shown later (pp. 22–31) that, if the pressure of the foreign gas in the atomic vapour is increased, the measured line absorption will correspond to the absorption of monochromatic light by a region of the line profile displaced from the centre by Δv_s.

It was pointed out earlier that, with low absorbances across the line profile ($k_v l \ll 1$), the total absorption is given by the integrated absorption coefficient for the line [equation (1.29)]. As the absorbance increases, the total absorption increases less rapidly than the integrated absorption coefficient, so that as $k_v l$ increases the discrepancy between them also increases.

The ratio of the total absorption A_v to the integrated absorption coefficient $\int_0^\infty k_v l \, dv$ is a function of the product $k_0 l$ (where k_0 is the absorption coefficient at the centre of the line), or is simply related to it in a function of the absorbance A_s. The form taken by the functions

$$\phi(A_s) = \frac{A_v}{\int_0^\infty k_v l \, dv} = \frac{\int_0^\infty (1 - e^{-k_v l}) dv}{\int_0^\infty k_v l \, dv} \tag{1.39}$$

depends on which factors govern the form of the absorption line. Ladenburg and Levy[51] calculated the function $\phi(A_s)$ for a pure Doppler profile ($a=0$), while Elsasser[52] calculated the function for a pure Lorentz profile ($a=\infty$).

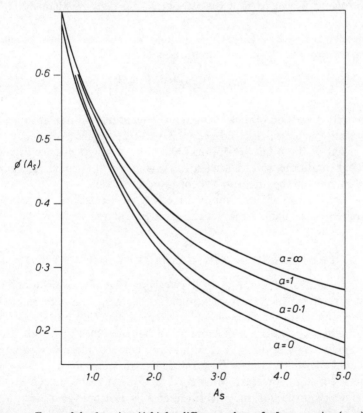

FIG. 1.5. Form of the function $\phi(A_s)$ for different values of a [see equation (1.24)].

The form taken by this function is represented graphically in Fig. 1.5 for A_s between 0·5 and 5·0. When the curves were plotted it was assumed that, in the case of a Doppler profile, the measured line absorption corresponds to the absorbance of the line at its centre, i.e.

$$A_s = (\log_{10} e) k_0 l \simeq 0.434 k_0 l \qquad (1.40)$$

while in the case of a Lorentz profile, in which the absorption peak is shifted relative to the unaffected emission line (see pp. 23–31), the measured line absorption corresponds to the absorbance in the line profile region displaced by Δv_s from the maximum, i.e.

$$A_s = (\log_{10} e) k_0 l \frac{\Delta v_L^2}{\Delta v_L^2 + 4\Delta v_s^2} \qquad (1.41)$$

or, using (1.20)

$$A_s \simeq 0.286 k_0 l \qquad (1.42)$$

Lebedev and L'vov calculated the function $\phi(A_s)$ for intermediate cases ($a = 0.1$ and $a = 1.0$). The calculations were made by numerical integration, using Simpson's rule, with an electronic computer. In the calculations, the limits of integration were chosen so that the variables assumed less than 0·01 times their maximum values.

From equations (1.27), (1.37) and (1.39), we get the following equation for measuring the product Nfl by the combined method:

$$Nfl = \frac{mc}{\pi e^2} \frac{\Delta v}{k\phi(A_s)} \left(1 - \frac{I}{I_0}\right) \qquad (1.43)$$

The combined method of measuring total absorption and line absorption enables us to associate the simplicity and accuracy of measurements made by the photoelectric variant of the total absorption method with independence of results from experimental conditions and line structure, this independence being inherent in the method of measuring the integrated absorption coefficient.

Examples of the use of the combined method for measuring line half-widths and absolute values of oscillator strengths are given in Chapter 6.

THE WALSH ATOMIC ABSORPTION METHOD

The principle of the method proposed by Walsh is that of measuring the absorption coefficient directly at the centre of a line. To do this, an atomic vapour must be subjected to a monochromatic beam of light with a wavelength corresponding to the centre of the absorption line. As sources, Walsh therefore proposed low-pressure hollow-cathode lamps, which afford sufficiently narrow emission resonance lines of the elements being determined. To produce the atomic vapour he used a flame, of which the absorption lines are broadened by Doppler and Lorentz effects. The emission line required for the measurements is isolated by means of a monochromator or other spectral device and its intensity is recorded photoelectrically.

The absorption, measured in absorbance units, is a measure of the concentration of the atoms being determined in the volume examined.

We shall consider later the extent to which the assumptions underlying this method of measuring absorption are true under natural conditions. Here it must be noted that, in principle, the direct measurement of the absorption coefficient at a specific point on the profile of a line greatly simplifies the procedure for determination as well as the calculation of analytical results, since in this case the absorption coefficient is proportional to the concentration of atoms in the atomic vapour, whatever the reasons for the shape of the profile. For analytical purposes, therefore, the method proposed by Walsh is without doubt preferable to the methods of measuring the integrated absorption coefficient or the total absorption described above.

In fact, determination of the integrated absorption coefficient necessitates the use of instruments with high resolving powers to record the line profiles; photometric determination of the profile of an entire line is a prerequisite of graphical integration, so that the procedure is complex and lengthy. The determination of total absorption does not require apparatus with high resolving powers and the measuring process is simple. However, the relationship between total absorption and the concentration of atoms depends on the range of absorbance measured, while the form of the relationship (the growth curve) is governed by the experimental conditions, in particular the magnitude of $a = (\Delta v_L / \Delta v_D) \sqrt{(\ln 2)}$. In addition, as we shall see later, the sensitivity of the method is two orders lower when a continuous source is used than when a line is used.

As regards comparison with line absorption methods, it should be noted that the method suggested by Walsh is, in fact, a particular case for $\alpha \ll 1$ of the more general method of line absorption with different lamps (α is the ratio of emission line-width to absorption line-width). The distinguishing feature of the Walsh method is the different type of measurement: instead of measuring the line absorption $A_L = 1 - (I/I_0)$, Walsh suggests measuring the absorbance $A = \log(I_0/I)$. Fig. 1.6 gives the relationship of A_L and A to $k_0 l$ (i.e. ultimately to the concentration of atoms in the atomic vapour) for $\alpha = 0$. The linear relationship to concentration is extremely important in practice and a definite advantage of absorbance measurements.

If the shape of the absorption line in the atomic vapour is due entirely to Doppler broadening, then the absorption coefficient measured with a source emitting a line at the frequency of the centre of the absorption line and having negligible half-width is given by equation (1.12), i.e.,

$$k_0^{(D)} = \frac{2}{\Delta v_D} \sqrt{\left(\frac{\ln 2}{\pi}\right)} \frac{\pi e^2}{mc} Nf$$

Walsh [15, 53] showed that this technique could be expected to yield useful results, even in practice it gives only an approximate measure of peak absorbance for the following reasons:

1. In an atomic vapour produced in a flame the frequency of the absorption line maximum is slightly displaced from that of the emission line maximum.

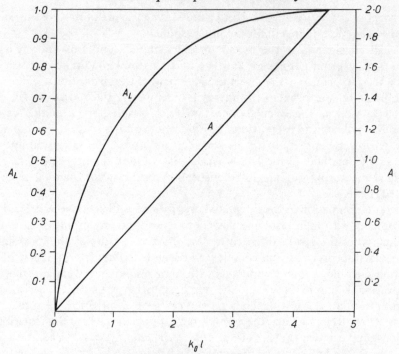

FIG. 1.6. Relationship of line absorption A_L and absorbance A to $k_0 l$.

2. The half-width of the emission line from the source is not always so small that the signal can be considered monochromatic.

3. The absorption line profile of an atomic vapour produced in a flame is determined not only by the Doppler effect but also by Lorentz and possibly other types of broadening.

The magnitude of these departures from the simple theory of equation (1.12) has been examined by the author[43]. The pressure of the foreign gas (in flames, this consists principally of nitrogen and oxygen) not only causes Lorentz broadening, but also shifts the maximum of a line by Δv_s relative to its undisturbed position (pp. 10–15). The maximum of the absorption line from a flame is therefore shifted by an amount Δv_s relative to the maximum of the emission line from the source (which, owing to the low pressure of the foreign gas, is not shifted). The expression for the absorption coefficient must accordingly allow, not only for Doppler and Lorentz broadening, but also for the line shift. It is therefore best to use equation (1.23), which allows for all these factors:

$$k_s = k_0^{(D)} \frac{a}{\pi} \int_{-\infty}^{\infty} \frac{e^{-y^2} dy}{a^2 + (\omega_s - y)^2} \tag{1.44}$$

where

$$a = \frac{\Delta v_L}{\Delta v_D} \sqrt{(\ln 2)}, \qquad \omega_s = \frac{2\Delta v_s}{\Delta v_D} \sqrt{(\ln 2)}, \qquad y = \frac{2\delta}{\Delta v_D} \sqrt{(\ln 2)}$$

δ is frequency displacement from $v - v_0$.

The allowance for the shift is particularly important with high values of a, when the Lorentz line-width is greater than the Doppler line-width. Values of $k_s/k_0^{(D)}$ are compared, in Table 1.6, for different values of a; the values of $k_s/k_0^{(D)}$ were calculated from equation (1.44), with and without an allowance for line shift (we assumed the shift $\Delta\nu_s$ to be, as proposed by Lindholm, equal to $0.36\,\Delta\nu_L$).

With $a = 0.5$, the difference between the results is 8 per cent, and with $a = 2.0$ it is about 34 per cent.

According to measurements made by Hinnov and Kohn,[41] the values of a at 2500°K vary between 0.46 and 3.0 according to the element and resonance line (Table 1.7). With the Walsh method, therefore, it is more correct to estimate the coefficient for the mixed line profile in the region displaced by the amount of the shift from the centre.

TABLE 1.6 Results of calculating coefficients of absorption with and without allowance for line shift

a	$k_s/k_0^{(D)}$ (allowing for $\Delta\nu_s$)	$k_s/k_0^{(D)}$ (no allowance for $\Delta\nu_s$)
10	0.0368	0.0561
2	0.192	0.257
1.5	0.248	0.322
1.0	0.356	0.428
0.5	0.569	0.616

TABLE 1.7 Relationship of the Doppler and Lorentz widths for the resonance lines of certain elements[41]

Line (Å)	$\Delta\nu_D$ (cm^{-1})	$\Delta\nu_L$ (cm^{-1})	a	Line (Å)	$\Delta\nu_D$ (cm^{-1})	$\Delta\nu_L$ (cm^{-1})	a
Li 6708	0.20	0.13	0.55	Ag 3281	0.11	0.13	1.03
Na 5890	0.13	0.13	0.86	Ti 3776	0.066	0.24	3.0
K 7665	0.074	0.24	2.7	Cr 4254	0.12	0.08	0.54
Ca 4227	0.13	0.07	0.46	Mn 4030	0.12	0.12	0.80
Sr 4607	0.083	0.11	1.1	Fe 3720	0.13	0.08	0.53
Ba 5535	0.053	0.09	1.4	Co 3526	0.13	0.12	0.78
Cu 3248	0.14	0.08	0.46	Ni 3524	0.14	0.09	0.54

In cases in which $a > 1$, the Doppler broadening can be ignored and the profile can be assumed to be solely a Lorentz profile. The coefficient of absorption can then be expressed by the equation:

$$k_s^{(L)} = \frac{2e^2}{mc}\frac{\Delta\nu_L}{\Delta\nu_L^2 + 4\Delta\nu_s^2} Nf \qquad (1.45a)$$

According to Lindholm, the line half-width $\Delta\nu_L$ and the line shift $\Delta\nu_s$ are linked by equation (1.20); it follows from this that

$$\frac{\Delta\nu_L}{\Delta\nu_L^2 + 4\Delta\nu_s^2} = \frac{0.237}{\Delta\nu_s} \qquad (1.45b)$$

It is interesting to note that this relationship remains true, with an accuracy of up to 5 per cent, for values of the ratio $\Delta v_L/\Delta v_s$ between 1·2 and 3·2. At a point displaced by Δv_s relative to the centre, therefore, the coefficient of absorption for a Lorentz profile is practically independent of variations in the ratio $\Delta v_L/\Delta v_s$ such as are encountered in practice (Table 1.4), and the coefficient can be expressed as follows:

$$k_s^{(L)} \simeq \frac{0 \cdot 474\ e^2}{mc\Delta v_s} Nf \tag{1.46}$$

The equations given here for the absorption coefficient are true only for monochromatic signals and single absorption lines. Let us consider the extent to which these assumptions are true for actual measurement conditions.

In principle, the emission lines in low-pressure gas-discharge lamps must be narrower than the absorption lines in flames. In fact, with an inert-gas pressure of a few millimetres of mercury, the Lorentz broadening is two orders less than with a gas pressure of 1 atm (in a flame), while at 500°K (the temperature of an uncooled hollow cathode with low currents passing through a discharge lamp) the Doppler half-width is 2·2 times less than in a flame at 2500°K. When these broadening factors alone are taken into account, therefore, the assumptions made above are sufficiently justified.

When hollow-cathode lamps are used as sources, however, the effect of self-absorption of the resonance lines within the lamp must be taken into account; sometimes this effect appreciably broadens the lines. Moreover, with many elements there is considerable hyperfine splitting of the resonance lines. Table 1.8 gives the hyperfine splitting data for certain resonance lines.[42]

We can see from the results given in this table that, for many resonance lines, hyperfine splitting effectively broadens the lines until their widths are comparable with the widths of the absorption lines (Table 1.7). For such instances, graphical methods have to be used for plotting the overall profiles of absorption lines. In addition, monochromatic emission from the source is unobtainable, since an emitted resonance line consists of a set of hyperfine components with a variable distribution of intensity, depending on the discharge conditions.

The use of equation (1.44) or approximation (1.46) for calculating absorption coefficients in general, for any element and any arbitrary source conditions, is therefore unjustifiable.

Let us now consider how the non-monochromaticity of the emission lines must affect the form taken by the relationship of absorbance to the concentration of atoms.[43] For purposes of simplicity we shall assume that the absorption line profile is of Lorentz form and that the emission line consists of two components, of equal intensity, separated from each other by a distance Δv_{spl} (Fig. 1.7). This last assumption is of real significance in the case of self-reversed emission lines. We shall further denote all the quantities relating to the different components by one or two strokes, while the quantities with no strokes will relate to a fictitious single-component line.

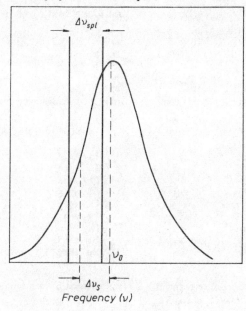

Fig. 1.7. Absorption and emission lines.

The true value of the absorbance A_x will represent the result of absorption of both components, i.e.

$$A_x = -\log \frac{I' + I''}{I'_0 + I''_0} = \log 2 - \log (e^{-k'l} + e^{k''l}) \tag{1.47}$$

where

$$I'_0 = I''_0 \tag{1.48}$$

$$k'l = \frac{Ak'}{0.434k} = \frac{A(\Delta\nu_L^2 + 4\Delta\nu_s^2)}{0.434[\Delta\nu_L^2 + 4(\Delta\nu_s + \Delta\nu_{\text{spl}}/2)^2]} \tag{1.49}$$

$$k''l = \frac{Ak''}{0.434k} = \frac{A(\Delta\nu_L^2 + 4\Delta\nu_s^2)}{0.434[\Delta\nu_L^2 + 4(\Delta\nu_s - \Delta\nu_{\text{spl}}/2)^2]}$$

If we substitute $0.36\Delta\nu_L$ for $\Delta\nu_s$ according to equation (1.20), and if we also substitute $x\Delta\nu_L$ for $\Delta\nu_{\text{spl}}$, we get

$$k'l = \frac{3 \cdot 5A}{1 + 4\left(0 \cdot 36 + \frac{x}{2}\right)^2} \tag{1.50}$$

$$k''l = \frac{3 \cdot 5A}{1 + 4\left(0 \cdot 36 - \frac{x}{2}\right)^2}$$

We shall take the ratio between the absorbances of the two-component (non-monochromatic) and single-component (monochromatic) lines as the criterion for the discrepancy in the results:

$$\beta = \frac{A_x}{A} \qquad (1.51)$$

TABLE 1.8 Hyperfine splitting of resonance lines

Line (Å)	$\Delta\nu_{spl}$ (cm^{-1})	Line (Å)	$\Delta\nu_{spl}$ (cm^{-1})
Au 2428	0.2	In 3039	0.4
B 2496	0.2	Li 6708	0.35*
Bi 3068	0.9	Mg 2852	0.2
Co 2407	0.5	Mn 2795	0.3
Cs 8521	0.3	Os 2909	0.3
Cu 3248	0.4	Pb 2833	0.45
Eu 4594	0.25	Rb 7800	0.2
Ga 2874	0.1	Sb 2068	0.75
Hg 2537	0.7	Sn 2863	0.25
		Ti 2767	0.7

* The isotopic displacement of 0.35 cm^{-1} is supplemented by the doublet structure so that the total line splitting is $\simeq 0.7$ cm^{-1}.

Table 1.9 contains the result of calculating the coefficients β, with relation to x, for different values of A.

TABLE 1.9 Relationship of the coefficient β to x, for different values of A

A	$x (\Delta\nu_{spl}/\Delta\nu_L)$					
	2:1	1:1	1:2	1:3	1:4	1:5
0.1	0.36	0.87	1.0	1.0	1.0	1.0
0.5	0.35	0.75	0.93	0.97	0.99	1.0
1.0	0.34	0.65	0.86	0.92	0.96	0.98
1.5	0.32	0.57	0.81	0.89	0.94	0.96
2.0	0.30	0.54	0.76	0.86	0.92	0.94

As an illustration, the relationship between the absorption of a non-monochromatic (two-component line) to that of a monochromatic (single-component) line has been plotted, for different degrees of emission-line non-monochromaticity, in Fig. 1.8. The curves are essentially true calibration curves, since the values of A for a monochromatic line are proportional to the concentration of atoms in the absorbing medium.

The following conclusion can be drawn. Increasing the degree of emission-line non-monochromaticity must reduce the absorption (i.e. must finally reduce the sensitivity of measurements) and must also cause loss of proportionality between the absorbance and the concentration of atoms in the atomic vapour.

While discussing the effects of line non-monochromaticity on the results of measurements, it will be useful to consider the ultimate case in which the source emits a continuous spectrum. Let us estimate the difference between the sensitivity

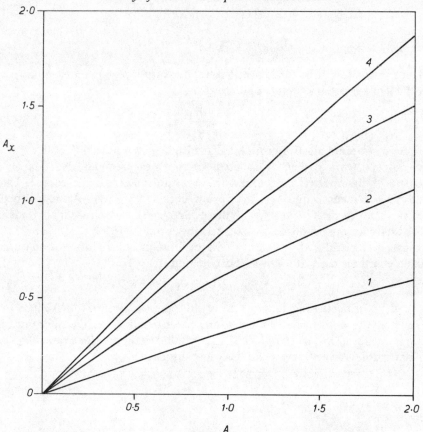

FIG. 1.8. Calculated calibration curves for different degrees of emission line non-monochromaticity: 1. $\Delta v_{spl} = 2\Delta v_L$. 2. $\Delta v_{spl} = \Delta v_L$. 3. $\Delta v_{spl} = 0.5\ \Delta v_L$. 4. $\Delta v_{spl} = 0.2\ \Delta v_L$.

of measurements made using sources of continuous and monochromatic light. We shall use equations (1.43) and (1.45a). Let us assume that, with low absorptions,

$$1 - \frac{I}{I_0} \simeq \ln \frac{I_0}{I} = 2 \cdot 3 A_{cont}$$

where A_{cont} is the absorbance when a continuous source is used. Therefore

$$A_{cont} = 0 \cdot 434 \frac{\pi e^2}{mc} \frac{k\phi(A_s)}{\Delta v} Nfl \qquad (1.52)$$

On the other hand, using Lindholm's equation (1.20) and equation (1.4), equation (1.45a) gives us

$$A_s = 0 \cdot 434 \frac{2e^2}{mc} \frac{0 \cdot 658}{\Delta v_L} Nfl \qquad (1.53)$$

When we compare equations (1.52) and (1.53), we find that

$$A_{\text{cont}} = 2\cdot 4 k\phi(A_s)\frac{\Delta\nu_L}{\Delta\nu}A_s \tag{1.54}$$

When the smallest possible absorptions are measured, $\phi(A_s)$ is close to unity and the coefficient k is equal to unity. Therefore

$$A_{\text{cont}} \simeq 2\cdot 4 \frac{\Delta\nu_L}{\Delta\nu} A_s \tag{1.55}$$

The mean Lorentz absorption line half-width $\Delta\nu_L$ is 0·1 cm^{-1} (Table 1.7). The continuous spectrum region $\Delta\nu$ isolated for the measurements depends on the particulars of the spectral device and the emitting power of the source. Under normal measurement conditions, the region does not exceed 1Å; at $\lambda = 3000$ Å, this corresponds to $\Delta\nu = 10$ cm^{-1}. With these parameters, the sensitivity of measurements with a continuous source is two orders lower than with a line source.

Let us now consider the nature of the calibration curves when a continuous spectrum emitter is used. It follows from equation (1.54) that

$$A_{\text{cont}} \propto \phi(A_s)N \tag{1.56}$$

Since $\phi(A_s)$ is a decreasing function of A_s, or (the same thing) of N, it is quite obvious that the absorbance does not increase as rapidly as the concentration. Fig. 1.9 is a calibration curve in which A_{cont} is plotted against A_s for a ratio $\Delta\nu_L/\Delta\nu = 10^{-2}$. The graph is a curve, even with very low absorbances.

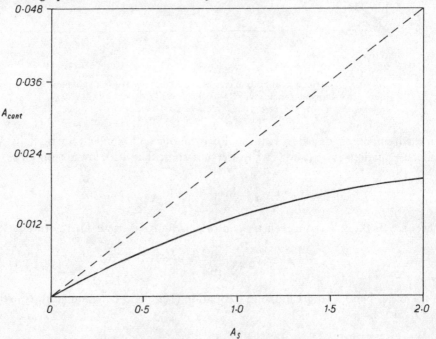

FIG. 1.9. Relationship between the absorption from continuous and line sources, with $\Delta\nu_L/\Delta\nu = 10^{-2}$

The calculation of sensitivity and the shape of calibration curves for continuous emitters has so far been considered for a rectangular output by the monochromator (entrance slit much narrower than exit slit). With a triangular distribution of intensity in the isolated region of the spectrum, the measurements prove to be about twice as accurate as with a rectangular distribution (s being assumed identical). All the other conclusions remain valid, whatever the shape of the continuous spectrum interval isolated.

To sum up the contents of this section, we can note the following important results:

1. The Walsh method for measuring atomic absorption is simpler and more sensitive than the methods previously used in absorption research.

2. When the method is interpreted theoretically, it is necessary to allow for the shift of the absorption line relative to the emission line when a foreign gas at atmospheric pressure is present in the atomic vapour.

3. In general, the theoretical calculation of absorption coefficients is complicated by the hyperfine structure of the lines and by self-absorption in the source.

4. The fact that emission lines are not monochromatic affects the sensitivity of the measurements and the curvature of calibration curves.

5. When a continuous source is used, measurements are much less sensitive than with a line source. The calibration graphs are non-linear even for the very lowest absorbances.

ANALYSIS LINES

Absorption lines resulting from transitions from the ground state are generally the most sensitive. The absorption coefficient denoting the strength of a line produced by absorption from a monochromatic beam must be proportional to the oscillator strength for the transition concerned and to N_i, the population of absorbing atoms at the lower level of the transition:

$$k \propto N_i f$$

When the ground level is single and is, moreover, far enough from the excited levels, we can assume that N_i is equal to the total number N of atoms, since the population of the excited levels at temperatures below 3000°K is negligibly small (Table 1.1).

When, however, the ground level is multiplet, i.e. split into several sub-levels close together, the population of the lowest sub-level (the normal state of the atom) may be smaller than that of any of the neighbouring sub-levels. In equilibrium conditions, the distribution of atoms between two energy states obeys Boltzmann's law:

$$N_i = N_0 \frac{g_i}{g_0} e^{-E_i/kT}$$

If E_i is small,

$$N_i \simeq N_0 \frac{g_i}{g_0}$$

In the normal arrangement of sub-levels in multiplet terms, the lowest sub-level has the lowest quantum number **J**. The statistical weight $g_i = 2\mathbf{J} + 1$ for a higher sub-level exceeds the statistical weight of the lowest sub-level. It may therefore be that N_i exceeds N_0 and that a transition from the higher sub-level is more sensitive than one from the lower. The resonance lines of aluminium and gallium (see Table 1.10) afford examples.

TABLE 1.10 Resonance lines of aluminium and gallium

Line (Å)	Transition	Lower level energy (cm^{-1})	f	Sensitivity[51,44] (p.p.m.)
Al 3082	$3\,^2P_{1/2}-3\,^2D_{3/2}$	0	0.22	2.4
Al 3093	$3\,^2P_{3/2}-3\,^2D_{3/2}$	112	0.23	1.7
Ga 4033	$4\,^2P_{1/2}-5\,^2S_{1/2}$	0	0.13	7
Ga 4172	$4\,^2P_{3/2}-5\,^2S_{1/2}$	826	0.14	5

The oscillator strengths for the corresponding line pairs are about the same, but at 3000°K the population of the $3\,^2P_{3/2}$ level is 90 per cent above that of the ground $3\,^2P_{1/2}$ level, while the population of the $4\,^2P_{3/2}$ level is 30 per cent above that of the $4\,^2P_{1/2}$ level. The lines absorbed by transitions from the $P_{3/2}$ level therefore prove to be the most sensitive, as has been confirmed by experiments in which sensitivities were measured.

Another important characteristic of lines is the oscillator strength. The oscillator strengths of the spectral lines have been measured for many elements, and there is therefore no difficulty in selecting the most sensitive lines. In cases in which the oscillator strengths (absolute or relative) in the spectrum of an element are unknown, or have not been sufficiently accurately measured, resort can be made to selecting the most sensitive lines by experimenting. The simplest method is that proposed and used by Allan.[45] This method consists of photographing the spectrum of a source while the beam of light is passing through a flame, and comparing the spectrograms obtained when a solution of an element and the pure solvent are introduced into the flame. Microphotometry of the spectrograms is used for estimating the absorption values. The most convenient sources are line spectrum emitters, but Allan[46] and later Fassel[47] proved that continuous sources can also be used, although sensitivity is not so good with them. In addition, high resolving powers are necessary with continuous backgrounds, since without them absorption lines may simply not appear on the spectrogram.[48] Obviously it is also possible to use photoelectric methods of recording, as was done by Robinson.[49]

It was established in the experiments conducted by the authors mentioned above that the most sensitive absorption lines do not necessarily correspond with the most sensitive emission lines used in emission analysis, including flame photometry. This fact has sometimes been misunderstood, although it has a simple explanation. We know that the emission intensity I of resonance lines, in equilibrium conditions, is given by

$$I = \frac{hc}{1 \cdot 51} \frac{f}{\lambda^3} N_0 e^{-h\nu/kT} \qquad (1.57)$$

and it follows that the emission intensity of resonance lines depends not only on N_0 and f, but also on the region of the spectrum in accordance with the expression

$$I \propto \nu^3 e^{-h\nu/kT} \qquad (1.58)$$

The variation in intensity at flame or arc temperatures is governed almost entirely by the exponential term, and emission intensity therefore falls rapidly in the short-wave region of the spectrum. The distribution with intensity of the resonance lines in the emission spectrum can therefore be different from the corresponding distribution in the absorption spectrum. The strongest resonance lines in absorption, if they do not coincide with the strongest resonance lines in emission, are always shorter in wavelength than the emission lines. All experimental results, without exception, support this conclusion. As illustration, Table 1.11 compares the most sensitive absorption and emission lines, with the same lower levels, of several elements.

A list of the most sensitive absorption resonance lines for seventy elements is given, together with the oscillator strengths, in Appendix 4.

TABLE 1.11 The analysis lines which are most sensitive in emission and absorption

Element	Absorption line (Å)	Transition	Emission line (Å)	Transition
Al	3093	$3\,^2P^o_{3/2}-3\,^2D_{5/2,\,1/2}$	3961	$3\,^2P^o_{3/2}-4\,^2S_{1/2}$
Co	2407	$a\,^4F_{9/2}-x\,^4G^o_{11/2}$	3527	$a\,^4F_{9/2}-z\,^4F^o_{9/2}$
Fe	2483	$a\,^5D_4-x\,^5F^o_5$	3720	$a\,^5D_4-z\,^5F^o_5$
In	3039	$5\,^2P^o_{1/2}-5\,^2D_{3/2}$	4101	$5\,^2P^o_{1/2}-6\,^2S_{1/2}$
Mn	2795	$a\,^6S_{5/2}-y\,^6P^o_{7/2}$	4031	$a\,^6S_{5/2}-z\,^6P^o_{7/2}$
Mo	3133	$a\,^7S_3-y\,^7P^o_4$	3798	$a\,^7S_3-z\,^7P^o_4$
Ni	2320	$a\,^3D_3-y\,^3G^o_5$	3415	$a\,^3D_3-z\,^3F^o_4$
Ti	3653	$a\,^3F_4-y\,^3G^o_5$	3999	$a\,^3F_4-y\,^3F^o_4$

ADVANTAGES OF THE ATOMIC ABSORPTION METHOD OF SPECTROCHEMICAL ANALYSIS

Walsh[15] was the first person to point out that, in principle, absorption methods have certain advantages over emission methods of spectrochemical analysis.

1. In atomic absorption measurements it is almost impossible for the resonance lines of different elements to be superimposed. We can assess this assertion for the interesting case of the determination of an impurity element. In such a determination by emission methods, lines of foreign elements, with any appreciable concentration, may well be superimposed on the analysis line.

If we assume that the total number of levels in an atom is equal to n, the number of possible transitions between levels, and consequently also the number of atomic emission lines, must be proportional to $n^2/2$.* At the same time, the number of

* No. of transitions $= \dfrac{n!}{(n-2)!\,2!} = \dfrac{n(n-1)}{2} \simeq \dfrac{n^2}{2}$ for n large.

transitions from the ground state must be proportional to the total number of levels, i.e. to n. Hence,

$$\text{Number of absorption lines} = \sqrt{(2 \times \text{number of emission lines})} \qquad (1.59)$$

The total number of lines observed in the arc or spark spectra of substances containing the elements being determined in the form of impurities is governed in practice by the total number of lines in the spectra of the one or two principal elements of the substance. On average, the number of emission lines from neutral atoms and singly charged ions for any element in the 2000–10 000 Å region, i.e. within a 40 000 cm^{-1} interval, is about a thousand.[55]

The line-width in arc or spark spectra, allowing for instrumental width and assuming ideal recording conditions, is about 1 cm^{-1}. The probability $\mathbf{P}_{em}$ that any of the thousand foreign emission lines will be superimposed on the analysis line (within its 1 cm^{-1} width) is therefore

$$\mathbf{P}_{em} = \frac{\text{line-width} \times \text{number of lines}}{\text{spectral range}} = \frac{1 \times 1000}{40\ 000} = 0.025 = 2.5 \text{ per cent}$$

In absorption measurements, the probability of superimposition is also governed by the width and total number of absorption lines in the spectrum range considered. Unlike emission line-widths, absorption line-widths are governed only by broadening in the absorbing layer; their average is not more than 0.5 cm^{-1}. When allowance is made for the fact that atomic lines comprise about half the total number of lines in emission spectra, the number of absorption lines given by equation (1.59) is $\sqrt{(2 \times 500)} \simeq 30$. The probability $\mathbf{P}_{abs}$ of superimposition, when atomic absorption measurements are made under the same conditions, is therefore

$$\mathbf{P}_{abs} = \frac{0.5 \times 30}{40\ 000} = 4 \times 10^{-4} = 0.04 \text{ per cent}$$

As a result, therefore, of the much smaller number of lines and their smaller width, the probability that lines will be superimposed is on the average two orders less in atomic absorption measurements than in emission spectroscopy under the very best conditions. Thus, the atomic absorption method almost entirely eliminates the superimposition of interfering lines on the analysis line of the element being determined, and so avoids an important problem in the emission analysis of complex specimens.

2. Hollow-cathode lamps and high-frequency discharge lamps, which are used as light sources, provide a sufficiently simple spectrum of any element being determinated, with little background. The required resonance line can therefore be isolated by instruments with low resolving powers.

3. In emission spectrochemical analysis, the radiations of excited atoms are recorded; the concentration of these atoms at a level with an energy E_i is determined by Boltzmann's law:

$$N_i = N_0 \frac{g_i}{g_0} e^{-E_i/kT}$$

Small variations in the temperature of the source have a great effect on the concentration of excited atoms, and consequently also on the intensity of the analysis lines. In absorption analysis the number of atoms in the *un*excited state is important, and this number is not greatly affected by the temperature. This fact considerably reduces inter-element effects as well as the effects of changes in the conditions under which the atomic vapour is produced. Herein lies the main analytical advantage of absorption methods over emission methods.

4. The sensitivity of emission methods depends greatly on the energy of the levels corresponding to the analysis lines, especially with low-temperature excitation sources, such as flames. Consequently, there is a great reduction in the sensitivity of determining many elements whose analysis lines lie in the short-wave ultra-violet region of the spectrum (<3000 Å). Such elements are, for example, gold, cadmium, mercury, magnesium, lead and zinc. In atomic absorption spectroscopy, the absorption is governed by the population of the ground state and, as a result, the energy of the transition producing the most sensitive resonance line is unimportant.

5. Since, in absorption methods, the quantity measured is the ratio of an unabsorbed signal to an absorbed signal, and not an absolute magnitude, the requirements of the recording apparatus, e.g. the adjustment of exit slits, the quality of radiation detectors, etc., are much less exacting than in emission methods. In fact, in emission methods systematic errors may result from even small variations in the sensitivity of detectors or the parameters of electronic circuits, as well as from shifts in the positions of analysis lines relative to the monochromator exit slits, any of which may occur in the interval between calibration and analysis. In absorption measurements, the same small variations in the parameters of a spectrometer do not affect the accuracy; they are compensated by the method of measuring the relative magnitude of two signals through the same measuring channel. For the same reason, there is no need to allow for the numerous parameters of the apparatus when estimating absolute values of absorption and the concentration of atoms in an atomic vapour. In emission measurements, the aperture ratio, slit-width, and dispersion of the monochromator, and the sensitivity of the detector, are parameters of this type.

6. One further circumstance which differentiates absorption from emission methods, and to which no attention has yet been paid, relates to the dimensions of the emitting and absorbing clouds. In emission spectrographic analysis, the dimensions of the emitting cloud do not affect the sensitivity of determination, since the sensitivity is governed by the ratio of the useful signal to the background; whatever the size of the emitter, this ratio is practically constant and beyond improvement, although it may in fact *decrease* owing to self-absorption as the length of the emitting layer is increased. In atomic absorption analysis, increasing the thickness of the atomic vapour is accompanied by a proportional increase in the sensitivity of determination.* This circumstance provides even greater opportunities for increasing the sensitivity of absorption methods in comparison with emission methods.

* If the sensitivity of emission measurements is limited by statistical fluctuations in the photoelectric current, increase in the brightness of the emitter and in its dimensions will tend to increase sensitivity. Even in this case, however, the relationship of sensitivity to the dimensions of the medium will be less apparent than when absorption measurements are made.

Chapter 2

ATOMIC ABSORPTION SPECTROPHOTOMETERS

This chapter will describe the main components of the spectrophotometers used for recording atomic absorption, which are essentially similar to the main components of the spectrophotometers used for absorption measurements in molecular spectroscopy—a light source, a monochromator, and a recording device.

Owing to the special features of atomic absorption, these units have special features too. In particular, line sources are normally used, although there are certain atomic absorption fields for which continuous sources are better. During atomic absorption measurements, the monochromator is set to a particular wavelength corresponding to a resonance line; the spectrum is not scanned as in molecular spectroscopy. In atomic absorption measurements, the atomic vapour superimposes a line or a continuous spectrum on the emission spectrum of the source; the atomization system is therefore located, not after the monochromator, as when measuring molecular absorption, but before it.

DESIGN AND MANUFACTURE OF HOLLOW-CATHODE LAMPS

The luminescence in a hollow cathode during a d.c. glow discharge was discovered in 1916 by Paschen.[1] When the spectrum of the discharge was investigated, it was found to contain lines, not only of the gaseous medium, but also of the elements in the cathode material, or of materials applied to the internal surface of the cathode. The mechanism by which vapours of the cathode material enter the emitting plasma is twofold: first, cathodic sputtering of the material results from collisions between positive ions of gas and the surface of the cathode; second, thermal evaporation of the material results from the heating of the cathode during the discharge.

The principal reason for the broadening of the non-resonance lines from a hollow-cathode lamp is the Doppler effect, so that lamps with cooled hollow cathodes have been used extensively in research into the hyperfine structure of lines.

Lamps have undergone various improvements in design, according to their various purposes. Some of the most typical designs were described in references 2 and 3. Hollow-cathode lamps are usually filled with inert gases—helium, argon or neon—at pressures between tenths of a torr and a few torr, according to the type of gas and the shape of the cathode.

Until recently, it was assumed that lamps could be used for long periods only if impurities from metal parts of the lamp were continually removed from the gas. To achieve this removal, it was usual for a special pumping system to draw a current of inert gas through the lamp; the gas, in most cases, circulated continually through cleansing traps in a closed loop.[2] The necessity for such a special cleansing system made the lamps inconvenient in spectroscopic research. In 1955, Crosswhite, Dieke and Legagneur,[4] gave the first description of sealed lamps with hollow iron cathodes (Fig. 2.1); in these lamps, liberated gaseous impurities are removed from the inert gas by a trap brazed to the tube and containing activated uranium. Later, Walsh and his colleagues[5] found that it was possible to use an anode made of a metal such as zirconium, which actively absorbed gaseous impurities (called a 'getter').

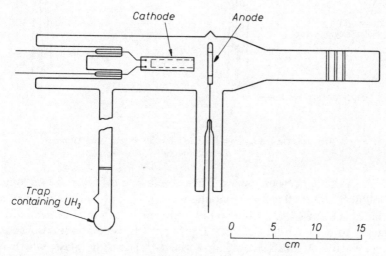

FIG. 2.1. Sealed hollow-cathode lamp with a trap containing activated uranium.[4]

Further use of lamps showed that their lives are governed, not by gaseous impurities leaking in or being liberated from the cathode, but by reduction in the pressure of the inert gas through its adsorption by particles dispersing from the cathode. Later designs by Jones and Walsh[6] therefore omitted a getter.

The type of lamp now most used in atomic absorption analysis is the sealed lamp with uncooled hollow cathode, since this lamp is easier to use than one in which there is a current of gas.

Sealed hollow-cathode lamps for atomic absorption analysis are now being produced by a number of Soviet and Western firms. Some workers use hollow-cathode lamps of their own design. Making lamps for oneself offers no special difficulty and can be done with a minimum of experience in the manufacturing of vacuum equipment. In view of this, we shall now briefly consider the procedure for manufacturing the lamps described by Bodretsova, L'vov, Pavlovskaya and Prokof'ev.[7]

The shape and principal dimensions of the lamps are shown in Fig. 2.2.* The lamp-bulb is a thin-walled (1 mm) round-bottomed molybdenum-glass flask 85 mm in diameter and with extensions to which are sealed the electrodes and the tube

* Commercial lamps of this type were first made by Hilger & Watts Ltd in 1957.

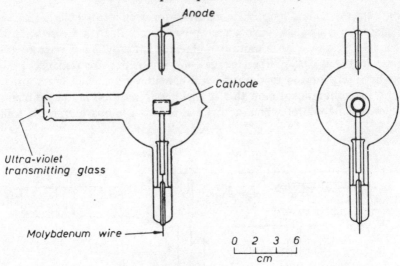

FIG. 2.2. Sealed hollow-cathode lamp for atomic absorption measurements.

through which the light beam passes. The cathode is a cylinder 20 mm long with a central hollow 8 mm in diameter throughout its full length and walls 2·5 mm thick. This cylinder is screwed to a stem 60 mm long and 5 mm in diameter, which is in turn secured to a molybdenum wire 1·5 mm thick. The anode also consists of molybdenum wire. A thin piece of ultra-violet-transmitting glass, which is sufficiently transparent up to 2100 Å, is fixed as a window to the end of the light-beam tube.

When the flask has been made it is baked, washed with a chromic mixture and running water, and finally rinsed out with distilled water. The electrodes are successively washed with benzene and dilute acids and then boiled in distilled water. Electrodes made of non-volatile elements are baked in a high-frequency vacuum furnace. When the electrodes have been sealed in position, the lamp is connected to the vacuum system by means of the extensions and then evacuated; it is degassed by means of a tubular furnace (400°C) or a gas burner (for low-melting-point elements). The lamp is then conditioned with argon or xenon—whichever is the gas with which it will ultimately be filled—since these gases enable the cathode to be well heated with quite low currents. After pretreatment for 20–30 minutes, the lamp is evacuated and the entire pretreatment cycle repeated a further four or five times. The conditioned lamp is finally filled to the optimal pressure with spectrally pure inert gas and then sealed.

The system by which lamps are filled with gases includes an initial vacuum pump and an oil diffusion pump for purposes of evacuation, a U-shaped oil manometer for measuring the pressure of the inert gas, a vacuum gauge for measuring the vacuum, and glass cylinders of spectrally pure inert gases (Ar, He, Xe and Ne) containing not more than 10^{-3} per cent of nitrogen or oxygen.

In the case of low-melting-point elements (indium, gallium, tin, etc.) it is best to use vertically mounted cylindrical cathodes with the aperture for the beam of light in the cathode side wall (Fig. 2.3). The cathode is made of some substance which is not easily atomized. The best substance is obviously aluminium, the spectrum of which contains very few lines. The low-melting-point element, in the form of lumps of metal, is inserted in the cathode.

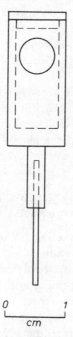

FIG. 2.3. Form of cathode for low-melting-point elements.

When precious metals (silver, gold, or platinum) are used, the cathode is made of a base metal and lined with foil of the precious metal. It is effective to apply certain metals (chromium and manganese) to the surfaces of cathodes by electrolysis. Elements which are easily oxidized (sodium, potassium, etc.) can be applied to the cathode in the form of compounds, but a lengthy conditioning process is then necessary in order to decompose the salts fully and to eliminate gaseous impurities; lamps for such elements have shorter lives than the types normally used, since the small amount of metal is removed rather quickly from the hollow within the cathode.

When lamps with hollow cathodes of calcium are manufactured, they must be conditioned with particular care since the surface of metallic calcium becomes oxidized while the cathode is being treated, and it takes a long time to reduce the calcium oxide.

Hollow-cathode lamps are used with either a.c. or d.c. The voltage drop in the lamp electrodes is 200–300 V and discharge strikes at a voltage of about 300–400 V. The power source for the lamps must therefore provide a voltage not lower than 500 V

Used with an a.c. supply, hollow cathodes emit light throughout the half-cycle in which they are truly cathodes. Usually half-wave or full-wave rectified, but not smoothed, current is used. The signal from the lamp is therefore interrupted, with interruptions at the mains frequency (50 c/s for half-wave current and 100 c/s for full-wave current). Box and Walsh[8] recommended a normal rectifier without a smoothing filter for this purpose; it could be used as either a half-wave or a full-wave rectifier. The open-circuit voltage was 800 V. The current passing through the lamp could be adjusted between 5 and 100 mA. Power was supplied to the rectifier from a standard stabilizer providing a constant voltage of 220 V with fluctuation not exceeding 0·25 per cent.

Stabilized power supplies are used for supplying d.c. to lamps. Currents of up to 100 mA can be provided with maximum voltages of 500 V or more. Of the standard rectifiers produced in the Soviet Union, the UIP-1 universal power supply is suitable for this purpose; it provides a smooth d.c. voltage of up to 600 V with fluctuations of only 0·5 per cent and with a load current of up to 600 mA.

SPECTRAL CHARACTERISTICS OF HOLLOW-CATHODE LAMPS SUPPLIED WITH D.C.

Although the procedure for constructing hollow-cathode lamps presents no particular difficulties, the optimal operating conditions (type of inert gas, its pressure, and the electrical conditions) require special investigation. Until recently, no details of systematic research into the most important parameters of sealed lamps had been published. When experimenters constructed the lamps, they achieved success largely by accident or by intuition rather than by observing the results of systematic preliminary research. Consequently some types of lamps have proved entirely unsatisfactory. For instance, when lamps constructed by different laboratories were used in atomic absorption measurements for manganese, the results varied in a ratio of 3:1.[9]

The most important characteristics of light sources for atomic absorption spectrochemical analysis are life-time, intensity of emission, line-width, signal-to-background ratio, and stability of illumination. In hollow-cathode lamps these characteristics depend on several parameters: the type of gas, its pressure, the electrical conditions, and the design of the cathode.

Lamp life

The life of a lamp is shortened by reduction in the pressure of the inert gas because of its adsorption by metal particles sputtered from the cathode. Of the different types of inert gas, neon is adsorbed more rapidly than the others. Nevertheless, this factor is not of overriding importance in the selection of the gas.

To lengthen lamp life, it is advisable to use lamps with large bulbs. Increasing the volume from 75 to 250 cm^3 lengthens lamp life from 3 to 50 ampere hours.[9] Since the current passing through a lamp seldom exceeds 50 mA, the life of a lamp with a volume of 250 cm^3 will be not less than 1000 hours. The figures given above are

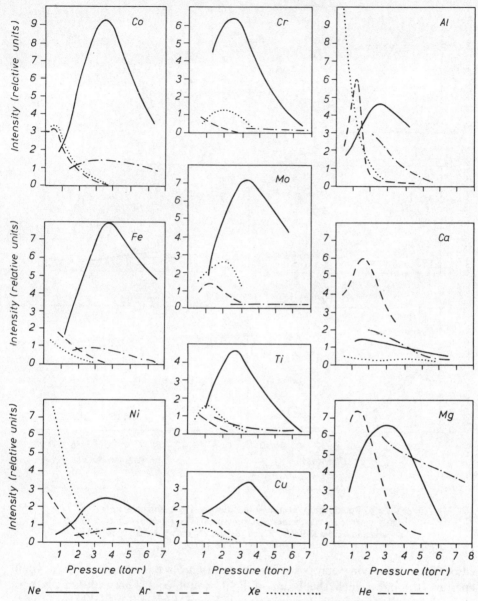

FIG. 2.4. Relationship of the intensities of the resonance lines for low-volatility metals to the pressures of inert gases. The following lines were investigated: Co 3453·5 Å; Fe 3719·9 Å; Ni 3492·9 Å; Cr 4254·3 Å; Mo 3902·9 Å; Ti 3653·5 Å; Cu 3247·5 Å; Al 3961·5 Å; Ca 4226·7 Å; Mg 2852·1 Å.

obviously approximate, since the degree of adsorption depends on the type of gas and the element atomized.

In order to reduce the amount of metal sputtered and consequently to reduce

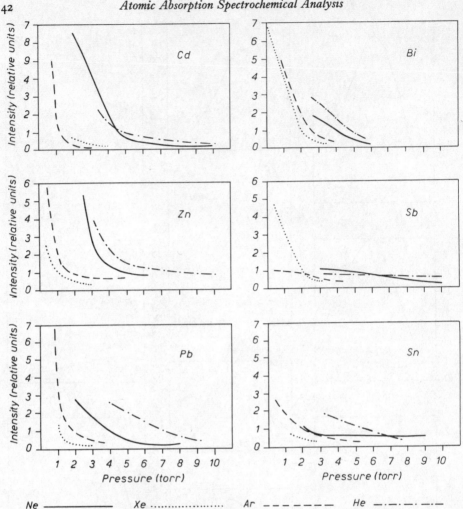

FIG. 2:5. Relationship of the intensities of the resonance lines for highly volatile metals to inert gas pressures. The following lines were investigated: Cd 2288·0 Å; Zn 2138·5 Å; Bi 3067·7 Å; Sb 2311·4 Å; Pb 4057·8 Å; Sn 2839·9 Å.

adsorption, Robinson[10] proposed using semi-enclosed cathodes with a small aperture through which the beam of light would pass. Our experiments with aluminium cathodes proved, however, that lamps of this type are highly sensitive to variations in the pressure of the gas surrounding the cathode; their range of working pressures is narrower and their lives are consequently much shorter. Using semi-enclosed cathodes, the starting potential is higher than with normal cathodes.

Relationship of resonance lamp emission intensity to the type of gas and its pressure

Measurements were made by Crosswhite, Dieke and Legagneur[4] for lamps with iron cathodes and by Bodretsova, L'vov, Pavlovskaya and Prokof'ev[7] for lamps with cathodes of Al, Bi, Ca, Cd, Co, Cr, Cu, Fe, Mg, Mo, Ni, Pb, Sb, Sn, Ti and Zn. The first of these references concerned research with semi-enclosed cathodes 40 mm long and 8 mm in diameter, while the second concerned research with the lamps described above. We shall dwell in greater detail on the results given in the latter reference.

Intensities were recorded photoelectrically over the entire range of inert gas pressures with which there was luminescence in the hollow cathodes. The discharge currents were kept the same throughout the experiments and were 15 mA for the highly volatile elements and 30 mA for low-volatility elements. With each metal, the measurements were made with the same experimental lamp connected to a vacuum system. After each change in pressure, the lamp was conditioned for 10–15 minutes in order to stabilize the intensity of the emission.

As a result of the experiments it was established that:

1. The intensity of the resonance lines of low-volatility metals (Al, Ca, Co, Cr, Cu, Fe, Mg, Mo, Ni and Ti) passes through a maximum as the pressure of argon or neon varies, while the resonance line intensity for highly volatile metals (Bi, Cd, Pb, Sb, Sn and Zn) has no maximum, but increases continuously as the pressure decreases. Figs. 2.4 and 2.5 show how resonance line intensities for the above groups of metals vary with the pressure of the inert gases helium, neon, argon and xenon. The emission maxima correspond to an argon pressure of 1 torr and to a neon pressure of 3 torr.

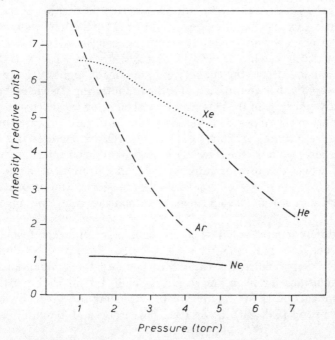

FIG. 2.6. Relationship of the intensity of the lines for atoms of inert gases to the pressure.

When interpreting the effects of the type of gas and its pressure on the line emission of metals in hollow cathodes, allowance must be made for the effects of the gas on the sputtering of the cathode material, on the diffusion of atoms from the cathode, and on the excitation of the metal vapours in the plasma.

The effect of gas pressure on the excitation process can be determined from the variation in the emission intensity of the inert gases in lamps. The graph in Fig. 2.6 shows the results of measuring the intensities of inert-gas arc lines. It follows from this graph that, as the pressure is reduced, emission either remains constant (Ne) or increases steadily (Ar, He, Xe), but that in no case does it decrease.

The loss of atoms of cathode metal by diffusion is inversely proportional to the pressure of the foreign gas. Other conditions being equal, therefore, the concentration of atoms within the cathode must decrease as the pressure drops. On the other hand, reduction in the pressure causes increase in the energy of the ions which bombard the cathode and consequently also causes the temperature of the cathode to increase. Sputtering of the cathode and thermal evaporation of the metal therefore increase. Reducing the pressure thus reduces the concentration of atoms of metal in a hollow cathode owing to diffusional losses but also causes increase in the concentration as a result of the increasing amount of metal entering the discharge. The resultant change in the concentration of atoms within the cathode depends on which of these processes exerts the greater effect.

It should at the same time be taken into account that the number of atoms of high-volatility metals entering the discharge is governed by thermal evaporation, while the number of atoms of low-volatility metals entering it is governed by cathode sputtering. Evidently, the intensity varies in different ways because, as the pressure of the gas in a lamp drops, the thermal evaporation of highly volatile metals increases more rapidly than the amount of cathode sputtering. In the first case, therefore, the relative increase in the amount of metal entering the discharge exceeds the relative increase in the losses by diffusion, while in the second case the reverse is true; starting from some particular pressure, the relative increase in the losses by diffusion exceeds the relative increase in the amount of metal entering the discharge. Accordingly, when the pressure drops, the concentration of atoms of highly volatile metals in a hollow cathode increases, but the concentration of atoms of low-volatility metals increases to some maximum concentration and then begins to decrease.

2. Of the different gases tested—helium, neon, argon and xenon—neon provides the highest resonance line intensity for Co, Cr, Cu, Fe, Mo, Ni, and Ti, argon provides it for Ca, while in the case of Al and Mg the difference between the maximum intensities for neon and argon is small, but in favour of argon. The strongest resonance lines for highly volatile elements are given by lamps filled with argon or neon (Cd, Zn), argon (Sn, Pb), argon and xenon (Bi), and xenon (Sb).

3. The relative intensity of a particular metal's resonance lines in different regions of the spectrum varies according to the inert gas. Table 2.1 gives the

relationship of the intensities (on an arbitrary scale) of the resonance lines Fe I 3720 Å and Fe I 2483 Å.

TABLE 2.1 Ratio of the intensities of the Fe I 3720 Å and Fe I 2483 Å lines, for different inert gases

Gas	Ionization potential (V_i)	Pressure (torr)	Current (mA)	$\dfrac{\text{Fe I 2483 Å}}{\text{Fe I 3720 Å}}$
He	24·58	6·6	100	25
Ne	21·56	3·0	30	14
Ar	15·76	1·2	60	5

The variation is due to the effects of the inert gas on the electron temperature of the discharge plasma. On changing from, say, helium to xenon, the plasma temperature decreases with the ionization potential of the gas and this is bound to result in redistribution of emission in the different regions of the spectrum.

This circumstance explains why, when the resonance lines of highly volatile elements are excited in the far ultra-violet region of the spectrum (Cd I 2288 Å, Zn I 2139 Å), the intensity of the lines in neon is almost the same as in argon, although in view of the sputtering mechanism, the heavy gases should be certain to provide higher emission intensities.

4. Unlike magnesium hollow-cathode lamps filled with argon, those filled with xenon emit an intense Mg II 2796 Å spark line and an extremely weak Mg I 2852 Å arc line. With a current of 20 mA, the ratio of the intensity of the Mg I 2852 Å line to that of the Mg II 2796 Å line is 20 in an argon lamp and 0·038 in a xenon lamp. Thus, the spark line in a xenon-filled lamp is about 500 times more intense than in an argon lamp.

The amplification is associated with the resonance nature of excitation of the Mg$^+$($^2S_{1/2}$) level, due to collisions between unexcited atoms of magnesium and ions of xenon, thus:

$$Mg + Xe^+ \rightarrow Mg^+(^2S_{1/2}) + Xe$$

Actually, the ionization energy of xenon is 12·127 eV, while the total ionization and excitation energy for the Mg II 2796 Å line is equal to 7·644 eV + 4·434 eV = 12·078 eV, i.e. the difference between the energies is only 0·049 eV.

The phenomenon of the amplification of ionic lines by second order collisions was investigated with copper and aluminium in neon as examples.[11] The maximum amplification recorded for the Al (5 ^3P) level was 30.

The effect noted here confirms the substantial part played by second order collisions in the excitation mechanism of metal vapours within a hollow cathode.

Relationship of emission intensity to the current

Similar research has been conducted for sealed lamps with uncooled hollow cathodes.[7] The results of measurements made for certain types of lamp are plotted

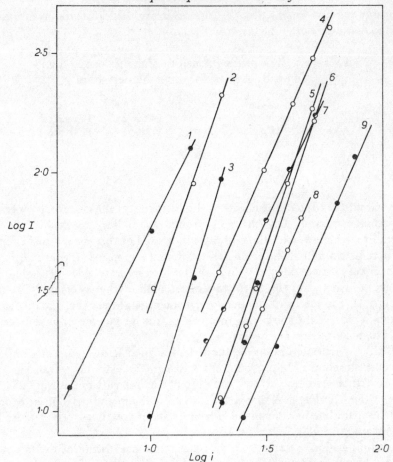

FIG. 2.7. Relationship of resonance line intensity to the current (in mA). 1. Pb 2833 Å (Ar).   2. Zn 3076 Å (Ne).   3. Cd 2288 Å (Ne). 4. Mo 3132 Å (Ne).   5. Ti 3653 Å (Ne).   6. Cr 3578 Å (Ne).   7. Ni 3414 Å (Ne). 8. Bi 3067 Å (He). 9. Ca 4226 Å (Ar).

in Fig. 2.7 on a logarithmic scale. This graph shows that the relationship of emission intensity I to the current i can be expressed as follows:

$$I = ai^n \qquad (2.1)$$

where a and n are specific constants for each combination of cathode and gas.

It should be noted that the power index n in equation (2.1) is much higher on average for neon than for argon (2·7 for neon and 2·1 for argon). Similar results were obtained by Crosswhite et al.[4] The power index n increases for krypton, argon, neon, and helium in that order, the figures being 1·3, 1·9, 2·5 and 2·9 for these gases.[4] These results must be considered in conjunction with the process by which metal enters the discharge, for instance the value of n for the Ne I line is only about 0·5.

Where there is a great deal of self-absorption in a lamp, there are deviations in the relationship of I to i; these show up in the form of deflections of the lines towards the horizontal axis. Such deflections, in turn, are an indication of self-absorption.

Relationship of the width of the resonance lines to the current and the type of gas[7]

Measurements were made by recording line profiles with a type IT-28-30 Fabry–Perot interferometer combined with a ZMR-3 monochromator. Fig. 2.8 is a block diagram of the apparatus. The central spot of the interference pattern was isolated by a diaphragm and its varying intensity was recorded as the pressure was varied in the pressure chamber housing the etalon. The pressure was altered at the constant rate necessary to ensure a linear frequency scale by allowing gas to enter the exhausted pressure chamber from a compressed gas cylinder through a special reducing valve. For pressures between 0 and 0·5 atm, the recording was linear with an accuracy of about 1·5–2·0 per cent. The interferometer mirrors had reflection factors of 0·81 ($\lambda=2852$ Å), 0·83 ($\lambda=3187$ Å), and 0·85 ($\lambda=3885$ Å).

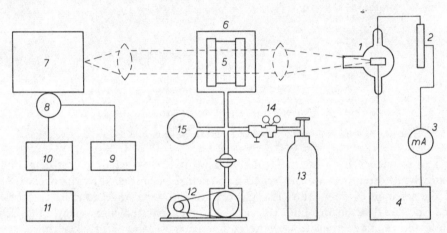

FIG. 2.8. Diagram of the interferometric apparatus. 1. Hollow-cathode lamp. 2. 5 k rheostat. 3. Milliammeter. 4. UIP-1 power supply. 5. IT-28–30 interferometer. 6. Pressure chamber. 7. ZMR-3 monochromator. 8. FEU-18 photomultiplier. 9. VS-22 rectifier. 10. UI-2 electrometer amplifier. 11. EPP-09 automatic recorder. 12. VN-461 pump. 13. Compressed gas cylinder. 14. Reducing valve. 15. Ballast cylinder.

The half-widths of resonance lines were measured for lamps with hollow cathodes of different metals, filled with inert gases providing the maximum emission intensity. The experiments for magnesium and nickel were made with lamps filled with different gases.

Table 2.2 contains the results of measuring half-widths with the minimum currents with which emission intensity could still be measured accurately enough with the interferometer. No allowance was made for the instrumental half-width; all the values given in the table are, therefore, higher than the true resonance-line half-widths and can thus be used only for gauging the general character of broadening.

TABLE 2.2 Half-widths of resonance lines in hollow-cathode lamps

Line (Å)	Gas	Pressure (torr)	Current (mA)	Half-width (cm^{-1})
Al I 3092·7	Ar	1·0	6	0·17*
Al I 3082·2	Ar	1·0	6	0·18
Bi I 3067·7	He	4·6	20	0·22*
Ca I 4226·7	Ar	1·6	25	0·13
Cd I 2288·0	Ne	3·0	10	0·35
Cd I 3261·1	Ne	3·0	6	0·10
Co I 2407·3	Ne	3·0	30	0·44
Cr I 3578·7	Ne	3·9	25	0·14
Cu I 3247·5	Ne	3·9	5	0·12*
Cu I 3274·0	Ne	3·9	5	0·14*
Fe I 2483·3	Ne	3·9	15	0·28
Fe I 3719·9	Ne	3·9	15	0·1
Mg I 2852·1	Ar	1·6	5	0·17
Mg I 2852·1	Xe	1·0	10	0·29
Mo I 3132·6	Ne	4·6	20	0·1
Ni I 3414·8	Ne	3·0	20	0·14
Ni I 3414·8	Xe	1·0	15	0·48
Ni I 3524·5	Ne	3·0	20	0·14
Ni I 2320·0	Ne	3·0	20	0·24
Pb I 2833·1	Ar	1·2	5	0·49
Sn I 2863·3	Ar	1·2	15	0·52*
Ti I 3653·5	Ne	3·9	20	0·14
Ti I 3642·7	Ne	3·9	30	0·14
Zn I 3075·9	Ne	3·3	10	0·12

* Half-widths were measured for the hyperfine components with the greatest intensities.

The resonance lines for most of the elements investigated consist of a single component (often because the hyperfine structure is concealed by the overlapping of the separate components). The Cu 3248 Å and 3274 Å, Al 3093 Å, Bi 3068 Å, and Sn 2863 Å resonance lines consist of several separated components (Fig. 2.9), while the relative intensities of the components depend largely on the current passing through the lamp. As an instance, the ratio of the strong to the weak component of the Al 3093 Å line as the current increases from 5 to 30 mA decreases from 4·8 to 2·0.

As the current passing through a lamp is increased, the line half-width increases. With the lines Ni 3524 Å and 3415 Å (in Ne), Ti 3653 Å and 3642 Å, Bi 3068 Å, Fe 3720 Å and Zn 3076 Å, an increase by two or three times in the current does not increase the line-width to more than 15 per cent above the figures given in the table; the increase is, however, 20–50 per cent for the lines Ni 2320 Å, Pb 2833 Å, Cd 3261 Å, Ca 4227 Å, Cr 3579 Å, Co 2407 Å, and Mo 3133 Å. The greatest increase in width, 100–200 per cent is for the Cd 2288 Å, Cu 3248 Å, Mg 2852 Å (in Ne), and Al 3093 Å.

The principal reason for line-width to increase with current is self-absorption. The Doppler effect cannot cause more than a 20–30 per cent increase in line-width when the temperature of the cathod increases between 500 and 800°K. Self-absorption also explains why the relative intensity of the components of hyperfine structure alters; this has been noted by a number of research experts, particularly by Turkin.[12] The

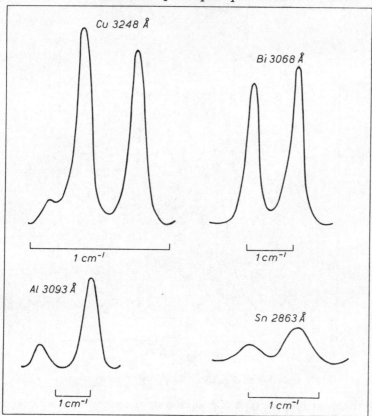

Fig. 2.9. Shapes of resonance lines with several components.

Cd 2288 Å, Cu 3248 Å, and Mg 2852 Å lines undergo appreciable self-reversal as the lamp current increases. Fig. 2.10 shows, as an example, the Mg 2852 Å line profiles with different currents. Self-reversal first becomes evident at 20 mA. With a current of 35 mA, self-reversal is so great that the line appears to consist of two components. Self-reversal occurs because there is, around the cathode, a cooler cloud of vapour from the element, the product of intensive cathode sputtering.

For the same current, the half-width of lines from xenon-filled lamps is greater than the half-width from argon-filled or neon-filled lamps.

For instance, in a xenon-filled lamp, the Mg 2852 Å line is 50 per cent wider than in an argon-filled lamp. The difference between the xenon-lamp and neon-lamp half-widths of the Ni 3415 Å line is even greater; with a current of 40 mA, these half-widths are in the ratio of 4·3 :1. This considerable difference obviously occurs because cathode sputtering is more violent in xenon than in argon or neon, particularly the latter. In xenon, the heating of the cathode is also much greater; in xenon a nickel cathode becomes red hot with a current of 40 mA but it requires a current of 80 mA to heat it equally in neon. The intensity of Ni 3415 Å line emission (at its maximum) is approximately the same, with the same current, for either

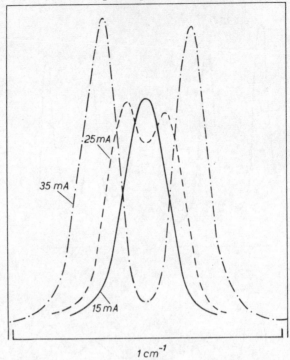

FIG. 2.10. Mg 2852 Å line profiles for different currents.

gas. When the most suitable type of lamp for atomic absorption is being selected, therefore, the line half-width should be used as the guide. Xenon is less suitable than neon or argon from the point of view of achieving the least self-absorption.

Signal-to-background ratio

The band-width isolated by a monochromator is much greater than the width of a resonance line. Background emission in the isolated band is therefore superimposed on the signal from the line.

The intensity of the continuous background, caused by the electron continuum is not great in hollow-cathode lamps. The principal danger is from foreign lines belonging to the cathode material or the gas filling of the lamp and lying in the isolated band.

Of the inert gases, helium, argon, and neon, helium has the simplest spectrum. The strongest lines in the ultra-violet spectra of these three gases, as observed in hollow-cathode emission, are listed in Table 2.3.

The spectrum of argon consists of a comparatively small number of lines, mostly concentrated above 3500 Å. In the ultra-violet region, neon has a much more complex spectrum, particularly between 3200 and 3700 Å. The neon lines are therefore much more likely to interfere with resonance lines of metals in the ultra-violet region of the spectrum than the argon lines. This conclusion is contrary to the conclusion drawn by Robinson[10] and David.[13]

TABLE 2.3 Ultra-violet spectra of gases in hollow-cathode lamps

Ne (Å)		Ar (Å)	He (Å)
II 2792·05	I 3369·91	II 2891·61	I 2945·10
II 2955·73	I 3375·64	II 2942·90	I 3187·74
II 3001·65	II 3392·78	II 2979·05	I 3613·64
II 3198·62	I 3417·90	II 3491·54	I 3819·61
II 3218·21	I 3447·70	II 3509·78	I 3888·65
II 3230·16	I 3450·76	II 3514·39	I 3964·73
II 3232·28	I 3464·34	II 3576·62	
II 3244·15	I 3466·58	II 3729·29	
II 3297·44	I 3472·57	II 3850·57	
II 3309·78	I 3501·22	I 3948·08	
II 3319·45	I 3520·47		
II 3323·75	II 3568·53		
II 3327·16	II 3664·11		
II 3334·87	II 3964·20		
II 3344·43	II 3713·08		
II 3355·05	II 3727·08		

In order to measure reliably large absorbances (up to 2) and to obtain a linear relationship between absorbance and the concentration of atoms in an absorbing atomic vapour, the intensity of the useful signal must be at least 100 times that of the background in the band-width isolated for measurements.

Table 2.4 gives ratios of the intensity of resonance lines to the minimum background on both sides of these lines, for several types of lamp.

TABLE 2.4 Signal-to-background ratio for certain resonance lines

Cathode	Gas	Pressure (torr)	Current (mA)	Resonance line (Å)	Spectral width of slits* (Å)	Signal/background ratio Short wavelength side	Long wavelength side
Iron	Ne	3	30	Fe 3719·93	3·0	48	16
			60	Fe 2483·27	1·0	50	25
Cobalt	Ne	3	40	Co 3526·85	1·5	11	1·9†
			60	Co 2407·88	1·5	18	12
Nickel	Ne	3	60	Ni 3414·77	2·1	80	4·3‡
			60	Ni 2320·03	1·7	21	6·7§
Aluminium	Ar	1	25	Al 3092·71	2·5	210	430
Brass	Ne	3	10	Cu 3273·96	6·0	270	270
Brass	He	7·7	70	Zn 2138·56	1·0	100	280

* The fact should be taken into consideration that the total band-width isolated by a monochromator, with equal entrance and exit slits, is double the spectral width of either slit.
† Superimposition of the Co 3529·03 Å line.   ‡ Superimposition of the Ni 3417·90 Å line.
§ Superimposition of the Ni 2321·38 Å line.

The figures in Table 2.4 show that there is an abrupt difference between the backgrounds on the two sides of the lines of cobalt and nickel; this is due to the superimposition of closely adjacent lines of these elements. The calibration curves must be distorted in consequence, even for medium values of absorbance.

Stability of emission intensity

Lamp stability depends on steady output by power sources and the individual lamp characteristics.

It was established earlier that intensity is related to current by equation (2.1), where $n = 2.5$; in order, therefore, to eliminate fluctuation in the intensity I (more than ± 0.5 per cent), steps must be taken to ensure that the current i passing through the lamp is stable to within ± 0.2 per cent.

On the other hand, unstable emission may be caused by internal processes of some sort in the lamp, associated with the shape of the cathode, the material of which it is made, the gas with which the lamp is filled, and thermal effects. Deviation from steady conditions represents a variation in the internal resistance R_l. With a constant voltage drop V_1 at the electrodes, this implies variation in the current passing through the lamp; fluctuations in the emission accordingly take place. The fluctuations in the current can be reduced if a ballast resistance R_b is connected in series with the lamp.

With V, the power supply voltage, constant,

$$\frac{V}{i} = R_b + R_l \qquad (2.2)$$

If we differentiate equation (2.2), we get:

$$\frac{V\,di}{i^2} = -dR_l$$

or for finite variations, and ignoring the sign,

$$\frac{V\,\Delta i}{i^2} = \Delta R. \qquad (2.3)$$

Hence we finally get

$$\frac{\Delta i}{i} = \frac{\Delta R_l}{R_b + R_l} \qquad (2.4)$$

Suppose a lamp resistance to fluctuate by 1 per cent, then the introduction of a ballast load equal to twice the lamp resistance ($R_b = 2R_l$) must reduce the fluctuations to 0.3 per cent. In order, therefore, to achieve maximum stabilization of the discharge conditions, the biggest possible ballast loads should be used, with the maximum power-supply voltages. For this reason, the current should be regulated not by altering the power-supply voltage, but by changing the ballast resistance.

Fluctuations in hollow-cathode emission caused by the internal properties of the lamps, must be measured under conditions in which no other sources of fluctuation play an important part, i.e. in which the lamps are powered from highly stable supplies, detector and recording circuit noise are much lower than the quantities measured, and any possible shifts of the lamp relative to the detector owing to vibration, etc., are eliminated.

Box and Walsh[8] and Gatehouse and Willis[14] conducted similar research, but their measurement conditions were arbitrary. The first two authors established that internal fluctuations take place in lamps, that they amount to about 0.25 per cent of the signal, and that they could not be eliminated by using more stable power supplies. It will, however, be shown on pp. 95–103 that fluctuations of this magnitude are most probably due to shot noise from the photomultiplier, and this assumption has been supported by experiment.[7]

Measurement of fluctuations in the intensity of the Fe 3720 Å line, with different luminous fluxes reaching the photomultiplier, has proved that the recorded noise is associated with shot noise in the detector, not with internal noise in the lamp. If the luminous flux, and consequently also the photo-current is increased 10^3 times by opening the monochromator slits, the relative signal fluctuations drop by about thirty times; this agrees satisfactorily with the statistical characteristics of shot noise.

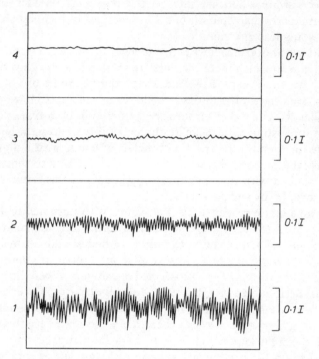

Fig. 2.11. Signal recordings with different levels of luminous flux. Frequency transmission band half-width 3 c/s. 1. $5 \cdot 10^{-15}$ amp; 2. $3 \cdot 10^{-14}$ amp; 3. $2 \cdot 10^{-13}$ amp; 4. $4 \cdot 10^{-12}$ amp.

Fig. 2.11 shows typical recordings of signals, made with different levels of luminous flux from a lamp with a hollow iron cathode; the lamp was neon-filled and the current passing through it was 40 mA. The time constant of the amplifier input circuit was 0·1 sec, and consequently the bandpass was 3 c/s. With the maximum luminous flux measured, which corresponded to a photo-current of $4 \cdot 10^{-12}$ amp, the fluctuation in the recording was $\simeq$0·1 per cent of the magnitude of the signal. It can therefore be assumed that lamp noise cannot exceed 0·1 per cent.

The higher level of noise (up to 1 per cent) recorded by Gatehouse and Willis[14] for certain hollow-cathode lamps was evidently due to shot effect, which was not considered. This opinion is confirmed by the fact that Gatehouse and Willis recorded different fluctuations in emission when recording different lines in the spectrum of the same source: the greatest fluctuations affected the weakest lines.

HOLLOW-CATHODE LAMPS SUPPLIED WITH PULSED CURRENT

The greatest limitation of hollow-cathodes lamps, supplied with direct current, is the self-absorption of resonance lines. For this reason, the discharge current through these lamps must be as low as possible (see Table 2.2), and often, as a result, the emission is insufficient for making precise measurements, or alternatively it is necessary to compromise and use intense, but purposely broadened, lines. This situation is particularly characteristic of easily vaporized elements such as cadmium, copper, magnesium, lead and zinc.

Self-absorption of resonance lines is due to vigorous sputtering and the relatively low effectiveness with which metal vapours are excited in a d.c. glow discharge. On the other hand, we know that the regions around the electrodes of a high-frequency glow discharge are very similar to the cathode regions of a d.c. glow discharge, and that in particular the internal electrodes are sputtered in both types of discharge.[15] It is therefore interesting to investigate the possibility of using a high-frequency glow discharge for exciting the spectra of metals in lamps with hollow electrodes, and to compare the spectral characteristics of these lamps with those of the same lamps when supplied with d.c. The first experiments of this nature were carried out by Bodretsova, L'vov and Mosichev.[16]

The first lamps investigated were sealed lamps with uncooled hollow cathodes, of the type described earlier. We investigated the characteristics of lamps filled with argon to a pressure of 1 torr and with hollow cathodes made of magnesium, and also of lamps filled with neon to a pressure of about 4 torr, with hollow aluminium cathodes. These two metals were not selected at random. When d.c. is supplied to lamps with magnesium cathodes, the degree of self-absorption is excessively great. Among the other metals, aluminium has the characteristic of being highly resistant to sputtering in a discharge. Doubts therefore arose regarding the possibility of exciting metal spectra in lamps supplied with high-frequency current.

We established, in the preliminary experiments, that lamps with semi-enclosed cathodes provide spectra which are many times more intensive than those provided by lamps with through hollow cathodes, and for further experiments we used lamps of the former type.

We found it best to compare the characteristics of lamps supplied with d.c. and high-frequency current by comparing the emission of resonance lines with the same degree of self-absorption in the discharges. The measurements were made with the interferometer apparatus described in the previous section.

The width of the resonance lines in hollow-cathode lamps is governed by the Doppler effect and self-absorption. Changes in temperature (between 400 and 600°K for uncooled hollow cathodes) have practically no effect on the overall width of lines. Variations in line half-width accompanying changes in the discharge conditions are therefore governed solely by the different degrees of self-absorption. In cases in which a line undergoes hyperfine splitting, it is best to gauge the degree of self-absorption by the recorded ratio of intensities of the components of the hyperfine structure. The ratio between these components is bound to approach unity as self-absorption increases.

The high-frequency discharge was maintained by means of a generator operating at a frequency of about 60 Mc/s.[15] The d.c. discharge was obtained from a universal UIP-1 power source providing a constant stabilized voltage of up to 600 V.

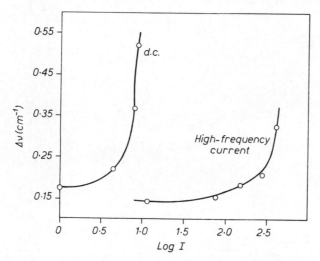

FIG. 2.12. Intensity of the Mg 2852 line, and its width, when d.c. and high-frequency current were supplied to the lamp.

Fig. 2.12 shows the relationship of the recorded half-width of the Mg 2852 Å line to the integral intensity for d.c. and high-frequency current discharges. In the former case the current through the lamp was varied between 10 and 40 mA, while in the latter case the power input into the high-frequency generator was varied between 10 and 50 W.

In these experiments the error in measuring line half-width, estimated by the reproducibility of the results, did not exceed 0·004 cm^{-1}, and the error in measuring intensity did not exceed 3 per cent. It should be noted that the true line half-width is less than shown in Fig. 2·12, since no allowance was made for the broadening of the lines by the apparatus. Since, however, the apparatus half-width remained constant during the experiment, the recorded variation in width was almost solely governed by the variation in self-absorption.

It follows from the curves in Fig. 2·12 that the use of a high-frequency discharge provides a gain in emission intensity, with the same line width, of more than 100:1 by comparison with a d.c. discharge. Even with the minimum discharge intensities suitable for making measurements in a d.c. discharge, the Mg 2852 Å resonance line is wider than in a high-frequency discharge, i.e. self-absorption is not eliminated in a d.c. discharge however low the current passing through the tube.

The later experiments proved that not only spectra of other highly volatile metals (In, Sn Pb, Cu, etc.), but also the spectra of the metals which are most difficult to sputter (Fe and Al), are intensively excited in high-frequency discharges. Fig. 2.13 gives the results of measuring the width of the Al 3093 Å line, and the ratio of the components of hyperfine splitting of this line, with relation to the integral intensity of emission I_Σ. The current was varied between 10 and 120 mA, the power input into the high-frequency generator between 16 and 60 W.

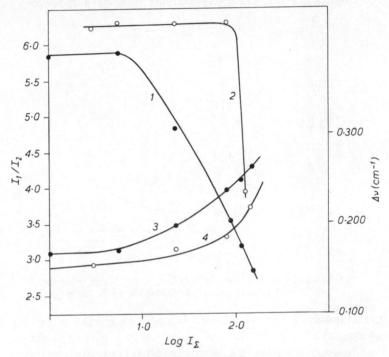

FIG. 2.13. Intensity, width and ratio between hyperfine structure components for the Al 3093 Å line in a d.c. discharge (full circles) and a high-frequency current discharge (open circles). 1 and 2 are the ratio of the hyperfine structure components. 3 and 4 are the half-widths of the more intensive component.

The Al I 3093 Å line ($3^2 P°_{3/2} - 3^2 D_{3/2,5/2}$) line consists of two components corresponding to transitions from the upper levels $3^2 D_{3/2}$ and $3^2 D_{5/2}$ to the lower level $3^2 P°_{3/2}$. The distance between the components of the line is 1·34 cm^{-1} (with an etalon constant of 1·0 cm^{-1}, these components are a distance of 0·34 cm^{-1} apart on the interferogram).

We can see from the curves in Fig. 2.13 that, whatever the current, the ratio between the components of the Al 3093 Å line in a d.c. discharge comes below the equivalent ratio for a high-frequency discharge. This fact confirms that self-absorption takes place in a d.c. discharge, even in the case of an element so difficult to sputter as aluminium.

With the same level of self-absorption, the maximum gain in intensity is about 25.

On the basis of the research described above, the simplified lamp designs shown in Fig. 2.14 were produced and tested. The distinctive feature of these lamps is that

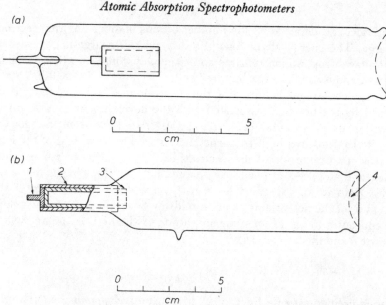

FIG. 2.14. High-frequency discharge lamps with (*a*) uncooled, and (*b*) cooled, hollow cathodes. 1. Copper bottom. 2. Kovar alloy (28 per cent Ni, 18 per cent Co, and 54 per cent Fe) cylinder. 3. Tip made of metal sputtered. 4. Ultra-violet-transmitting glass end.

they are smaller than the types normally used and there is no second electrode. Actually the comparatively large dimensions of the sealed hollow-cathode lamps normally used are largely associated with the adsorption of gas by the layer of metal deposited on their walls, and this in turn governs the life of the lamp. The fact that metal is not intensively sputtered in a high-frequency discharge enabled us greatly to reduce the dimensions of lamps without to any extent shortening their lives.

The high-frequency lamp with the electrode sealed into its envelope is particularly interesting. In these lamps the electrode can be cooled considerably in order to reduce Doppler line broadening.

Measurements of the Ne I 3369 Å line profile made with an interferometer and of the line excited in a lamp of this design show that, with a power input of 32 W into the high-frequency generator, the line half-width is 0·036 cm^{-1} less when liquid nitrogen is used for cooling, than with no cooling. With an uncooled electrode temperature of 400°K, this difference corresponds to a change of 250°K in temperature. The temperature within a lamp electrode cooled with liquid nitrogen ($T_{boiling} = 77$°K) is thus 150°K, i.e. it only differs by 70°K from the cathode wall temperature.

The most important result of the research described above is that, with the same emission, less metal is sputtered in a high-frequency discharge than in a d.c. discharge. Let us discuss this special feature of the high-frequency discharge on the basis of modern ideas regarding the mechanism by which electrode metal is sputtered and spectra excited.

As is the case with a d.c. discharge, the atomization of the cathode substance in high-frequency discharges is caused by positively charged ions striking the wall of the cathode, and depends on the number of ions bombarding the electrode and their energy. The energy of the ions in a d.c. glow discharge is governed by the cathode voltage drop, which remains approximately constant when the discharge power alters, and is usually a few hundred volts.

The energy of the ions bombarding the electrode in a high-frequency discharge is governed by the potential associated with the development of a positive space charge within the lamp. The formation of this charge is due to the fact that the mobility of electrons in a high-frequency field is greater than that of the ions, and for this reason a proportion of the electrons leave the discharge gap and enter the electrodes, and an excess of positively charged ions develops in the discharge.[17] A theoretical estimate of the effect leads to the following relationship of the d.c. component v_o of the potential for a point midway down the gap between electrodes d to the concentration n_p of ions, the amplitude of the high-frequency voltage v_m, the pressure P and the frequency f:[17]

$$v_o \alpha \frac{n_p v_m^2}{f^2 P^2 d^2}$$

The measurements made by Levitskii[17] qualitatively confirmed the effects of these parameters on the potential, and proved that, with gas pressures above 0.4 torr, the space potential is not above 20–30 V, even with high discharge powers. Thus the energy of the ions bombarding the electrode in a high-frequency discharge, and consequently the number of dispersed atoms per ion, must be at least an order lower than in a d.c. discharge. From this we are able to understand why a smaller amount of electrode metal is sputtered in high-frequency discharges than in d.c. discharges, with the same spectrum excitation effectiveness.

Another feature of the high-frequency discharge is the purely electronic mechanism by which spectra are excited. Actually, when high-frequency current is supplied to a lamp, filled with xenon, with a magnesium cathode, there is no anomalous 'amplification' of the ionic Mg 2796 Å line owing to second order collisions (see p. 45), and the relationship between the Mg 2852 Å and Mg II 2796 Å line intensities is almost the same as the ratio of the intensities for a lamp filled with argon.

As opposed to d.c. discharges, therefore, second order collisions play no part in the excitation of luminescence in high-frequency discharges. This conclusion is in agreement with the generally accepted point of view on the mechanism of spectrum excitation in high-frequency plasma.

Dawson and Ellis[18] proposed a method of supplying hollow-cathode lamps with current which was even more effective than high-frequency current, i.e. high-amplitude-pulse current, and which provided an improvement in resonance line strength in comparison with the optimum d.c. power supply conditions; the improvement was between 50 and 800 times, depending on the type of metal. At the same time, comparative measurements of analytical sensitivity in atomic absorption showed that there was little line broadening. The reason is that there is no thermal vaporization of the metal, since the effective current passing through the lamp does not, owing to the long intervals between pulses, exceed a few milliamperes.

It should be noted that the experimental procedure used by Dawson and Ellis[18] for investigating the spectral characteristics of lamps suffers from some shortcomings. Lamp luminosities were compared photographically, without taking into consideration the intervals between power pulses or the possible influence of the Schwartzschild effect on the results of assessing luminosity. There are also serious objections to assessing line-widths when measuring low values of absorbance, since the effect of line broadening is generally less for low values of absorbance than for medium or high values (see pp. 22–31).

In view of this, Katskov, Lebedev, and L'vov compared the spectral characteristics of lamps to which d.c. and pulsed currents were supplied, basing their comparison on photoelectric measurements of luminosity and the recording of resonance line profiles with an interferometer.

The pulsed system used in the work described by Dawson and Ellis [18] for power supply to hollow-cathode lamps included a square-pulse generator, a power amplifier, and an amplifier-power-supply unit. Owing to the special features of the optical system in the spectrophotometer used in our own measurements (see pp. 201–13), we used a photoresistor illuminated by an incandescent lamp instead of the square pulse generator. The beam of light was chopped by a rotating sector

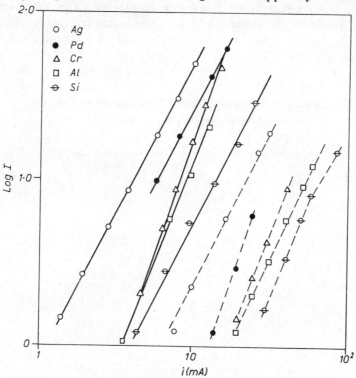

FIG. 2.15. Relationship of the luminosity of the Ag 3281 Å, Pd 2467 Å, Cr 3579 Å, Al 3092 Å and Si 2516 Å resonance lines to the effective value of the current passing through the lamp, for different methods of power supply. Pulsed current—the continuous lines; d.c.—the broken lines.

mounted alongside to act as pulse generator. The almost square pulses emitted from the photo-resistor as the sector-disk rotated passed to the power amplifier unit, which had the same circuit as in reference 18 and was supplied with power from a standard power source with an output of about 600 V. The pulse frequency was about 300 c/s. The pulse interval (the ratio of the repetition period to the pulse duration) was governed by the shape of the slots in the disk, and was equal to 5. The effective current passing through the lamp was measured with a milliammeter connected in series with the lamp. The amplitude of the pulse currents was calculated from the oscillograph-measured voltage drops in the 10 Ω resistance in the lamp circuit. As was to be expected, the amplitude proved to be five times the effective current. The signals were recorded using an amplifier with a synchronous detector.

Fig. 2.15 gives the results of measuring the resonance line strengths for some hollow-cathode lamps, under both power supply conditions, relative to the d.c. or effective (mean) values of the current.

As before (pp. 40–54), the line-widths of the same resonance lines were measured by recording the profiles by means of a Fabry–Perot interferometer, scanning the interference pattern by varying the pressure. The results are given in Table 2.5. It should be noted that, when the measurements were made, the apparatus half-width was not taken into account; all the given values are therefore greater than the true resonance line half-width and it is thus possible only to estimate the order of the change in line-width as the discharge conditions vary.

TABLE 2.5 Resonance line half-widths under different lamp power-supply conditions

Lines Å	D.C.		Pulsed current		
	i (mA)	$\Delta\nu$ (cm^{-1})	i_{mean} (mA)	i_{ampl} (mA)	$\Delta\nu$ (cm^{-1})
Ag 3281			5	25	0·130
	10	0·137	10	50	0·172
	20	0·196	20	100	0·246
	30	0·246			
Al 3093	20	0·160	20	100	0·164
Cr 3579	25	0·130	30	150	0·132
Mo 3133	40	0·119	44	220	0·172
	60	0·136			
	75	0·154			
Ni 3415	30	0·130	27	135	0·134
Rh 3435	10	0·090	10	50	0·105
	20	0·115	20	100	0·121
Si 2516	75	0·220	40	200	0·213

If we consider the figures in Table 2.5 we find that, with equal mean currents passing through the lamp, the resonance line half-width is slightly greater under the pulsed conditions than with d.c. This is bound to be the case, since the cathode material enters the discharge, not only as a result of the thermal effect, which is the same under the conditions concerned, but also owing to cathodic sputtering, which depends on the amplitude of the current.

In view of this, it is inadvisable to increase lamp luminosity by increasing pulse amplitude above some limiting value specific for each metal–gas combination. Appreciable resonance-line broadening takes place at currents of 200 mA in lamps with molybdenum cathodes, while the same occurs at 50 mA in lamps with silver cathodes. Dawson and Ellis[15] reached the same conclusion; they found that increasing to 600 mA the current through a lamp with a copper cathode halved the sensitivity of an atomic absorption measurement for copper.

Nevertheless, if we compare the quality of the resonance lines in both strength and width, with reasonable values of lamp current, it can safely be stated that, for the same or nearly the same mean current, line-width is practically independent of the method of power supply (i.e. whether d.c. or pulsed). That being so, it is legitimate to compare lamp luminosities for different power supply conditions, but with the same values of mean current. Table 2.6 gives the results of this type of assessment for several resonance lines. It can be seen that the improvement in strength varies widely according to the metal—between 20:1 and 130:1. Variation results from differences in the relationship between the intensity I of the lines and the current i. If, as before (p. 46), this relation is put in the form of the function

$$I = ai^n \qquad (2.1)$$

where a and n are constants characteristic of each combination of cathode and gas, we find that the exponent n (i.e. the slope of the lines in Fig. 2.12) varies, for the lamps investigated, from 1·8 for Ni to 3·0 for Pd (Table 2.6).

TABLE 2.6 Improvements in resonance line strength with pulsed power supply to the lamps

Lines (Å)	i_{mean} (mA)	i_{ampl} (mA)	$\left(\dfrac{I_{pulsed}}{I_{d.c.}}\right)_{meas}$	n	$\left(\dfrac{I_{pulsed}}{I_{d.c.}}\right)_{calc}$
Ag 3281	8	40	23	1·9	21
Al 3093	20	100	55	2·5	56
Co 2407	10	50	36	2·3	41
Cr 3579	20	100	102	2·9	107
Mo 3133	30	150	21	1·9	21
Ni 3415	15	75	20	1·8	18
Pd 2476	10	50	132	3·0	125
Rh 3435	10	50	44	2·3	41
Si 2516	30	150	26	2·0	25

The right-hand column in Table 2.6 contains the results of calculating the anticipated improvement in strength, with the same mean lamp currents, from the equation

$$\frac{I_{pulsed}}{I_{d.c.}} = \left(\frac{i_{pulsed}}{i_{d.c.}}\right)^n \qquad (2.5)$$

In general, the validity of this equation depends on the extent to which the constants a and n remain unchanged when different lamp power-supply conditions are used.

The fact that the results of the calculations agree with the observed improvements indicates that equation (2.5) is equally valid for both power-supply conditions and over a far wider range of current variation than is shown in Fig. 2.15 for each of the supply conditions separately.

It is interesting to compare the results of this assessment with the improvements measured for certain types of lamp by Dawson and Ellis. It should be taken into consideration that the improvement values given by Dawson and Ellis[18] relate not to line strengths but to the amount of energy reaching the photographic plate during a prescribed measurement time. Taking a pulse duration of 15 μs and a pulse frequency of 300 c/s, we find that, for pulsed power-supply conditions, the effective exposure time is $2 \cdot 2 \times 10^2$ times less than for a d.c. supply. The improvement in emission strength under pulsed conditions is therefore the same number of times greater than the improvement figures given by Dawson and Ellis.[18]

Improvements in strength derived from the figures given by Dawson and Ellis[18] are entered in the penultimate column of Table 2·7, and the results of calculating improvements using equation (2.5) are given in the right-hand column. In every case, the value of the index n was taken as 3.

TABLE 2.7 Comparison of the improvement in the amount of light measured by Dawson and Ellis[18] with the calculated improvement in luminosity

Lines (Å)	$i_{d.c.}$ (mA)	i_{ampl} (mA)	Increase in amount of light[18]	Improvement in luminosity[18]	Improvement in luminosity found using equation (2.5)
Ca 4227	15	620	800	$178 \cdot 10^3$	$71 \cdot 10^3$
Co 3527	30	570	50	$11 \cdot 10^3$	$6 \cdot 8 \cdot 10^3$
Cu 3248	15	50	400	$89 \cdot 10^3$	$37 \cdot 10^3$
Mg 2852	15	500	400	$89 \cdot 10^3$	$37 \cdot 10^3$
Mn 2798	30	600	50	$11 \cdot 10^3$	$8 \cdot 0 \cdot 10^3$
Sr 4607	15	590	600	$130 \cdot 10^3$	$61 \cdot 10^3$

Comparison of the observed and calculated values reveals a systematic mean discrepancy of a factor of about 2. In our opinion this is caused by not allowing for the Schwartzschild effect, namely the relationship of photographic emulsion sensitivity to exposure time. We know (see, for instance, Nagibina and Prokof'ev[19]) that the maximum sensitivity of photographic film corresponds to an exposure of $\simeq 10^{-2}$ sec, and that as the exposure time is increased the sensitivity steadily decreases. In the experiments conducted by Dawson and Ellis,[18] the exposure time when the d.c. conditions were used was $2 \cdot 2 \times 10^2$ times the exposure time with the pulsed conditions. It is therefore perhaps valid to attribute the discrepancy indicated above to excessive values of improvements given by photographic measurements.

Let us discuss the practical value of the pulsed method of power supply to hollow-cathode lamps as applied to atomic absorption measurements. This method of power supply improves the strength of resonance lines by two orders of magnitude, on the average, without affecting line-widths. The pulsed method has the further important advantages of simplicity, particularly in comparison with the more expensive and complex boosted output tubes (pp. 69–71), low noise level (in comparison with the high-frequency discharge), the high stability and, as has been pointed out,[18] less critical lamp sensitivity to gas pressure and consequently the possibility of longer lamp life. We should not forget in the end that it is necessary for the recording circuit to use a synchronous detector (which is, in any event, advisable in many other respects).

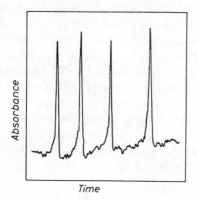
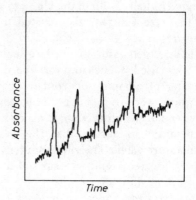

FIG. 2.16. Measurements of the atomic absorption of chromium, recorded under different lamp power-supply conditions: 1. Pulsed current ($i_{mean} = 30$ mA); 2. d.c. ($i = 30$ mA).

In conclusion, recordings of the atomic absorption determination of 1.10^{-10} g of chromium from the Cr 3579 Å line in a graphite cuvette are given in Fig. 2·16 to illustrate the advantages of pulsed power supply over the normal d.c. supply. Not only the considerably lower noise level attracts attention, but also the increased absorption, which results from the high monochromacity of the light (i.e. narrow lines and low noise).

RADIO-FREQUENCY AND METAL-VAPOUR LAMPS

Like hollow-cathode lamps, radio-frequency lamps are now extensively used in high-resolution spectroscopy.[2] Up to the present, they have been undeservedly neglected in atomic absorption spectroscopy. The great advantage of a radio-frequency lamp is ease of manufacture, since the lamp is usually no more than a glass or quartz vessel containing a small amount of metal. Depending on the method by which the lamp is energized, there are two types of discharge—a radio-frequency glow discharge (usually with external electrodes) and an electrodeless ring discharge. Radio-frequency glow discharges are started by bringing radio-frequency electrodes close to the tube-shaped lamp. The lamp will work with a wide range of pressures—between 10^{-3} and 10^2 torr.

Generators with working frequencies of 10^6–10^{10} c/s are used for exciting radio-frequency discharges. As has been proved,[15] frequencies at the higher end of the range are preferable, since less metal is absorbed by the lamp walls and emission is more intense when they are used.

A standard 125 W power supply generating at a frequency of 2450 Mc/s has been used[21] for exciting the spectrum of mercury when determining the isotopic state of the element by atomic absorption. To reduce the mercury-vapour pressure in the discharge, and thereby reduce self-absorption, the 8 mm diameter discharge tube, which contained the pure isotope Hg-202, was cooled with running water, the temperature of which was thermostatically controlled to 25°C. Under these conditions, the half-width of the Hg 2537 Å emission line was about 0·06 cm^{-1}.

The present author also excited the Hg 2537 Å resonance line by means of a radio-frequency glow discharge to determine mercury by the atomic absorption method. A generator with a frequency of 145 Mc/s and a useful power of up to 40 W was used.[20] The lamp was a vertical quartz tube 100 mm long, with an internal diameter of 2·6 mm and containing a small quantity of mercury. The external electrodes were 10 mm apart on opposite sides of the central part of the tube. The lower end of the tube, being at a distance from the discharge zone, was held at room temperature to ensure that the mercury-vapour pressure retained its room-temperature value. The 2·6 mm internal diameter of the capillary tube was ideal—a larger diameter would increase self-absorption, and a smaller would make the discharge more difficult to excite.

Interferometric measurement of the profile of the Hg 2537 Å line of monoisotopic Hg-198 proved that, allowing for instrumental broadening, the line half-width was about 0·09 cm^{-1}. With the heated part of the capillary tube at about 700°K, the Doppler half-width must have been 0·054 cm^{-1}. The broadening resulting from self-absorption is therefore less than the Doppler width.

Radio-frequency lamps for highly volatile elements (Cd, Zn, etc.) contain a small amount of the metal in the capillary tube. Lamps for low-volatility elements (rare earth elements, U, Ac, etc.) are charged not with the metals, but with their halides, the high volatility of which ensures that a discharge is excited and that the metal is vaporized.[25] Halide-filled lamps need to be preheated. They enable the atomic spectra of metals to be excited with an intensity two or three orders higher than hollow-cathode lamps supplied with d.c.

Richards[25] proved that intense atomic spectra can be excited for any element without additional heating of the discharge tube by using a standard microwave generator and an extremely simple vacuum system with pump for the initial exhaustion of the discharge tube prior to its filling with inert gas. For this, 10 mg of pure metal or its halide is sufficient. If the vapour pressure of the substance is not more than 1 torr at 800°C, an equal amount of AgCl or AlCl$_3$ is added to the specimen (when metal oxides are used). The helium pressure in the quartz discharge tube is kept at 3–10 torr. Usually the spectrum of the metal begins to appear a few seconds after the discharge has started, the heating period being as long as 2–3 minutes only with particular refractory substances. This method proved effective for exciting the spectra of tungsten and zirconium, when weighed amounts of metallic tungsten and zirconium oxide were placed in the discharge tube. The lamp is extremely stable and bright (comparable with a 3-amp arc). It is quite suitable for atomic absorption measurements.[26]

A radio-frequency electrodeless ring discharge is excited when the discharge lamp is placed in the magnetic field of the coil of a radio-frequency generator. Unlike radio-frequency glow discharges, the ring discharge can exist within only a narrow pressure range. The lamp must therefore hold inert gas at a specific optimal pressure. Lamps of this type are nowadays important in the optical pumping of lasers.

Bell, Bloom, and Lynch developed a simple type of lamp, which they tested with rubidium.[22] This lamp is a Pyrex sphere, 1 cm in diameter, with walls 0·2 mm thick and with a small branch tube 2–3 mm long leading from it. A vacuum system was used in charging the lamps with an alkali metal and an inert gas; a few milligrams of metal were injected into the sphere, which was then filled with krypton to a pressure of 1·6 torr, this pressure suiting the minimum discharge excitation potential; the tube was then sealed.

Emission was excited by a 3·3 W radio-frequency ($\simeq$100 Mc/s) generator. The lamp and the radio-frequency coil were enclosed in a glass cover to facilitate thermostatic control, which eliminated fluctuations in emission due to variations in metal-vapour pressure in the lamp.

Later, Brewer[23] increased the diameter of the sphere to 3 cm (volume $\simeq$10 cm^3) and used a more powerful generator (40 W, 20 Mc/s); by these means he increased the emissive power of the lamp by a factor of about 50.

Franz[24] used a similar design of lamp and generator, and achieved the same increase in emitting power with caesium lamps. He established that, of the inert gases tested (argon, krypton, and xenon), the best results were obtained with a xenon filling at 1·5 torr.

The effectiveness of spherical radio-frequency lamps for atomic absorption spectrochemical analysis was discovered in 1961 by Kirillova, Levikov, and L'vov, who established that such lamps can be produced for rubidium, caesium, potassium, sodium, zinc, cadmium, mercury, selenium, and tellurium. The bodies of the lamps for elements whose resonance lines fall in the ultra-violet region (Zn, Cd, Hg, Se, Te) were made of quartz. Lamp diameters were 2·0–2·5 cm. The lamps were filled with argon up to a pressure of 1·5 torr. The generator used with the GU-50 lamp (Fig. 2.17) was a compact unit mounted on the bar of the spectro-

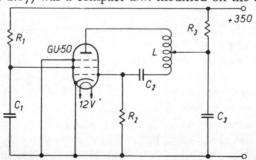

FIG. 2.17. Radio-frequency generator for GU-50 spherical lamp. $R_1 = 7\cdot5$ k; $R_2 = 5$ k; $R_3 = 100$; C_1, $C_3 = 1000$ pF; $C_2 = 200$ pF; L-coil of 2 mm diameter copper wire; coil diameter 20 mm, number of turns 8.

meter. Power was supplied to the generator from an universal UIP-1 power source.

To eliminate fluctuations in intensity caused by convection currents in the air surrounding it, the lamp, together with the coil, was enclosed in a protective cylinder with a quartz window. This precaution is particularly important when lamps for the less volatile elements are used; these elements, zinc, cadmium, etc., are those whose vapour pressure is particularly sensitive to temperature changes.

Research into the intensity and stability of emission and the profiles of the resonance lines established that spherical radio-frequency lamps are greatly superior to hollow-cathode lamps. The measurements were made with the interferometric apparatus used for investigating hollow-cathode lamps (Fig. 2.8).

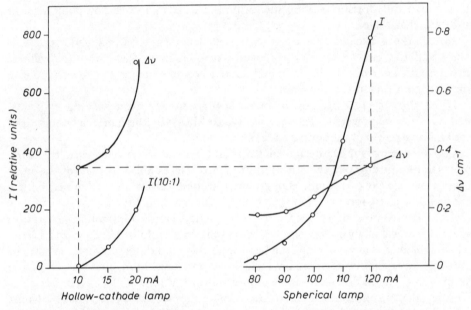

FIG. 2.18. Intensities and half-widths of the Cd 2288 Å line in a hollow-cathode and a spherical lamp.

In Fig. 2.18, the Cd 2288 Å line emitted by a cadmium hollow-cathode lamp filled with neon is compared in intensity and width with the line emitted by a spherical cadmium lamp 20 mm in diameter. When interpreting the diagram, it should be noted that the ordinates for the hollow-cathode lamp have been expanded ten times. The diagram shows that, with a current of 10 mA, the width of the Cd 2288 Å line from the hollow-cathode lamp is the same as the width of the line from a radio-frequency lamp with a generator taking 120 mA ($\Delta\nu = 0.35$ cm^{-1}), but there is a difference of about $10^3:1$ between the intensities provided by the lamps. Under these excitation conditions (120 mA), the Cd 2288 Å line in the spherical radio-frequency lamp is twice as wide as with the minimum current for a radio-frequency discharge (80 mA).

There is no self-absorption with the lower currents because, owing to the skin effect (the essential feature of which is that plasma tends to concentrate the radio-frequency field close to the surface of the lamp) the thin layer against the lamp walls, about 1·6 mm thick, takes part in the emission.[23] This theory has been confirmed by visual observation of the form taken by discharges in lamps. With low discharge powers, two zones differing in colour are to be seen; an outer zone gives the characteristic emission of metal resonance lines, and an inner shell emits the luminescence of the inert gas. The luminosity of the lamp is governed by the emission of the surface layer facing the observer; emission by the layer on the far side of the lamp is completely absorbed by the non-luminous core of the lamp.

The fact that there are differently coloured luminous zones in a lamp proves that the excitation conditions in spherical radio-frequency lamps are those required for minimum self-absorption. The difference in colour is particularly appreciable in cadmium and zinc lamps, since emission from these metals is bright blue and in sharp contrast to the lilac emission from argon. Increasing the discharge power causes the differently coloured zone to disappear (with a current of about 100 mA in the case of cadmium) and the line-width to increase. Fig. 2.19 contains Cd 2288 Å line recordings for different excitation conditions. The minimum observed half-width with a current of 80 mA is about 0·16 cm^{-1}, with 100 mA it is 50 per cent greater, and with 120 mA the line is self-absorbed and the true half-width increases to 0·35 cm^{-1}. In high-power discharges, therefore, spherical radio-frequency lamps are heterogeneous emitters. Obviously, the temperature drop between the hot core of the discharge and the colder peripheral regions is a contributory cause.

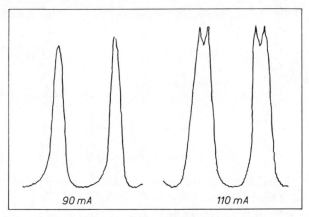

FIG. 2.19. Profiles of the Cd 2288 Å line in a spherical lamp with different excitation powers. Thickness of standard 0·5 cm.

Spherical radio-frequency mercury lamps emit the Hg 2537 Å line with great intensity but, as regards line-width, they are inferior to the radio-frequency glow lamps with external electrodes described earlier. This is evidently because of the high mercury-vapour pressure in the spherical lamps.

If the lamp temperature is stabilized and the generator working conditions are chosen correctly, the fluctuations in emission from spherical radio-frequency lamps can be reduced until recorded fluctuations are governed by the shot effect in the detector. When the lamp noise was measured experimentally,[22,23] it was found that the recorded noise level coincided with the theoretically calculated level and was independent of frequency.

The author's measurements of the stability of emission of the Cs 4555 Å line also proved that, up to luminous fluxes corresponding to a photocurrent of 4.10^{-11} amp, the fluctuation level did not exceed the shot noise, amounting to about 0·03 per cent of the signal (with a recorder time constant of 0·2 sec).

Spherical radio-frequency lamps have now been developed for a large number of elements: As, Bi, Cd, Cs, Ca, Cu, Fe, Ga, Hg, In, I, K, Mg, Mn, Mo, Na, P, Pb, S, Se, Sn, Te, Tl, Zn. Many of the lamps are being mass-produced by Soviet industry.[70,71] Pure elements were used in the lamps for all the elements listed above, apart from the least volatile, Cu, Fe, Mn and Mo. These low-volatility elements were introduced into the lamps in the form of compounds with adequate vapour pressures. In lamps for iron, manganese, molybdenum and copper, $FeCl_2$, $MnCl_2$, MoO_3, and CuI_2 were used.

In work carried out by Ivanov et al.,[70,71] xenon was used for filling the lamps. To judge by comparative measurements made in my laboratory, however, argon-filled lamps provide the same luminosity and narrower resonance lines with less continuous background. In my opinion, therefore, it is preferable to use argon. It is safe to assume that the anomalous line-broadening in radio-frequency lamps and hollow-cathode lamps when they are filled with xenon is of the same nature, and not associated with self-absorption.

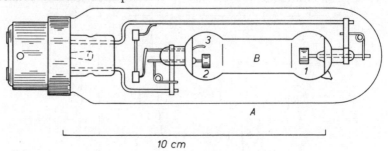

Fig. 2.20. Metal vapour lamp.[28]

Metal-vapour discharge lamps also are now used in atomic absorption spectrochemical analysis. Their design is illustrated in Fig. 2.20. The cylindrical flask A contains the discharge tube B, which has three electrodes, one of which (3) is for initiating the discharge. For thermal insulation, the space between A and B is exhausted to a high vacuum. The discharge tubes have an argon filling at a pressure of a few torr; they contain weighed amounts of the metal concerned. A glow discharge is excited between the electrodes 2 and 3, and this becomes an arc discharge between electrodes 1 and 2.

The lamps are supplied with a.c. at 220 V, and are connected in series with a ballast resistance of 200–220 Ω. To simplify striking, a 0·5–1·0 μF capacitor, discharging at 450–500 V, is connected in parallel with the tube.

Metal-vapour discharge lamps are produced for metals with sufficiently high vapour pressures—alkali elements, cadmium, zinc thallium and mercury. Soviet industry produces the spectral microtube metal-vapour lamps (for instance the SMK-2 for cadmium and the SMZ-2 for zinc).

The emissive intensity of metal-vapour lamps is very high (almost as high as that of spherical and other radio-frequency lamps). There is, however, considerable self-absorption and self-reversal of the lines they yield. The reason is that, unlike the discharge in a spherical lamp, the discharge in a metal-vapour lamp occurs throughout the entire volume of the discharge tube and not in a thin surface layer.

Since self-absorption increases as the current density through a lamp increases, experimenters used currents lower than recommended by manufacturers. The author's interferometer measurements made with a sodium lamp proved, however, that, even with a current as low as 0·75 amp, the Na 5890 Å resonance line was entirely self-reversed; a current of 1·1 amp had been recommended in the lamp running data.

Thus we can conclude that, from different line spectrum sources considered, hollow-cathode lamps and radio-frequency lamps are the most suitable for atomic absorption spectrochemical analysis. All of the elements determined by the atomic absorption method can be determined with them.

At this point, we should emphasize another fact. Usually, when the particulars of spectral sources are being assessed, stability of emission is considered the most important factor. This is because, with single-beam systems, fluctuations in source intensity reduce the accuracy of measurement and the sensitivity of the method. Nevertheless, emission stability is not the most important requirement of a source; for, in principle, source noise can always be eliminated by using double-beam or two-channel methods of measurement. Emissive power is far more important; for, if the signal is too small, detector noise will be the most important cause of measuring errors.

As was pointed out on pp. 54–63, the recorded fluctuations in emission from hollow-cathode lamps and spherical radio-frequency lamps are associated, not with internal noise in the lamps, but with detector shot noise. As signal intensity increases, shot noise steadily decreases. In particular, therefore, emission fluctuations are less significant in spherical radio-frequency lamps than in hollow-cathode lamps supplied with d.c.

LAMPS WITH SEPARATE VAPORIZATION AND EXCITATION OF ELEMENTS

All the light sources considered earlier—hollow-cathode lamps supplied with direct or high-frequency current and radio-frequency and metal-vapour lamps—have one point in common: in all of them, one and the same discharge vaporizes the elements and excites the emission. In these sources, it is impossible to regulate independently the entry of the element into the discharge and the excitation of its atoms. It is not surprising that increasing the input power to increase the luminosity is accompanied, in all of them, by an increase in vapour concentration (as a result of thermal vaporization and cathode sputtering) and consequently by self-absorption and broadening.

The main problem, where sources for atomic absorption spectroscopy are concerned, is to develop sources of resonance emission in which vaporization and excitation can be separately regulated. Sources of this type based on hollow-cathode lamps were proposed by Sullivan and Walsh.[72] Their special principle is the use of two independent (electrically separate) discharges: (1) the normal hollow-cathode discharge to vaporize the cathode material and (2) a discharge solely for exciting the vapour.

Fig. 2.20, which illustrates diagrammatically the arrangement of the electrodes in this type of lamp, shows that the lamps differ in construction from normal hollow-cathode lamps in having two auxiliary electrodes, the discharge between which excites the vapours diffusing from the hollow cathode. The auxiliary electrodes are in glass tubes with small diaphragms, which confine the core of the discharge and localize it close to the cathode. It is a glow discharge and the atoms are excited in its positive column. The tungsten electrodes are intensely heated and vaporized, for which reason, the authors recommended that they should be coated with a layer of oxide, to reduce the voltage drop in them. Later reports have shown that the authors found it better to use a low-voltage arc as the additional discharge, the cathode for which was an oxide spiral heated by a.c.

Boosted output lamps have proved extremely effective for atomic absorption measurements. In the first place, boosted output lamps are approximately 100 times more luminous than simple hollow-cathode lamps, while the increased intensity is not accompanied by line broadening, as is confirmed by the linearity of calibration curves up to absorbances of 1·0–1·5. In the second place, the beneficial increase in resonance-line intensity is not associated with a disadvantageous increase in the intensity of ionic lines or inert gas lines, which are practically unchanged. This is evidently because the electron temperature in an arc (or in the positive column of a glow discharge) is lower than in the negative glow region of a hollow-cathode discharge. The resulting simplification of the spectrum with its predominance of intense resonance lines makes it easier to isolate the analysis lines of elements with complex spectra, e.g. elements of the iron group.

Bodretsova, L'vov and Mosichev investigated some of the characteristics of boosted output lamps. The measurements were made with the interferometric equipment described on p. 47. A xenon-filled (0.5 torr) lamp with a nickel cathode and a neon-filled (4 torr) lamp with a copper cathode were used for the research. The lamp design is shown in Fig. 2.21.

Preliminary experiments established that, to achieve narrow resonance lines, it is best to isolate the emission of the additional arc discharge from the emission within the hollow cathode. The arc discharge was therefore supplied with a.c. at 50 c/s through a stabilizer and the hollow-cathode discharge was supplied with d.c. Further measurements proved that, whatever the ratio between the hollow-cathode current (5–25 mA) and the arc current (50–200 mA), the width of the resonance lines excited in the arc discharge is practically constant and generally less than that of resonance lines excited in a hollow cathode by the minimum usable current (5 mA). The additional arc discharge thus enables greater currents to be passed

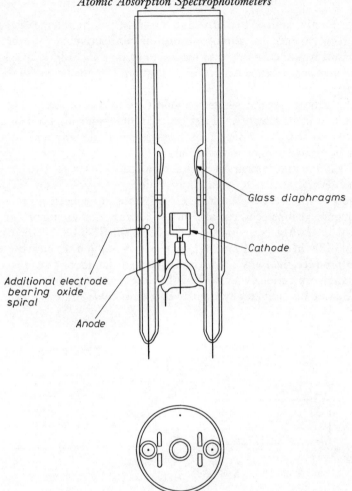

FIG. 2.21. Lamp with separate vaporization and excitation (boosted output lamps).[72]

through a hollow cathode, while producing lines of no greater and perhaps smaller line-width. Although, for the same current the ratio in emission strength with and without the additional discharge is only 3:1 for nickel and 7:1 for copper, there may be an increase in luminosity of two orders of magnitude if the current passed through the hollow cathode is increased. For instance, with a current of 25 mA passed through the hollow cathode and of 200 mA through the auxiliary electrodes, the magnitude of the signal for nickel is 30 times greater than from an ordinary hollow cathode with 5 mA passing through it; it is 100 times greater for copper.

MULTI-ELEMENT LIGHT SOURCES

There are two reasons why the problem of designing multi-element light sources is important. Firstly, unless the problem is solved, multi-channel devices cannot easily be constructed for atomic absorption measurements. Secondly, multi-element lamps would cure the ill of having to change and adjust lamps and would reduce the number of lamps needed in determining elements one by one in single-channel instruments.

There have been several suggested solutions to the problem, but all can be grouped in two main categories. First are solutions requiring the atoms of several elements to be excited in a single discharge; second are solutions requiring each element to be excited in a separate discharge.

Solutions in the first category include hollow-cathode lamps with cathodes made of alloys of several elements, or of rings of different metals, and radio-frequency lamps loaded with several elements. Descriptions have been given of spherical radio-frequency lamps containing the following combinations of elements: Bi-Sb-Zn-Cd, Ca-Mg-Na, Bi-In-Ga-Pb and Bi-Sb-Tl.[70,71] In construction, these lamps differ in no way from single-element lamps and they are therefore much simpler to produce than any of the multi-element lamps of the second category. However, they are seriously limited in that their emissive powers for individual elements cannot be individually regulated. Comparable luminosities for different

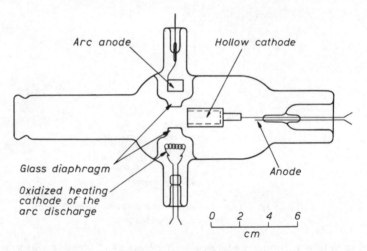

FIG. 2.22. Hollow-cathode lamp with an additional low-voltage arc discharge (boosted output lamp).

elements sometimes correspond to discharge powers differing by a whole order of magnitude (see Table 2.2, p. 48). This is due to differences between vapour pressures, amounts of cathode sputtering and resonance-line characteristics (oscillator strengths and transition energies). When elements with different characteristics are combined in a single lamp, therefore, the strength and width of the resonance lines may be far from optimum.

Lamps in this category are further limited because the combination of elements included in the cathode (or spherical lamp) cannot be altered to allow use of the lamp for other analytical purposes.

Lamps with hollow cathodes made of alloys have an additional disadvantage in that the intensity ratio of the resonance lines alters gradually because the volatile elements vaporize most readily, condense on the surface of the cathode, and suppress the sputtering of the less volatile components. To reduce this effect, Massmann[73] used cathodes built from rings of the different elements, with the rings inversely proportional in width to the sputtering rates of the metals concerned.

In the second category of solutions to the problem of exciting the spectrum of each element separately from those of other elements come all methods using optical means to combine in a common light path the emission from many single-element sources. Fig. 2.23 shows several methods of such optical combination.

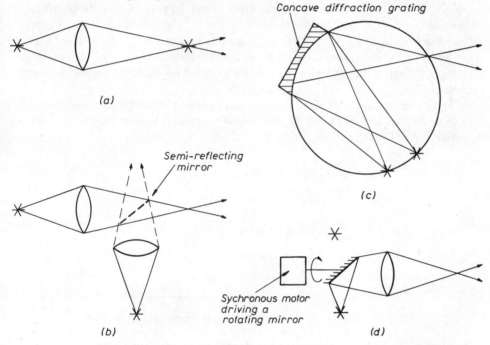

FIG. 2.23. Principal ways of optically combining the emission from several lamps into a common light path.

Systems (a) and (b) were used by Butler and Strasheim.[74] The main shortcoming of system (a) is the considerable loss of light through reflection by the lenses and through the partial masking of one cathode by another. Further, if more than three sources are to be combined, system (a) presents serious difficulties in adjustment. The main shortcoming of system (b) is the loss of light at the semi-reflectors needed to deflect some beams through a right angle.

System (c) was proposed by Walsh.[75] This system is effectively an 'inverse spectrograph'. The lamps are located in the focal plane of the dispersing element at the positions of the resonance lines being isolated in the spectrum; the total flux, consisting of pure resonance-line emission from all the lamps used, is combined at the slit—in its normal use, the entrance slit—of the spectroscope. Of course, either a prism or a diffraction grating can be used as the 'inverse-dispersing' element. The system is cumbersome and rather difficult to adjust. Light losses are considerable.

System (d), proposed by the present author, consists of alternately reflecting the emission from the independent lamps into the optical axis of the instrument by means of an oscillating or rotating mirror mounted in the optical axis. The shortcoming of this system is the shortening of the effective measurement times for the signals from each of the lamps. However, as regards utilization of the light, compactness, and simplicity of adjustment, system (d) is better than the others.

The systems in this second category have the common defect of requiring a large number of lamps, which naturally makes the apparatus more complicated and expensive (not to mention increasing its dimensions). In view of this, the possibility of combining several independent emitters within a single lamp is particularly interesting (Fig. 2.24). A lamp of this design was first used by Massmann.[76] The lamp has separate excitation for each ring cathode, but the individual cathodes can be supplied from a single power supply by connecting them in parallel through separate ballast resistances. If the cathodes are close together and small there is no need for special optical systems to combine the emissions from the separate sources. The diameters of the rings should successively increase, so that later cathodes do not mask earlier ones.

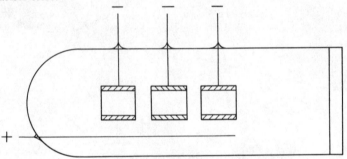

FIG. 2.24. Ring cathodes, with separate power supplies in a single tube.

In view of the success achieved in developing boosted output lamps (pp. 69–71), it should be effective to produce multi-element tubes with separate sputtering of elements from hollow cathodes and the simultaneous excitation of vapour spectra in an arc discharge. In the type of boosted output lamp described on pp. 70–71, the metals were sputtered in a hollow-cathode discharge. This is not, however, necessary in principle. Drabkin and Ivanov[77,78] proved that a cathode of any shape, in particular a metal rod or plate, can be used as the source of atomic vapours. One such possible lamp is shown in Fig. 2.25. It consists of the inert-gas-filled flask, with window. The oxide cathode 4, the anode 1, and several electrodes made of the metals whose spectra are required, are sealed into the flask.

The illustration also shows the electrical circuit for power supply to the lamp. The voltage from the supply E_1 is for maintaining an arc discharge between the anode 1 and the cathode 4, the current in which (0·5–2 amp) is determined by the resistance R_1. The potential of the source E_2, negative relative to the discharge

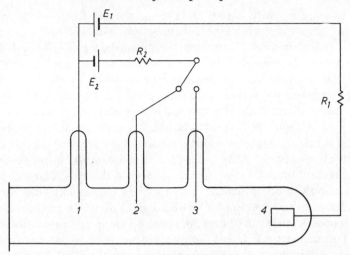

Fig. 2.25. Boosted output lamp.[77] 1. Anode. 2 and 3. Electrodes. 4. Cathode.

region, is fed to electrode 2 or 3. The current in the electrode circuit (5–50 mA) is determined by the resistance R_2. Ionic bombardment causes the materials of electrode 2 (or 3) to enter the arc discharge plasma, where the spectrum is excited. The distance between the electrodes is 25–40 mm, enough to prevent them from contaminating one another.

The experiments proved that, as regards luminosity, these lamps are tens of times superior to d.c. hollow-cathode lamps. The spectra of a group of elements can be emitted simultaneously from several rods made of different metals or alloys in the path of the arc discharge. These lamps have the great advantage of simplicity of design.

There are now, therefore, several different types of multi-element spectral source. When these different sources have been developed and compared in practice, it will be possible to decide how best to carry out further development.

CONTINUOUS SPECTRUM SOURCES

Continuous spectra can be emitted from lamps operated under either pulsed or continuous conditions.

High-power pulsed lamps provide bright emission for short periods (between 10^{-6} and 10^{-3} sec) and are of use in the observation of transient absorption phenomena. Pulsed lamps should be used for the photographic recording of spectra, for instance in qualitative analysis; for, if the recording time (exposure) is short, e.g. 10^{-3} sec, the emission of the flame atomic vapour, being faint, is not recorded. The lamps are pulsed by the discharge of a capacitor through the electrode gap. There are two principal types of pulsed lamps; spherical ones with small electrode gaps, filled with inert gas up to pressures of several atmospheres, and tubular ones with large electrode gaps, filled with inert gas at low pressures of 50–100 torr. Most of the tubular lamps filled with xenon or krypton provide flashes with continuous spectra in the ultra-violet and visible ranges and with colour temperatures of 5500–7000°K. Pulsed sources of light have been described very fully by Marshak.[27]

The pulsed photographic method of measurement was used by Nelson and Kuebler[29] for investigating atomic absorption by substances vaporized in a capacitive discharge lamp (see pp. 126–7). The source was a pulsed Lyman lamp, with an internal diameter of 3 mm, made of quartz and filled with argon at 15 torr. The pulses came from a 3·75 μF capacitor charges to 10 kV. The pulse duration was about 20 μs.

Continuous spectrum sources running continuously are best for photoelectric measurements of absorption spectra, when the emission must be constant throughout the measurement time. Of the different sources of this type, filament lamps are the most commonly used. Filament lamps emit a continuous spectrum in the visible and adjacent ultra-violet and infra-red regions of the spectrum, with a maximum at about 9000 Å; they can be used for spectrophotometric measurements down to 2600 Å. Tungsten strip-filament lamps are often used as sources with known spectral energy distributions. Soviet industry produces the following types of strip-filament lamp suitable for photometry: the 40-W SN 6-40, 100-W SN 6-100, and 200-W SN 8-200. The bulbs are of special design, so that there is no parasitic reflection from their walls. They have ultra-violet-transmitting windows. Fig. 2.26 shows the spectral energy distribution for a strip-filament lamp.

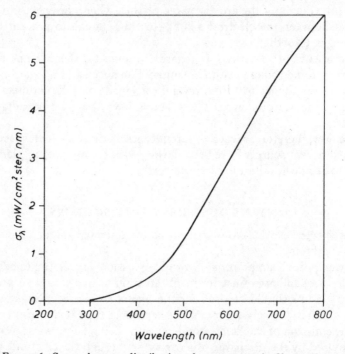

FIG. 2.26. Spectral energy distribution of a tungsten strip-filament lamp.

The emission intensity of filament bulbs drops abruptly at wavelengths shorter than 2600 Å. Below this wavelength, it is better to use gas-discharge sources, such as hydrogen and xenon lamps. Hydrogen lamps emit continuous spectra in the 3200–1700 Å range and run on low-voltage arc discharges. Of the types of hydrogen lamp produced in the Soviet Union, the 25-W VSFU-3 lamp, used for instance in SF-4 spectrophotometers, is the most common. These lamps are supplied with power from an EPS-86 electronic stabilizer, which maintains the discharge current constant to within 0·1 per cent. The more powerful VU-1 lamps (150 W) are also used.

Hydrogen lamps are now being supplanted by deuterium lamps[30] with the same parameters, these lamps provide more intense emission (Fig. 2.27).

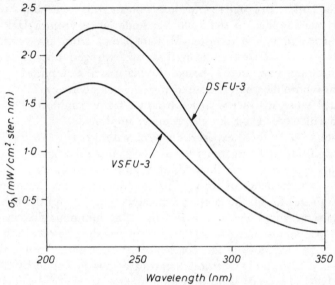

FIG. 2.27. Spectral energy distribution of hydrogen (VSFU-3) and deuterium (DSFU-3) lamps.[30]

Hydrogen lamps and tungsten-filament lamps were used by Allan in his work to determine the most sensitive lines to use in atomic absorption.[31]

High-pressure and ultra high-pressure xenon-filled lamps, used with high-current arc discharges, provide the most intense emission in the ultra-violet and visible regions of the spectrum. Emission from xenon lamps is less stable than emission from hydrogen lamps. Nevertheless, it is convenient to use these lamps as an alternative to pulsed lamps for photographic research, owing to the short exposure times. As an instance, a 150 W xenon-filled lamp was used by Fassel and Mossotti,[32] for investigating the absorption spectra of high-melting-point metals in flames. When the spectra were measured with a Jarrell–Ash 3·4 m grating spectrograph (order II), the exposure time was only 8 sec.

It should once more be noted that it is inadvisable to use continuous sources rather than line sources for photoelectric measurements of atomic absorption, except for certain tasks of a special nature (qualitative research into absorption spectra, measurement of total absorption, etc.). It was shown earlier (pp. 28–31 that, if continuous sources are used, there is a considerable loss of sensitivity and a substantial curvature of the graphs, even at low absorbances. These theoretical conclusions are confirmed by the author's experimental measurements (see pp. 278–82) and by other results.[79]

A number of research workers have nevertheless expressed the opinion that, since continuous sources are more stable than line sources, they allow elements to be determined with about the same detection limit. In view of this opinion, let us consider certain data obtained in the author's laboratory.

The frequency distribution of emission fluctuations in iron and chromium hollow-cathode lamps, hydrogen lamps and tungsten-filament lamps was investigated. Power was supplied to the hollow-cathode lamps from a UIP-1 voltage-stabilized source and to the tungsten-filament lamps from an accumulator; an EPS-86 current-stabilized source was used for the hydrogen tubes. The alternating component of the signals from the photomultiplier was first amplified by means of a wide-band linear amplifier, and measured at frequencies between 40 c/s and 10 kc/s with an AS-2M noise analyser with a constant relative bandpass. The d.c. signal was measured with a type U1-2 d.c. electrometer amplifier.

It was established by these experiments that, within the frequency range given above, the amplitude of fluctuations for all the lamps investigated is constant to within ±20 per cent, except for the hydrogen lamp, which gave an increased level of fluctuations at a frequency of 100 c/s (mains induction). The calculated fluctuations (see pp. 36–40) for all the lamps practically coincide with the mean levels of the fluctuations measured in the experiments. The luminous flux in every case corresponded to a photo-current of 10^{-12}–10^{-13} amp in the photomultiplier cathode.

With the normal luminosities for atomic absorption measurements, the fluctuations in emission from hollow-cathode lamps are thus governed by the detector shot noise and do not exceed the noise in continuous sources of the same luminosity. There are therefore inadequate grounds for the opinion that continuous sources are more stable. Clearly, the greater stability of continuous sources found in comparisons of continuous and line sources by Gibson et al.[80] was related to the different luminosities of the sources; in any event, it was hardly enough to compensate for the non-monochromaticity of the continuous source and for the consequent loss of sensitivity in using it.

MONOCHROMATORS

A spectral device is necessary to isolate the required resonance line from the light emitted by the source. It must be chosen according to the complexity of the lamp spectrum and the nature of the analytical problem. If an instrument is intended for determining only one element, or even several elements with simple enough spectra, it suffices to use the simplest possible equipment for isolating the analysis lines. If it is necessary to determine several elements with more complex spectra, monochromators of higher resolution should be used.

In the first case, narrow-band filters can be used for isolating the required resonance line. They are either absorption filters (for instance, coloured glass) or interference filters, which work on the Fabry–Perot etalon principle.

In transmittance and band half-width, combinations of several interference filters (multiplex light filters) and multi-layer dielectric coatings are particularly good. The theory and production technology of multiplex filters have been developed and described by Korolev.[81] The filters can be made for either the visible or the ultra-violet regions of the spectrum; they transmit band half-widths of the order of 1 Å with transmittances up to 50–70 per cent. The use of high-quality filters in simplified spectrophotometers for atomic absorption measurements is very interesting, since in comparison with slit-type monochromators, filters are easier to use and have larger aperture ratios.* Unfortunately, the production of filters with good properties is technologically complex. The filters are not, therefore, much used as yet in spectrochemical analysis.

Apparatus including filters is used for determining alkali metals and mercury by atomic absorption spectrochemical analysis. In the Malmstadt photometer,[33] which is intended for determining sodium, the Na 5890 Å and Na 5895 Å resonance lines are isolated by an interference filter with a transmission-band half-width of about 100 Å. Since the luminous flux is great, a photoresistive cell connected to a sensitive galvanometer (200 μA) is used for photometric recording.

Another simple and compact photometer with filters for determining sodium and potassium was developed by Box and Walsh.[82]

Poluektov and Vitkun[34] used a combination of absorbing filters and a phosphor, which isolate the emission of the Hg 2537 Å resonance line, for determining mercury by the atomic absorption method. The light passes through an UFS-1 filter, which isolates the ultra-violet region of the spectrum, and falls on a glass plate covered with a layer of the phosphor, which is only sensitive to the short-wave ultra-violet region of the spectrum. Beyond the plate coated with the phosphor there are ZS-17 and SZS-10 filters for eliminating the mercury red and ultra-violet emissions (the UFS-1 filter transmits the red region of the spectrum). The FEU-20 photomultiplier positioned beyond the combination of filters thus records the emission from the phosphor, which is excited by the Hg 2537 Å line alone.

The Hg 2537 Å line is much more easily isolated in the IKRP-445 and IKRP-446 devices, which are intended for the continuous automatic measurement and recording of concentrations of mercury vapours in industrial atmospheres.[35] These devices do not contain filters. The source of emission is a BUV-OP mercury lamp supplied with power from a high-frequency (30 Mc/s) generator. More than 90 per cent of the lamp emission energy is concentrated in the Hg 2537 Å line. A magnesium photocathode is used as detector; it is sensitive to ultra-violet radiation and has a long-wave absorption edge at 3000–3500 Å. Owing to this combination of source and detector spectral characteristics, the recording system receives only the Hg 2537 Å resonance line.

A device working by a similar system was used by Osborn and Gunning[21] for

---

* An extremely full review of the latest developments in the construction of interference light filters has been given in the book by Bochkova and Schreider.[15] See also H. A. Macleod, *Thin-Film Optical Filters* (Adam Hilger Ltd, London, 1969).

determining the isotopic composition of mercury by the atomic absorption method (see pp. 260–3). The source was a high-frequency discharge (2450 Mc/s) and the detector was a WL-775 photocell, sensitive over the 2000–3000 Å range.

Equipment as simple as this can be used mainly for isolating the resonance lines of alkali elements, cadmium, zinc, and mercury, the spectra of which are extremely simple, particularly from radio-frequency and metal-vapour lamps. For instance, in spherical rubidium lamps, more than 30 per cent of the emitted energy goes into the resonance lines.[23]

For most other elements, the spectra emitted by hollow-cathode lamps are too complex for standard filters, with their quite wide bandpass, to be usable for isolating spectral lines. The resonance lines of a number of elements (for instance, the iron group) are close to non-resonance lines. Thus, the highly sensitive Fe 2483·27 Å resonance line is alongside the Fe 2484·19 Å line, and the most sensitive Co 2407·25 Å line is alongside the Co 2406·27 Å and Co 2408·75 Å lines. In general, therefore, resonance lines must be isolated by monochromators capable of separating spectral regions about 1 Å wide.

We shall investigate the characteristics of a monochromator suitable for atomic absorption analysis. The slits should not be narrower than 0·01 mm if the band width of the isolated region is to be set reproducibly; a width of 0·01 mm can be set accurately enough with slits graduated at intervals of 0·001 mm. We know[3] that if the entrance and exit slits have equal spectral widths (the optimum condition for maximum optical conductance with a fixed bandwidth), the total bandwidth of the emergent flux is twice the spectral width of each slit. Slits with a spectral width of 0·5 Å are therefore necessary for isolating a 1 Å region of the spectrum. With a mechanical slit width of 0·01 mm, the monochromator should have a reciprocal linear dispersion of 50 Å/mm.

The first requirement of the monochromator, therefore, is that the linear dispersion should be 0·02 mm/Å or better. In addition, it must be possible to scan the spectrum reproducibly and accurately for consistent results.

Furthermore, instruments for atomic absorption measurements must be adequately powerful, in the energy sense. We shall now consider the parameters on which depends the power of an atomic absorption spectrometer. The power ratio L of a monochromator, multiplied by the spectral bandwidth, is equal to the ratio of the emitted flux Φ to the energy output of the source B, the latter being in terms of unit bandwith, unit solid angle, and unit area.

$$L = \frac{\Phi}{B \Delta \nu} \tag{2.6}$$

If the entrance and exit slits have the same spectral width, the power ratio as specified in (2.6) can be written[3]

$$L = T \frac{h}{F} S \frac{d\phi}{d\lambda} \tag{2.7}$$

where T is the transmittance of the optical system, h the height of the illuminated part of the slit, F the focal length of the objective element (collimator), S the area of the dispersing element, and $d\phi/d\lambda$ is the angular dispersion.

In comparing different instruments for their suitability for atomic absorption work, it should be noted that, providing the illuminating system is matched to the monochromator in terms of relative aperture, that is, as long as the dispersing element of the monochromator is fully illuminated, the only factor having a major effect on the power ratio is the angular dispersion of the dispersor. This follows from considerations of the Lagrange–Helmholtz law. For the case of diffraction-grating instruments, the angular dispersion rarely differs by more than 2:1 with the various gratings used. (Gratings with either 600 or 1200 lines/mm are used for the ultra-violet and visible regions of the spectrum in the mass-produced instruments of the Soviet Union.)

Referring to the above and also pp. 85–91, it is seen that spectroscopic instruments with a particular type of dispersor are roughly equivalent so far as atomic absorption purposes are concerned.

It would seem, at first glance, that the aperture ratio of a monochromator used for atomic absorption measurements must depend on the focal length of the objective. This conclusion, however, is true only when the illuminated height of the slit (h) is constant. Actually, the illuminated height h of the slit depends on the cross-section of the beam passed by the illuminating system of the spectrophotometer and on the relative aperture of the instrument. It will be proved later (p. 90) that if the illuminating system is correctly adjusted, so that the incident beam fills the aperture of the dispersing element, which remains constant from instrument to instrument, the diameter of the image of the source on the slit is directly proportional to the focal ratio of the instrument (equation 2.15). If the beam from the source is finite in cross-section (as must always be the case in atomic absorption measurements owing to the finite dimensions of the absorbing medium) the ratio F/h must remain constant.

All spectroscopic devices of any particular type are thus about equivalent as regards aperture ratio. From this point of view, the SD-2 or SFD-2 short-focus monochromators have no advantages over large grating instruments such as the DFS-13 or DFS-8. Moreover, since the gratings in the DFS-13 device have 1200 lines per mm, and those in the SFD-2 spectrophotometer have 600 lines per mm, the aperture ratio of the DFS-13 device is twice that of the SFD-2 spectrophotometer (we assume that the gratings have the same blaze angle and that the quality of their manufacture is the same).

In general, grating instruments have a larger power ratio than prism instruments in the visible region of the spectrum, mainly because of their greater angular dispersion. This has another beneficial effect in that the increased linear dispersion increases slit-width and other instrumental tolerances.

Foreign research workers use both prism devices (examples are the Hilger Uvispek, the Beckman DU, and the Zeiss PMQ-II) and grating devices (the JACO 0·5-metre model-82000 monochromator, the Perkin–Elmer model 290 monochromator, etc.) as monochromators.

Of the devices produced in the Soviet Union, we are able to use the UM-2 prism monochromators for the visible region of the spectrum, the ZMR-3 mirror monochromator with its changeable optics, the quartz monochromator from an SF-4 spectrophotometer, the monochromator from an SFD-2 spectrophotometer with a replica grating of 600 lines/mm, and the SD-2 (NIFI LGU) monochromator with 600 and 1200 lines/mm gratings. The relative apertures in these devices are between 1:7 and 1:10; the dispersions are given in Fig. 2.28.

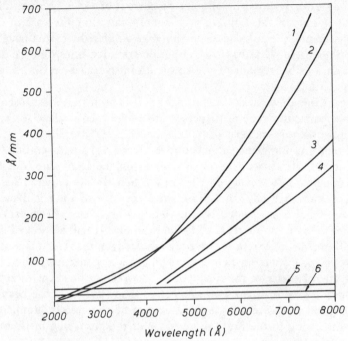

FIG. 2.28. Reciprocal linear dispersions of certain monochromators produced in the Soviet Union. 1. ZMR-3 (quartz); 2. SF-4; 3. UM-2; 4. ZMR-3 (glass); 5. SFD-2; 6. SD-2 (1200 lines/mm).

An original type of two-channel spectrophotometer, based on a standard grating monochromator with a plane grating was used by Yang and Legallais, also by Rikmenspoel.[83] Two half-size diffraction gratings mounted side by side are rotatable individually about vertical axes and each grating can be set to isolate a particular spectral region. The device was intended for measuring the molecular absorption of biological specimens. The incident beam of radiation illuminated both diffraction gratings. The two spectral bands emergent from a single slit were recombined by a split-mirror arrangement.

It should be taken into consideration that the real aperture ratio of this type of monochromator is reduced by 75 per cent, since the grating area used is halved and so is the time for measuring each of the luminous fluxes.

Some research workers use spectrographs as monochromators, spectrographs in which the spectrum can be traversed in the focal plane being readily adaptable for the purpose. The exit slit is firmly secured in place of the plate-holder. Suitable instruments are the Hilger E 492 spectrograph,[36] and the Soviet ISP-51 and KSA-1 prism spectrographs and DFS-8 and DFS-13 grating spectrographs.

Some equipments for atomic absorption analysis have included medium quartz spectrographs, which are fixed adjustment instruments. For instance, Allan used the Hilger medium spectrograph with a movable slit.[37]

Fuwa and Vallee used the same spectrograph, but with slits cut in interchangeable aluminium plates to correspond to the locations of different resonance lines.[38] The slit plates took the usual place of the photographic plates in the plate-holder. The positions for the slits were found by photographing the spectrum on the emulsion-coated plates. The plates were blackened to reduce scattered light. Initially, these authors had used standard glass photographic plates, applying a phosphorescent layer of sodium salicylate at the positions of the resonance lines after the emulsion at these positions had been removed.

Butler and Strasheim[74] described a four-channel attachment to the medium Hilger spectrograph. Each of the channels included a slit of fixed length (8·5 mm) and width (60 μm) and a lightpath with flat and concave mirrors for conveying the light to a vertically mounted 13-stage EMI photomultiplier in a light-tight casing. The lightpaths were arranged in pairs in two different horizontal planes, so that closely adjacent lines, or even one and the same line in regions of the focal plane at different heights, could be measured. The two neighbouring lightpaths were so designed that the minimum distance between the slits of a pair was 4 mm. The channels were moved along two guide rods by means of a fine steel cable, running through blocks and rigidly fixed to the housing of the photomultipliers. The gears and anti-backlash device provided two different cable speeds (1:150). A complete revolution of the handle (6 cm diameter) moved a channel 530 μm, affording scope for fine adjustment. Each had a calibrated wavelength scale. All scales were projected optically on a single screen. When the outlet slits in the two different planes were completely filled, the illuminated height of the entrance slit was 18 mm. In spite of the curvature of the lines in the focal plane of the spectrograph, the 40-μm entrance slits ensured that the luminous flux was practically utilized in full and provided resolution adequate for analytical purposes.

It should be noted that, in spite of the previously mentioned general shortcomings of prism instruments in comparison with grating instruments, there are definite advantages in using medium quartz spectrographs in multi-channel atomic absorption instruments. In the first place, the variable dispersion of the spectrograph is compensated by the fact that resonance lines are not uniformly distributed in a spectrum, so that the actual distribution of resonance lines in the focal plane of the spectrograph is fairly uniform. This makes it possible to utilize the comparatively short (about 22 cm long) focal plane of the spectrograph very effectively for isolating lines in the entire working range of the spectrum—from 2000 to 8500 Å. It must also be noted that, in the region up to 3000 Å, in which almost half the resonance lines of elements lie, the dispersion of the instrument device is no less than that of grating instruments of the same size. Secondly, the plane focal surface of the instrument substantially simplifies the construction of movable channels. Thirdly the instruments are now widely used in spectro-analytical laboratories. The possibility of using them in atomic absorption apparatus therefore simplifies and cheapens the atomic absorption method.

The principal difficulty in designing multi-channel photoelectric attachments for medium spectrographs is that of separating the light from closely adjacent lines and feeding it to independent detectors. This problem can be effectively solved by means of fibre optics,[84,85] which depend on the propagation of a beam of light by successive total internal reflections along a transparent cylindrical rod. The light pipe, made up of separate fine cylindrical fibres a few microns in diameter, is capable, not only of transmitting practically all of a beam of light from one point to any other (the losses of light are small if the curvature of a fibre is small relative to its diameter), but also of reshaping the image of an object by altering the positions of the fibres at the ends of the light pipe. It is possible to flatten out the entrance to the light pipe so that it simultaneously serves as the exit slit of the spectrograph. The advantage of such light pipes is that the number of exit slits can be greatly increased (with the same focal plane dimensions) and the feeding of light to the detectors greatly simplified.

A simple and effective way to separate resonance-line emissions from white light has been proposed by Sullivan and Walsh.[86] This method consists of using the resonance emission from atomic vapours excited by light containing resonance frequencies. A hollow-cathode lamp takes the place of a monochromator (the authors call it a 'resonance monochromator'); the function of the lamp is to produce a cloud of unexcited atoms of the element. As the beam of light passes through this cloud, the resonance frequencies are absorbed and re-emitted uniformly in all directions. Only the re-emitted radiation reaches the detector, which is mounted perpendicular to the direction of the beam.

This method of monochromation was used in a simplified atomic absorption spectrophotometer for determining magnesium and copper (Fig. 2.29).[86] The light from the high-intensity source (a boosted output lamp) (1) passed through the absorbing medium (2) into the resonance monochromator (3). The re-emitted resonance radiation was recorded by the photomultiplier (4) at the outlet of the branch tube perpendicular to the axis of the tube. The natural emission of the lamp serving as resonance monochromator was screened by a cylindrical screen round the hollow cathode. Preliminary experiments for magnesium and copper were successful.

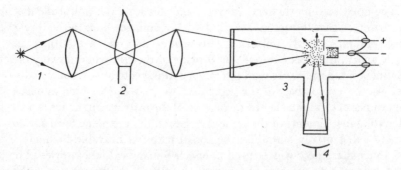

FIG. 2.29. Diagram of resonance monochromation. 1. Source of light; 2. Flame; 3. Tube with hollow cathode; 4. Detector.

Simpler methods of producing a fluorescent cloud of vapours, such as heated cells containing the elements to be determined, can also of course be used in resonance monochromators. Experiments using simple hollow-cathode lamps and heated cells containing calcium and magnesium were also extremely successful.[86]

The advantage of resonance monochromation is its extremely high resolving power, which is governed by the width of the absorption lines in the re-emitted cloud of vapour (of the order of 0·01 Å). This makes it possible to isolate the central part of the wide emission line in a source of light. For this reason, resonance monochromators may greatly improve the linearity of calibration graphs whose curvature is due to line broadening in the source. The high resolving power may also prove extremely useful for increasing the signal-to-background ratio when strong continuous emission from a flame is superimposed on emission from the light source.

Sullivan and Walsh[86] suggest that resonance monochromatizing opens up extremely interesting prospects in the design of simple polychromators for the simultaneous determination of several elements. At the same time, this method of monochromation for atomic absorption measurements has a number of limitations in comparison with other simple methods, such as the use of filters:

1. Resonance monochromators have aperture ratios about two orders of magnitude smaller than interference filters with transmittances up to 50–70 per cent. The reason is that the resonance monochromator fails to utilize all the re-emitted radiation, only the radiation confined by the solid angle at the detector being recorded. The luminous flux is therefore also two orders of magnitude less than when filters are used. As we shall prove later (see pp. 95–114), this restricts the possibility of improving detection limits.

2. The spectra of most elements (apart from elements in the second group of the periodic table) contain several resonance lines of comparable strength. The absorption recorded beyond a resonance monochromator is the combined absorption of all the resonance lines. It is easy to see that the calibration graphs, as in the case of non-monochromatic lines, must be curved, and the sensitivity will be slightly less accurate. This shortcoming was confirmed when copper was determined;[86] the resonance monochromator simultaneously isolated the Cu 3248 Å and Cu 3274 Å resonance lines, which differ 2:1 in sensitivity. Of course, additional monochromation of the re-emitted radiation, using filters can be employed, but the resonance monochromator then to a large extent loses one of its principal advantages, simplicity.

3. We should also consider such disadvantages of the resonance monochromator, in comparison with conventional monochromators, as the necessity for pretreating the lamp to give steady cathode sputtering, the possibility of fluctuations due to lamp instability, and the short lamp life due to the absorption of gas, and the condensation of sputtered cathode material on the window.

ILLUMINATING SYSTEMS

The function of the illuminating system in a spectrophotometer is to pass light from the source through the atomic vapour to the monochromator. In atomic absorption spectrochemical analysis, it is particularly important that the illuminating system should be well chosen, for the following reasons.

Firstly, atomic absorption is unlike molecular absorption spectrophotometry in having an absorbing medium that is strongly emissive. Owing to partial excitation of atoms in the flame, the spectrum may contain atomic and molecular line emissions as well as continuous emission from, for instance, incandescent particles of sample mist and excess carbon (particularly in reducing flames).

Secondly, the dimensions of the atomic vapour in atomic absorption measurements are often very limited; we refer, in particular, to long narrow flames from slot burners and to graphite cuvettes. The dimensions of the beam of light passing through the atomic vapour must be in accord.

Let us consider the optical systems used for atomic absorption measurements, taking the above special features into consideration. Fig. 2.30 shows (*a*) a two-lens system forming an image of the source in the centre of absorption region, (*b*) a single-lens system with the source focused on the slit with the absorption region close to the condenser, and (*c*) a two-lens system with the absorbing region positioned in a collimated beam.

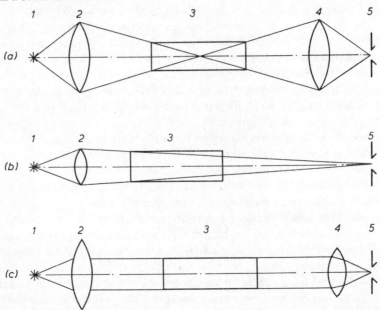

FIG. 2.30. Illuminating systems used for atomic absorption measurements.
1. Source of light. 2 and 4. Lenses;
3. Absorption region. 5. Monochromator slit.

The first system focused the combined image of the source and the centre of the absorption region on the monochromator slit. The second system differs in that the absorption region is at a distance from the slit. This has given some workers grounds for thinking that the luminous flux reaching the monochromator from the atomic vapour must be less in the second case than in the first and that, for this reason, given equal luminous fluxes from the source, the relative contribution of emission from the atomic vapour will also be less in the second case. It is not difficult to prove that this opinion is mistaken. The luminous flux passing through the slit (*b*) can be represented, in simplified form, by two truncated cones, in which the section S_1 corresponds to the emitter (the flame), S_2 to the slit of the spectroscopic device, and S_3 to the collimator of the monochromator (Fig. 2.31).

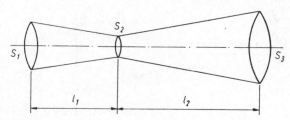

FIG. 2.31. Illumination of the monochromator slit in system (*b*).

We know that the luminous flux in a beam from a light source is:

$$\Phi = \frac{BS_1 S_3 \cos \theta}{l^2} \qquad (2.8)$$

where B is the brightness of the light sources, S_1 is the area of the emitter, S_3 is the area on which the beam falls, l is the length of the beam, and θ is the angle between the axis of the beam and the normal to the plane S_1.

For the adjoining cones (Fig. 2.31),

$$\frac{S_1}{l_1^2} = \frac{S_3}{l_2^2} \qquad (2.9)$$

On the other hand, according to the Lagrange–Helmholtz law, if there is no attenuation as a pencil of rays passes through an optical system, its brightness is constant along its entire path, so that the brightness of a pencil in the S_2 plane is equal to its brightness in the S_1 plane, i.e.

$$B_1 = B_2 \qquad (2.10)$$

Using equations (2.9) and (2.10), it follows from equation (2.8) that

$$\Phi_1 = \Phi_2 \qquad (2.11)$$

i.e. provided that the lens is filled to the same extent in each case, the luminous flux reaching the collimating lens of a monochromator is independent of the position of the source of emission relative to the slit.

System (*b*) therefore does not provide a lower relative flux from the atomic vapour than system (*a*).

At first sight, system (*c*) with its collimated beam, is superior to system (*a*) [Fig. 2.32 (*a*)]. It would seem that, by having the focusing lens at a distance from the absorbing cell, we shall at the same time reduce the angular aperture of the beam of light from the absorbing volume without losing any of the luminous flux

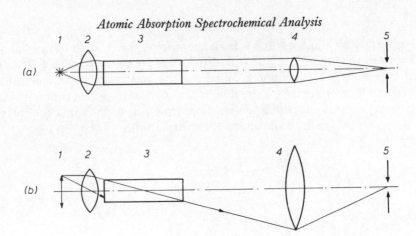

FIG. 2.32. (a) Hypothetical and (b) actual course of rays in system (c).
1. Source of light; 2 and 4. Lenses; 3. Absorbing medium. 5. Slit of monochromator.

from the light source, i.e. we shall achieve the best possible signal-to-background ratio. In fact, however, this is not so. Owing to the finite dimensions of light sources, the aberrations of optical systems, and diffraction, no present-day optical system is capable of providing a perfectly parallel beam.[39] The actual course of the rays in system (c) is shown in Fig. 2.32 (b). The beam from the first lens is a bundle of rays at varying angles to the optical axis of the system. The purpose of the second lens is to direct all the oblique rays passing through the absorbing volume to the slit of the monochromator. The angular aperture of the beam of light passing from the absorbing volume through the second lens is therefore governed, as in system (a), by the cross-section of the absorbing volume.

The flux passing from the absorbing volume to the slit is therefore the same in system (c) as in system (a). On the other hand, according to the Lagrange–Helmholtz law, given a beam of finite angular aperture (defined in our case by the cylindrical absorbing volume) it is impossible for the luminous flux that fills an aperture of definite size to exceed the flux obtained when the aperture is filled by the image of the source. Therefore, the luminous flux passing through a cylindrical absorbing volume in a 'parallel' beam system (c) cannot exceed the flux passing through the volume with the source imaged at an intermediate point (system a).

The ratio of the fluxes from the light source and the volume cannot therefore be better in system (c) than in system (b). This conclusion arises out of radical considerations, and can evidently be generalized as follows: no illuminating systems can provide better ratios between the fluxes from the source of light and the absorbing volume than systems with an intermediate image of the source in the absorbing volume and with the aperture completely filled by the beam passing through that volume.

It was stated earlier that emission from the absorbing volume consists of superimposed continuous and line spectra. To increase the ratio of the useful signal from the (line) source to the background noise from the absorbing volume, it is possible to reduce the width of the monochromator slits, since the flux for a line spectrum is proportional to the slit-width, while for a continuous spectrum it is proportional to the square of the slit-width. This method, however, involves a decrease in the absolute luminous flux from the light source and, in consequence (see p. 101), a relative increase in the shot noise at the detector. For this reason, the absolute luminous flux reaching the monochromator from the light source must not be decreased. In view of this, the possibilities afforded by proper choice of illuminating systems and monochromators should be fully heeded.

1. It follows from the Lagrange–Helmholtz law, formulated above in its more general form, that the luminous flux Φ which can pass through a volume with cross-sectional area S and angular aperture θ cannot exceed

$$\Phi = \pi B S \sin^2 \theta \qquad (2.12)$$

With the brightness B of the source constant, therefore, the amount of light passing through the volume cannot be increased, whether by increasing the dimensions of the source and the lenses, by altering the relative positions of the component parts of the illuminating system, or by any other optical means.

Of the types of illuminating system considered above, system (a) has the greatest illuminating power, and the single-lens system (b) has the lowest illuminating power (with an absorption volume of the same dimensions).

2. For the luminous flux from a source to be most fully utilized, the angular aperture of the beam reaching the slit must be such that the collimating lens of the monochromator is filled.

To illustrate the arguments given above, we shall design an illuminating system for a particular case. The source of light is a hollow-cathode lamp with a bright-spot diameter of 5 mm; the atomic vapour is within a 3 mm diameter graphite cuvette 50 mm long; the monochromator is a ZMR-3 instrument with a relative aperture of 1:7. The principal symbols required for the calculation are shown in Fig. 2.33.

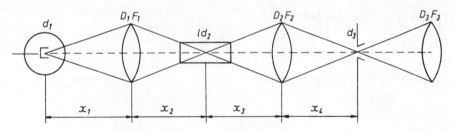

FIG. 2.33. Symbols for design of an illuminating system.

From equation (2.12), the maximum luminous flux passing through the atomic vapour is:

$$\phi_2 = \pi B \frac{\pi d_2^2}{4} \frac{\pi d_2^2}{4} \frac{4}{l^2}$$

The flux ϕ_1 passing through the first lens cannot be less than ϕ_2, i.e.

$$\phi_1 = \pi B \frac{\pi d_1^2}{4} \frac{\pi D_1^2}{4x^2} \geqslant \pi B \frac{\pi d_2^2}{4} \frac{\pi d_2^2}{4} \frac{4}{l^2}$$

Hence

$$D_1 \geqslant \frac{2d_2^2 x_1}{l d_1} \quad (2.13)$$

The equation for a thin lens

$$\frac{1}{x_1} + \frac{1}{x_2} = \frac{1}{F_1}$$

and the relationship

$$\frac{x_1}{x_2} = \frac{d_1}{d_2}$$

give us:

$$\left.\begin{aligned} x_1 &= \frac{F_1(d_2 + d_1)}{d_2} \\ x_2 &= \frac{F_1(d_2 + d_1)}{d_1} \end{aligned}\right\} \quad (2.14)$$

If, for instance, we take $F_1 = 150$ mm and substitute the other parameters given above, we get: $x_1 = 400$ mm, $x_2 = 240$ mm, $D_1 \geqslant 21 \cdot 6$ mm.

If the angular aperture of the beam reaching the slit of the monochromator is in accordance with the aperture ratio of the instrument, the condition for equality between the fluxes leaving the absorption volume and reaching the monochromator slit can be written as follows:

$$\pi B \frac{\pi d_2^2}{4} \frac{\pi d_2^2}{l^2} = \pi B \frac{\pi d_3^2}{4} \frac{\pi D_3^2}{4 F_3^2} \quad (2.15)$$

It follows from this that the diameter of the image of the source at the slit is:

$$d_3 = \frac{2 F_3 d_2^2}{D_3 l} \quad (2.16)$$

Using the thin-lens equation

$$\frac{1}{x_3} + \frac{1}{x_4} = \frac{1}{F_2}$$

and the relationship

$$\frac{x_3}{x_4} = \frac{d_2}{d_3}$$

we get:

$$\left.\begin{aligned} x_3 &= F_2 \left(1 + \frac{D_3 l}{2 F_3 d_2}\right) \\ x_4 &= F_2 \left(1 + \frac{2 F_3 d_2}{D_3 l}\right) \end{aligned}\right\} \quad (2.17)$$

The diameter of the second lens will be:

$$D_2 = F_2 \frac{2d_2}{l}\left(1 + \frac{D_3 l}{2F_3 d_2}\right) \quad (2.18)$$

With the parameters given above and $F_2 = 100$ mm, we get: $x_3 = 219$ mm, $x_4 = 184$ mm, $D_2 = 26\cdot3$ mm, $d_3 = 2\cdot5$ mm.

MODULATION OF LIGHT SOURCE EMISSION

The previous section has shown that the relative magnitude of the light flux reaching the monochromator slit from an atomic vapour cannot, with the brightness of the source and the volume of the atomic vapour constant, be reduced by any optical means. It is true that, if the flux from the atomic vapour is continuous emission, the relative magnitude of this flux reaching the detector can to some extent be reduced by isolating narrower spectral regions. This method does not, however, make it possible to reduce the relative strength of superimposed line emission from the atomic vapour.

The superimposition of foreign emission introduces appreciable error into the results. For example, the superimposition of emission comprising only 1 per cent of the intensity of the signal from the source will reduce the readings by 4 per cent (with an absorbance of about 1·0).

A common method for isolating a useful signal from background signal consists of modulating the useful signal and so isolating it by electronic means. The luminous flux from a lamp can be modulated either by mechanically interrupting the beam of light with a rotating sector or an oscillating or rotating mirror or by modulating the current supplied to the lamp. If the foreign emission is constant, isolating the modulated signal amounts to separating the variable signal from the constant one. This can be done with any modulation frequency and, if the shot effect is ignored (see pp. 95–103), with any a.c. amplifier bandpass.

It must, however, be taken into account that the superimposed emission may include a perceptible variable component, caused for instance in flames by poor mixing of the gases, uneven scattering by the nebulizer, external currents of air, etc. These fluctuations may have a frequency distribution that is not uniform. Fluctuations in emission from d.c. or a.c. arcs, for example, may decrease by almost two orders of magnitude as the frequency is increased from 2 to 10^3 c/s.[40] One would expect a similar non-uniform distribution in emission from a flame.

The usual method of limiting the effect of these fluctuations is to use selective amplifiers and modulation frequencies which are high in comparison with the frequency of fluctuations from other sources. The shot effect cannot be eliminated. When the modulation frequency is selected, we should also bear in mind the possibility of fluctuations in the amplifying system at frequencies of 50 or 100 c/s due to the mains a.c. supply to the valve filaments and poor rectification and inadequate filtering in the anode circuits. Modulation frequencies between 200 and 1000 c/s are normal in spectrophotometers.

A general method of counteracting interference is to restrict the bandpass Δf of the recording device. At low frequencies (up to 10^4 c/s), selective a.c. amplifiers, with T-shaped RC filters included in the negative feedback circuit of one or more of the stages, are used for this purpose. The selective properties of the amplifier are given by the quality factor Q:

$$Q = \frac{f}{\Delta f} \qquad (2.19)$$

This in turn will depend on the amplification G of the stages in the negative feedback:[41]

$$Q = \frac{G+1}{4} \qquad (2.20)$$

It is advisable to use the highest possible Q in order to achieve the smallest bandpass. Increasing Q to over a few hundred, however, involves using multi-stage circuits, which can provide the required amplification G. On the other hand, the possibility of constructing amplifiers with high values of Q depends on the stability of the modulation frequency, i.e. usually on the stability of the mains frequency f_m supplied to the mechanical modulators. If the mains frequency varies by Δf_m, the amplification factor G of the selective amplifier alters. The quantities Q, $\Delta f_m/f_m$, and $\Delta G/G$ are linked by the following approximation.[41]

$$Q \simeq \frac{\Delta G}{G} \frac{f_m}{\Delta f_m} \sqrt{2} \qquad (2.21)$$

If the variation Δf_m in the mains frequency is 0·01 per cent f_m, and fluctuations $\Delta G/G > 1$ per cent are not permissible, then approximation (2.21) shows that it is inadvisable to increase Q to above 140. Serious difficulties are therefore involved in obtaining high values of Q.

Another method of achieving minimum values of Δf is to use lower modulation frequencies f. However, it is necessary to note that the frequency distribution of noise in the measured and foreign emissions may not be uniform, particularly in the low-frequency range.

The signals amplified by the selective amplifier are diode detected, together with the noise, in the Δf range of frequencies and measured by means of an indicating and recording device. Recording devices also have their particular bandpass Δf_1, governed by the oscillation period T (in seconds) of the galvanometer system, thus:

$$\Delta f_1 \simeq \frac{1}{T} \qquad (2.22)$$

or by the time constant of the RC circuit which shunts the output device (low-frequency filter):

$$\Delta f_1 = \frac{1}{\pi RC} \qquad (2.23)$$

R is expressed in ohms, C in farads.[41]

The effective half-bandwidth Δf_{eff} of the entire recording system is determined by the equation:

$$\Delta f_{eff} = \frac{\Delta f \cdot \Delta f_1}{\Delta f + \Delta f_1} \qquad (2.24)$$

It follows from equation (2.24) that the magnitude of Δf_{eff} (within a factor of 2) depends in practice on the lower of the quantities Δf and Δf_1.

Either non-linear circuits (for instance diode circuits) or synchronous detecting circuits are used as detecting systems. In the former, the useful signal and the noise, of either polarity and whatever their phase, pass together through the detector. The output device therefore records not only the rectified component of the useful signal, but also the d.c. component of the unipolar noise. The superimposition of the d.c. component of noise on the useful signal greatly complicates the measurement procedure, particularly when the noise may vary during the analysis.

The synchronous detector has an electronic or mechanical switch synchronized with the modulation frequency. In the case of electric modulation synchronization is achieved by supplying power to the source of light and the switch from the same generator, while in the case of mechanical modulation the switch is synchronized with the modulation frequency by means of an additional element (a photoresistor, a system of relays, etc.) included in the modulation circuit. The moment when the synchronized detector is switched on, signals of any polarity pass through it. The opposite polarity pulses for all the signals differing in phase from the modulation frequency compensate one another, so that the d.c. component, integrated for a specific period and recorded by the output device, is almost zero. The synchronized detection system thus makes it possible to eliminate the superimposition of the d.c. components of noise on the useful signal.

The bandpass of a synchronized detector, and consequently also the level of fluctuations of the output signal, is governed, as with diode detectors, by the time-lag in the recording device or the time constant of the RC circuit; owing, however, to filtration of the signals as to phase, the synchronized detection system provides a slightly better signal-to-noise ratio. This last advantage of synchronized detection becomes particularly appreciable if the noise level is close to or greater than the useful signal.[41]

Examples of the use of different electrical and mechanical modulation circuits and of the isolation of signals will be given later (pp. 107–14).

In addition to the methods of modulation described above, resonance emission can also be modulated by an optical method, in which an atomic vapour is interposed in the lightpath periodically. Optical modulation has an advantage over mechanical and electrical methods in that the lamp emission is not all modulated, but only its resonance frequencies. With the optical method of modulation, therefore, not only can the lamp emission be isolated from foreign emission, but the resonance lines can be distinguished in the spectrum of the lamp.

It should be borne in mind that optical modulation can be used effectively only when emission from the atomic vapour is weaker than emission from the light source. If this is not so, the periodical emission from the absorbing medium may obscure the simultaneous reduction in intensity of light from the source.

Although in principle the optical method of modulation makes it possible to dispense with additional equipment for monochromating the light reaching the detector, in fact the presence of several resonance lines in the spectra of most elements must be taken into consideration, as well as the increase in recording circuit noise resulting from exposure of the detector to unmodulated emission from the light source and the atomic vapour used in the analysis. In order, therefore, to restrict the spectral region, any of the usual monochromating equipment is employed—filters or monochromators.

The optical method of isolating atomic resonance lines from a light source spectrum was first used by Alkemade and Milatz.[63] As the absorbing medium in their experiments, these authors used a flame into which the appropriate element was introduced. Metal-vapour lamps were the light sources. A double-beam system, in which a rotating sector alternately sent the beam of light through the flame or past the flame was used. The recording device, with a filter, made it possible to measure the difference between the intensities of the beams which were compared; this difference corresponded to the intensity of resonance emission from the source (see Fig. 47).

Bowman, Sullivan and Walsh[87] proposed a different method of optical modulation. The modulating cell was the discharge in a open ended hollow cathode. A pulsed voltage supplied the discharge, so that the beam of light passed through the hollow cathode was absorbed periodically. Boosted output lamps were used as sources. A small grating monochromator was an additional means of monochromation. The detector recorded the alternating component of the signal; this corresponded to the intensity of the resonance emission absorbed in the modulating cell.

This method of optical modulation has an advantage over the method proposed by Alkemade and Milatz in that its resolving power is higher, since the width of absorption lines in a hollow-cathode discharge is far less than in a flame. In view of this, the system of optical modulation using a hollow-cathode discharge can be used for monochromating resonance emission from sources which emit broadened resonance lines.

The system described by Bowman, Sullivan and Walsh[87] is a single-beam system simple to design; for the modulating hollow cathode can be positioned within the envelope of the source.

Optical modulation systems share the advantage of facilitating the isolation of the strongest resonance line from nearby non-resonance lines and weak resonance lines, for which reason simpler than normal means of monochromating can be used with larger aperture ratios. For example, Bowman *et al.*[87] showed that, when optical modulation is used for isolating the Fe 2483 Å, Ni 2320 Å and Co 2407 Å resonance lines, the superimposition of closely adjacent lines in the spectrum of the source can be entirely eliminated; the monochromator isolated bandwidths of 5 Å, 7 Å and 10 Å respectively.

ERRORS IN SPECTROPHOTOMETRIC MEASUREMENTS

Errors in spectrophotometric measurements are indicated by fluctuations in the readings of the recording instrument. The magnitude of these fluctuations depends on the stability of the light source, on the stability of attenuation of the light beam between the source and the detector, and on noise arising in the detector, amplifier, and recorder. Let us consider the effects of these various sources of error, remembering that, in absorption measurements, the important factor is the accuracy with which the absorbance A is determined, not the measured intensities of the absorbed and unabsorbed signals. We recall that

$$A = \log(I_0/I) \qquad (2.25)$$

Differentiating, we get:

$$dA \simeq 0.434 \left(1 - \frac{I_0 dI}{I dI_0}\right) \frac{dI_0}{I_0} \qquad (2.26)$$

If we substitute finite increments for the infinitesimals, and allow for the possibility that ΔI and ΔI_0 may have opposite signs, we get:

$$\Delta A = 0.434 \left(1 + \frac{I_0}{I} \frac{\Delta I}{\Delta I_0}\right) \frac{\Delta I_0}{I_0} \qquad (2.27)$$

Hence we get the following equation for the ratio between the relative error in measuring A and the relative error in measuring I_0:

$$\frac{\Delta A}{A} \Big/ \frac{\Delta I_0}{I_0} = \frac{0.434}{A}\left(1 + \frac{I_0}{I}\frac{\Delta I}{\Delta I_0}\right) \qquad (2.28)$$

Equation (2.28) indicates the relationship between the relative error in measuring A and the measured magnitude of the absorbance. The exact form of the relationship will differ according to the reasons for the fluctuations ΔI_0 and ΔI. There are three possible cases:

1. The fluctuations are constant whatever the intensity of the signals, i.e. $\Delta I =$ constant. This form of fluctuation is caused by internal noise in the read-out (for instance, oscillations of a galvanometer needle) or by recorder errors.[43,44] Amplifier noise, discussed on pp. 91–4, and noise due to superimposed foreign emission also come under this heading. The following equation gives the relative error:

$$\frac{\Delta A}{A} \Big/ \frac{\Delta I_0}{I_0} = \frac{0.434}{A}\left(1 + \frac{I_0}{I}\right) \qquad (2.29)$$

2. The fluctuations are proportional to the signal intensities, i.e. $\Delta I \propto I$. This type of fluctuation is characteristic of lamp emission, mechanical displacement of a beam of light relative to a detector, and refraction of light by its passage through atmospheric media of differing temperature. With these fluctuations,

$$\frac{\Delta A}{A} \Big/ \frac{\Delta I_0}{I_0} = \frac{0.868}{A} \qquad (2.30)$$

3. The fluctuations are proportional to the square root of the intensity, i.e. $\Delta I \propto \sqrt{I}$. This type of fluctuation is caused by statistical variation in the illumination and the current resulting from the discrete nature of light and electricity (shot effect). In this case,

$$\frac{\Delta A}{A} \bigg/ \frac{\Delta I_0}{I_0} = \frac{0 \cdot 434}{A} \left[1 + \sqrt{\left(\frac{I_0}{I}\right)}\right] \tag{2.31}$$

The relationship between the absorbance and the relative error in its determination is plotted in Fig. 2.34 for the three cases mentioned above. The curves show how, for equal relative fluctuations in I_0, the error in measuring absorbance depends, not only on the absorbance itself, but also on the nature of the fluctuations in intensity.

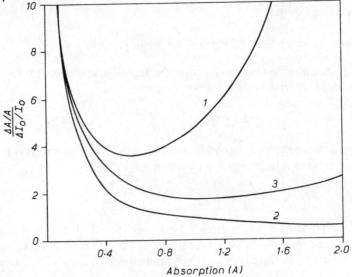

FIG. 2.34. Relationship of the photometric error to the absorbance for different types of fluctuation.
1. I = constant. 2. $\Delta I \propto \sqrt{I}$. 3. $\Delta I \propto I$.

The difference between the three types of fluctuation is particularly evident for medium and high absorbances where errors of the first case are most important.

It must once more be emphasized that the implied relationships between the values of $\Delta A/A$ for the three different types of fluctuation are valid provided the relative errors $\Delta I_0/I_0$ are the same in the three cases. In practice, the relative importance of the different sources of error in the overall error depends on a number of conditions associated with the nature of the fluctuations.

Let us estimate the magnitudes of the types of error described above and consider the principal measure that can be taken to reduce them. The errors in readings from the recording device, and its fluctuations, are practically independent of the measured signal, being the same when either strong or weak signals are measured. When signals are recorded, the measurement error is usually one scale division if the divisions amount to 0·01 times the complete scale. $\Delta I_0/I_0$ is then equal to 0·01. According to equation (2.29), the error in measuring an absorbance $A = 2·0$ is 22 per cent. When low absorbances are determined, the error is also great: for $A = 0·05$ the error is 18 per cent, and for $A = 0·01$ it is about 90 per cent.

To increase the accuracy with which small absorbances are measured, the instrument scale is expanded. The essentials of this procedure are that most of the signal is balanced against output from a stable power supply, while the remaining unbalanced part of the signal ($1/s$) is expanded over the entire scale. Since each division is now, not 0·01 of the signal, but 0·01/s, the accuracy with which intensities are measured is increased s times. This is usually referred to as a scale expansion of s times. The error in measuring absorbances is reduced, according to equation (2.29), in the same proportion. As an instance, with an absorbance $A = 0·01$ and $s = 10$, the measurement error is about 9 per cent.

A more convenient method for improving the accuracy of either low-absorbance or high-absorbance measurements requires the direct measurement of signals in terms of absorbance rather than intensity. If the instrument records a signal in units of $\log I$ rather than of I, the determination of absorbances amounts to measuring the difference $\log I_0 - \log I$ and not the actual values of both $\log I_0$ and $\log I$. The value of $\log I_0$ can be balanced in advance (backed off), so that readings correspond directly to the magnitude of A. Adjusting the sensitivity of the recording device makes it possible to record any absorbance with the same relative error.

Let us now consider errors associated with lamp instability and fluctuations in the attenuation of the beam between the source and the detector. It is characteristic of this type of fluctuation that the relationship between the intensity I and the fluctuations ΔI is linear. The principal means of eliminating or reducing the fluctuations consist of stabilizing and improving the light sources and arranging the different units (light source, atomic vapour and detector) more effectively. An indirect method of eliminating the effects of these fluctuations on the results, without reducing the fluctuations themselves, is simultaneously to measure the I_0/I ratio by means of double-beam ratiometric systems in place of the normally used method of successive measuring.

Light source fluctuations are eliminated or reduced by using more stable power supplies or a current stabilizer, by initially warming up the lamps to equilibrium discharge conditions, by stabilizing the thermal conditions in the lamps, by improving the design of the lamps, by preventing the lamp from vibrating, etc. These methods are generally known; some of them were discussed when light sources were described.

The measurement of signal ratios by means of ratiometric systems is based on using for comparison a light source signal which is sufficiently homologous with the signal being measured (resonance line intensity). In principle, a beam of white light from the source can be used for this signal, or any line (resonance or non-resonance) in the source spectrum.

Different possible ratiometric systems are shown diagrammatically in Fig. 2.35. When white light is used for comparison (system a), a proportion of the beam is reflected to an additional detector, the remainder passing through the absorbing medium. When the analysis line is used for comparison (system *b*), the signals corresponding to the two beams of light are received by a common detector and separated electronically, the frequency modulation or phase modulation of the beams being different. When a non-resonance line is used for comparison (system *c*), it is best to use the same monochromator, but with two independent measuring channels for isolating the comparison beam from the main beam; since the non-resonance line is not absorbed as it passes through the atomic vapour, there is no need to isolate the comparison signal before its passage through the atomic vapour.

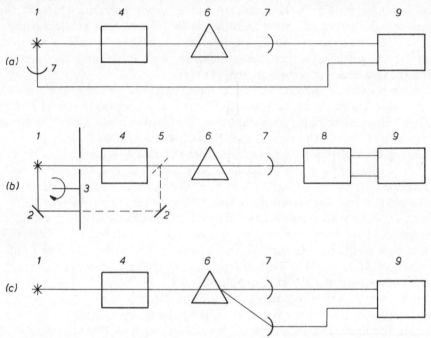

FIG. 2.35. Possible ratiometric systems. 1. Light source; 2. Mirror; 3. Rotating disk; 4. Absorption medium; 5. Semi-reflecting mirror; 6. Monochromator; 7. Detector; 8. Discriminator; 9. Recording unit.

It is obvious that system (*b*) provides the most homologous relationship between the main signal and the comparison signal, and that system (*a*) is the worst in this respect. On the other hand, system (*a*) is the simplest to produce and (*b*) the most complicated. When the same resonance line is used for comparison, the system becomes much more complicated owing to the double-beam optical system and the necessity for using a discriminator for separating the signals. The system in which white light is used requires only an additional detector. The system in which a non-resonance line is used requires a two-channel monochromator.

The system selected depends on a number of factors. If, for instance, the lamp used as a light source emits a large proportion of its energy in the resonance lines (radio-frequency lamps for alkali metals), system (*a*) is the best, since the white beam and resonance line must be highly homologous. To allow for fluctuations associated with the passage of the beam of light through an atomic vapour, it is best to use system (c), since in this system the intensities of the analysis and reference lines are subjected to the same effects.

Let us now dwell on internal fluctuations in photodetectors due to the discrete nature of light and electricity. While the types of fluctuation considered above can be reduced to some extent, or totally eliminated, by more or less complex measures, this last type of fluctuation cannot in principle be eliminated, though the effects of noise on the results can be reduced by using suitable measurement conditions.

Photodetector noise is caused firstly by statistical scatter about the mean N of the number of photons received in given increments of time and secondly by scatter in the emission of photoelectrons $(n - \bar{n})$. If photons are emitted at random, it can be shown that emission of photoelectrons is also random.[45] Detector fluctuations can therefore be regarded as a single source of noise.

In addition to shot noise in the detector, we must allow for thermal noise due to the random movement of electrons in the ballast resistor of the detector. Since photomultipliers are the detectors most commonly used in modern spectrophotometers, let us consider certain theoretical conclusions concerning them. Similar results have been obtained for photocells with valve amplification.*

A theoretical relationship can be established between the signal-to-noise ratio and the photocurrent i_ϕ in the detector. If V_ϕ is the signal voltage and $\sqrt{\bar{V}^2}$ the root mean square noise voltage due to all causes, then the signal-to-noise ratio is given simply by

$$\rho = \frac{V_\phi}{\sqrt{\bar{V}^2}} \tag{2.32}$$

To express the ratio in decibels, we may write:

$$\rho_{dB} = 20 \log \rho \tag{2.33}$$

The theoretical relationship of ρ_{dB} to i_ϕ is plotted in Fig. 2.36. Curve 1 is for the case in which there is no photomultiplier dark current ($i_d = 0$). The bandpass of the recorder is 1 c/s ($\Delta f = 1$ c/s). Under normal operating conditions for photomultipliers (gain 10^6, ballast resistance $\simeq 10^6 \Omega$), the shape of the graph is governed solely by the shot effect. The slight curvature for low values of i_ϕ is due to the superimposition of the thermal effect in the ballast resistance. Curves 2 and 3 allow for the existence of photomultiplier dark current: 10^{-14} amp for antimony-caesium, and 10^{-12} amp for oxygen-caesium photocathodes. This graph shows that, if the photocurrents are higher than the dark currents ($i_\phi > i_d$), the effects of the dark currents can be ignored.

In absorption spectrophotometry, we are usually dealing with luminous fluxes sufficiently large for the dark current to be unimportant. The relationship of ρ to i_ϕ then corresponds to the upper region of curve 1 and can be expressed thus:

$$\rho \simeq 1 \cdot 8 \cdot 10^9 \sqrt{(i_\phi/\Delta f)} \tag{2.34}$$

here i_ϕ is the photocurrent in amps and Δf is the bandpass in c/s.

* The application of the theory of electrical noise to light detectors (photomultipliers and photocells) is considered by Chechik et al.[42]

In precision spectrophotometric measurements, the shot noise is perceptible. As an example, let us consider the atomic absorption spectrophotometer described by Box and Walsh[8] (a diagram of the amplifier for this spectrophotometer is given on p. 110). Type IP 28 photomultipliers were used as detectors for the ultra-violet and visible regions and IP 22 photomultipliers for the infra-red region. With a dynode voltage of 100 V the photomultiplier amplification was $2 \cdot 10^5$ and a full-scale deflection of the microammeter was given by an anode current of $5 \cdot 10^{-9}$ amp. With this amplification and anode current, the photocurrent i_ϕ was $2 \cdot 5 \cdot 10^{-14}$ amp. The amplifier bandpass was governed by the RC circuit, with a 200 μF capacitor and a 4 kΩ resistance. Equation (2.23) gives us $\Delta f = 0 \cdot 4$ c/s and expression (2.34) then gives $\rho \simeq 2 \cdot 5 \cdot 10^2$. In other words, under the specified conditions, the photomultiplier shot noise comprises about 0·4 per cent of the signal. If the luminous

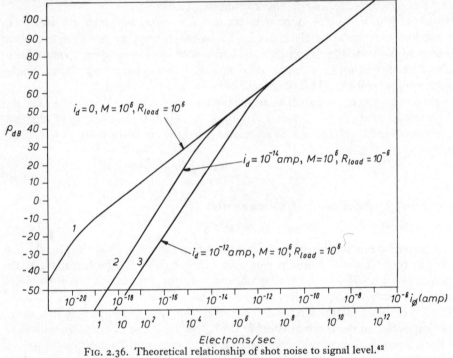

FIG. 2.36. Theoretical relationship of shot noise to signal level.[42]

flux is increased four times, the shot noise will drop to 0·2 per cent of the signal.

Box and Walsh, when investigating the stability of hollow-cathode lamps, concluded that there are lamp emission fluctuations, which cannot be eliminated, amounting to about 0·25 per cent of the signal. However, when the measurements were made with photocurrents i_ϕ less than $1 \cdot 10^{-13}$ amp, the fluctuations should have been ascribed, not to the lamp, but to the shot effect.

Fluctuations related in magnitude to the photocurrent in the photomultiplier were also found when noise was investigated in the measuring circuit of the spectrophotometer developed by Yang and Legallais.[46] An IP 28 photomultiplier was used. The measured fluctuations corresponded to the calculated shot noise values (Table 2.5).

TABLE 2.5 Comparison of calculating and measured values of ρ[46]

i (amp)	Calculated	Measured
$1 \cdot 10^{-10}$	$0{\cdot}73 \cdot 10^4$	$0{\cdot}7 \cdot 10^4$
$0{\cdot}25 \cdot 10^{-10}$	$0{\cdot}37 \cdot 10^4$	$0{\cdot}35 \cdot 10^4$
$0{\cdot}10 \cdot 10^{-10}$	$0{\cdot}23 \cdot 10^4$	$0{\cdot}23 \cdot 10^4$

It is interesting that, in this work, absorbances of about 0·0001 were measured for considerable luminous fluxes (corresponding to $i_\phi > 1 \cdot 10^{-10}$ amp). An incandescent lamp was the light source. With these high luminous fluxes, it proved possible to reduce the voltage in the IP 28 photomultiplier to 500 V.

Finally, it was also established by Bodretsova et al.,[7] the results of which were given on p. 37, that fluctuations in the signals recorded with a spectrophotometer from a hollow-cathode lamp, with photocurrents of up to $4 \cdot 10^{-12}$ amp, corresponded to photomultiplier shot noises.

When precision measurements are being made, therefore, the internal noise of the recording apparatus cannot be ignored. To reduce it, the luminous flux should be increased, i.e. brighter sources of light should be used, and efficient illumination systems and monochromators with large aperture ratios. The bandpass of the recording device should also be reduced (see p. 91).

It has already been pointed out that equation (2.34) is valid on condition that the photomultiplier dark current is less than the photocurrent from the useful signal. If the dark current, or in general the current due to the superimposition of a background, is comparable with or greater than the useful signal photocurrent, equation (2.34) should be rewritten as follows:

$$\rho = 1 \cdot 8 \cdot 10^9 \frac{i_\phi}{\sqrt{[(i_\phi + i_d) \Delta f]}} \qquad (2.35)$$

This means that the shot noises are governed by the total flux recorded by the receiver. The superimposition of foreign emission shows up, not only in increased noise caused by the fluctuation in this emission (p. 91), but also in increased shot fluctuations. In view of this, the following circumstances should be noted. As was pointed out in p. 91, the spectral region isolated by the spectroscopic device can be narrowed to reduce the superimposition of a continuous background from the absorbing medium on the line spectrum of the source. The signal-to-background ratio is then bound to increase in proportion to the reduction in slit-width. In the case of shot noise, however, this method is not always appropriate.

Actually if the width of the slits is reduced, the magnitude of i_ϕ decreases more rapidly than $\sqrt{(i_\phi + i_d)}$, and as equation (2.35) indicates, the relative magnitude of the shot noise must increase. When the accuracy with which the signal is measured is governed by the detector shot effect, wider slits should therefore be used (assuming there are no other reasons for using narrow slits, such as a complex light-source spectrum).

We considered all the principal causes of error in spectrophotometric measurements and methods of reducing them or of preventing them from affecting analytical results. Our conclusions are summarized in Table 2.8. In addition to the tabulated methods, there is another general means of making determinations more precise; it consists of averaging the results of several measurements. According to mathematical statistics, the mean square error $\sigma_{\bar{x}}$ for n independent measurements is $\sqrt{n}$ times less than the error of a single measurement, σ_x, i.e.

$$\sigma_{\bar{x}} = \frac{\sigma_x}{\sqrt{n}} \tag{2.36}$$

TABLE 2.8 Classification of errors in spectrophotometric measurements

$\Delta I = f(I)$	Types of error	Methods of eliminating or reducing the error
ΔI = constant	Errors in taking reading	Use of scale expansion. Use of logarithmic amplifiers
	Vibration of galvanometer needle or recorder pen	Use of scale expansion. Use of logarithmic amplifiers.
	Amplifier noise (at mains frequencies)	Selection of suitable modulation frequency.
	Superimposed emission noises (with non-uniform frequency distribution)	Selection of suitable modulation frequency. Limitation of recorder bandpass.
	Photomultiplier shot noise resulting from superimposition of foreign emission	Limitation of recorder bandpass. Reduction in brightness of foreign emission.
$\Delta I \propto I$	Variable lamp emission	Use of double-beam systems. Use of two-channel single-beam systems.
	Mechanical displacements of the beam of light	Use of two-channel single-beam systems.
	Refraction of light	Use of two-channel single-beam systems.
$\Delta I \propto \sqrt{I}$	Photomultiplier shot noise arising during measurement of light source signal	Limitation of recorder bandpass. Increase in flux from light source: increase of brightness, use of efficient optical system, use of monochromator with larger aperture ratio.

When constant quantities are to be determined, the result is meaned by integrating the current over a certain length of time. The method is as follows: the photocurrent amplified by the photomultiplier i_A charges a capacitor with a capacity C. After a period of time τ the capacitor potential is:

$$V = \frac{i_A \tau}{C} \tag{2.37}$$

The measurement amounts to determining the voltage V. When integrating circuits are used, the measurement error is reduced in proportion to $\sqrt{\tau}$.

SYSTEMS FOR MULTIPLE REFLECTION OF LIGHT BEAMS THROUGH ATOMIC VAPOURS

The repeated passage of a beam of light through an atomic vapour is one of the methods of making measurements more sensitive. This method is extensively used in molecular spectroscopy. Different combinations of mirrors are used for passing the beam of light through the atomic vapour with the minimum of reflection losses.

One of the most widely used systems was proposed by White.[47] This system consists of three spherical concave mirrors, all with the same radius, cut in the form of squares. Two of them (M_1 and M_2) are side by side at one end of the atomic vapour, while the third (M_3) is at a distance equal to the radius of curvature of the mirrors at the other end (Fig. 2.37). This system, with its conjugate foci, ensures

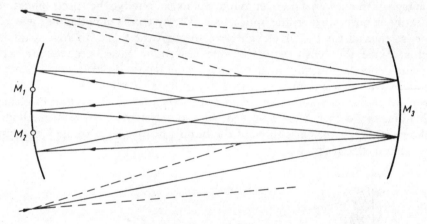

FIG. 2.37. White's mirror system.[47]

that the beam of light passes repeatedly across the space between the mirrors, without any restriction on the angular aperture of the beam. The number of crossings by the beam of light (a multiple of four) is regulated by means of mirrors M_1 and M_2 (Fig. 2.37 shows the path followed by the central beam in eight passages). Reflection losses depend on the quality of the mirror coatings. In the apparatus tested by White, the intensity of the light flux after twelve passages was 20–30 per cent of the initial intensity. White's system, with the beam passing twelve times, was used by Russell, Shelton and Walsh[5] for increasing sensitivity of atomic absorption measurements using the concentric flame from a Méker burner. Sensitivity was improved by about ten times.

Tabeling and Devaney[48] established that light losses in multiple-reflection systems so increase the detector noise that it is unwise to use a system traversed more than five times by the beam. That being so, a system was used through which the beam passed five times (Fig. 2.38) instead of White's multi-pass system. The

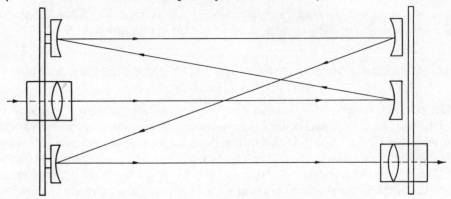

FIG. 2.38. Vertical system of mirrors for five passages of a beam of light through a flame.[48]

advantage of the new system over White's is its simplicity; the spectrophotometer can easily be positioned on the optical axis. The system is mounted in the vertical plane, so that all the beams of light pass through the flame, though at different heights. Since the sensitivity of atomic absorption measurements for certain elements depends critically on the region of the flame, the true improvement in sensitivity with this system of mirrors is much less than might be thought. For this reason, when Fassel and Mossotti[32] measured the atomic absorption of elements with refractory oxides, they mounted the same system of mirrors horizontally, as shown in Fig. 2.39. Only three of the beam's five passages through the system were utilized effectively.

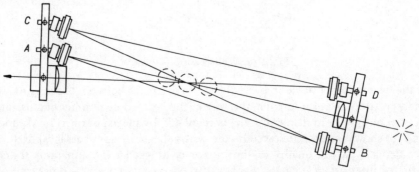

FIG. 2.39. Horizontal system of mirrors for three passages through a flame by a beam of light.[32]

Millikan[49] described how, in order to increase the pathlength of a beam from a tungsten-filament bulb in an air-ethylene flame, he used a system of several flat and spherical mirrors (Fig. 2.40). The diameter of the flame was 7 cm. The beam passed nine times through the flame, but not always centrally, and the effective length of the flame was 50 cm.

All the light beam multi-pass systems described above were used with concentric

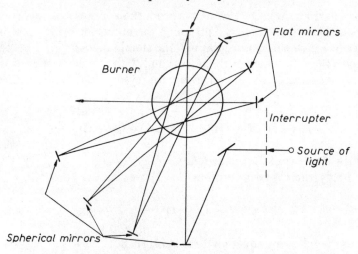

FIG. 2.40. Horizontal system of mirrors for passing a beam of light nine times through a flame.[49]

flames. These systems cannot be used with absorbing mediums severely limited in angular aperture (e.g. flames from slot burners) since the repeated passages by the beam of light require a medium of considerable angular aperture. For this reason, a system of two semi-reflecting mirrors, mounted perpendicular to the light path through the absorbing medium has excellent prospects. Systems of this type have been used for studying self-absorption in light sources by Frish and Bochkova[50] and by Podmoshinskii and Kondrasheva.[51] Either flat or concave mirrors can be used (Fig. 2.41). When self-absorption is measured one of the mirrors is opaque and the other is semi-transparent.

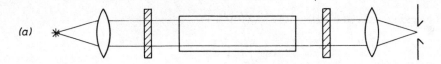

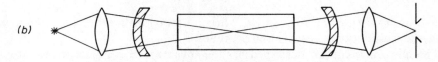

FIG. 2.41. Standard Fabry–Perot mirror systems; (a) with flat mirrors, (b) with concave mirrors.

Let us consider the effects of a system of two semi-reflecting mirrors on the degree of absorption. The beam from a light source undergoes repeated reflections within the system of mirrors. At each reflection, a proportion of the beam leaves the system. Thus the total beam of light passing through the mirrors is made up of individual separate beams which have passed through the atomic vapour 1, 3, 5, 7, etc., times.

If r denotes the reflectance of the mirror and T the transmittance of the atomic vapour, the intensity of the flux leaving the system can be expressed by the equation:

$$I = I_0(1-r)^2 (T + T^3 r^2 + T^5 r^4 + \ldots) = I_0(1-r)^2 \frac{T}{1 - T^2 r^2} \qquad (2.38)$$

where I_0 is the intensity of the incident flux.

If, accordingly, there is no absorption ($T = 1$), we get:

$$I_1 = I_0(1-r)^2 \frac{1}{1 - r^2} \qquad (2.39)$$

This gives us the following equation for the absorbance.

$$A = \log \frac{1 - T^2 r^2}{(1-r^2)T} = \log \frac{1 - T^2 r^2}{1 - r^2} - \log T = \log \left(1 + r^2 \frac{1 - T^2}{1 - r^2}\right) - \log T \qquad (2.40)$$

Since $-\log q$ corresponds to the absorbance when the beam has passed once through the medium, we shall write A_0 for $-\log T$. We must take into account the fact that mirror systems are most effectively used for absorptions below the limit of direct measurement by simpler means, so that $T > 0.99$. On the other hand, $r \leqslant 0.90$ for actual mirror coatings which are semi-transparent in the ultra-violet region. It can therefore be stated that $1 - T^2 \ll 1 - r^2$. Equation (2.40) can then be rewritten as follows:

$$A \simeq 0.434 \, r^2 \frac{1 - T^2}{1 - r^2} + A_0 \qquad (2.41)$$

The gain achieved by using the system is

$$a = \frac{A}{A_0} \simeq 1 + r^2 \frac{1 + T}{1 - r^2}$$

Since $T \simeq 1$, we finally get:

$$a \simeq \frac{1 + r^2}{1 - r^2} \qquad (2.42)$$

It follows from equation (2.42) that, when mirrors with a reflectance $r = 0.9$ are used, the sensitivity can be increased 9.5 times.

The mirror system just described has the advantage of being usable for atomic vapours with as restricted an angular aperture as is desired, since, as the light repeatedly passes from mirror to mirror, the individual beams are superimposed on one another and there is no need to increase the width of the medium.

Allowance should, however, be made for the fact that the losses of light during passage through a system consisting of two semi-reflectors are extremely great. In the ideal case in which no light is absorbed by the reflecting layer, the ratio of the incident luminous flux to the emergent flux is:

$$\frac{I}{I_0} = \frac{1-r}{1+r} \qquad (2.43)$$

With a reflectance $r = 0.9$, the intensity of the emergent flux will be about 5 per cent of the intensity of the incident flux; when absorption by the mirrors is allowed for, it will be even less. Thus, if a mirror absorbs 0·05 per cent of the light falling on it, only 1·3 per cent of the incident flux will pass through the system.

The gain of an order of magnitude in sensitivity of measurement is thus accompanied by a reduction amounting to two orders of magnitude in the intensity of the signal from the light source. The use of a mirror system of the Fabry–Perot interferometer type for improving the limit of detection may therefore not provide the desired result in cases in which the measurement of the signal is governed by the detector shot noises. If there is a large margin of light-source flux intensity, the use of this type of system may provide greater sensitivity and an improvement in detection limits.

It is interesting to note that mirror systems of the type mentioned above are now extensively used in laser beam amplifiers and generators, in which they play the part of the resonator as the emission power is being amplified.[52]

The atomic absorption spectrophotometer (AAS), with an atomic vapour within this type of mirror system, is a device working on a principle exactly the inverse of the laser beam amplifier (LBA). In the LBA (Fig. 2.42), the beam of light I_1 passing through the resonator is amplified. This amplification is caused by the stimulated transition of atoms from the excited into the unexcited state. The excess of atoms in the excited state is achieved by optical pumping from powerful emitters. If the amplification factor k_{LBA} is known we can determine the unknown signal intensity I_1 by measuring the amplified signal I_2.

In the atomic absorption spectrophotometer, the beam of light reaches the resonator and, as it passes repeatedly between the mirrors, is absorbed by the unexcited atoms in the absorbing medium; a proportion of the beam leaves the resonator. The atoms excited by absorption emit their energy in all directions and again become unexcited. By measuring the intensities of I_1 and I_2 of the incident and emergent beams, we can determine the absorption factor k_{AAS} of the system.

DESIGNS OF SPECTROPHOTOMETERS

Many descriptions have been published of the spectrophotometers used by different research workers for atomic absorption spectrochemical analysis. We shall consider a few of the designs and the recording systems used. We shall first describe simple designs, then the more complex ones.

The simplest recording circuit was used by Allan.[37] The Hilger medium quartz spectrograph is equipped with a moving exit slit, to which an IP 28 photomultiplier is secured in place of the plate-holder. The d.c. signal from the photomultiplier is measured directly by a galvanometer with a set of shunt coils for adjusting its

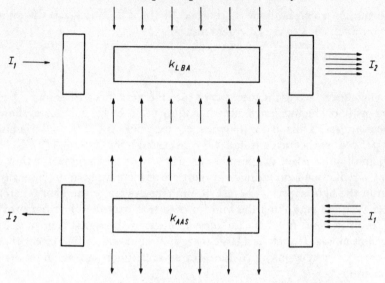

FIG. 2.42. Diagrams of (a) a laser beam amplifier, and (b) an atomic absorption spectrometer (with a standard Fabry–Perot type system of mirrors).

sensitivity. The photomultiplier dark current and flame background are compensated by means of a 1·5 V battery and a rheostat. Current is supplied to the hollow-cathode lamps through a stabilizer. The system for illuminating the spectrograph slit includes no lenses, and the distance from the burner to the slit is 45 cm. Beyond the burner there is a diaphragm which isolates the part of the flame required for making measurements.

Another simple spectrophotometer made up of standard units was described by

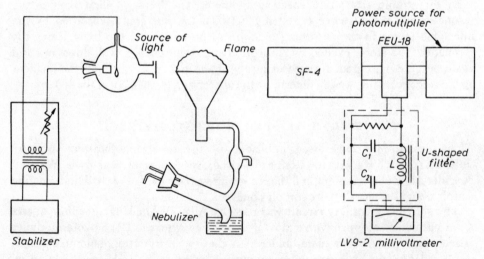

FIG. 2.43. Diagram of the apparatus using an SF-4 spectrophotometer.[53]

Poluektov and Grinzaid.[53] Fig. 2.43 is a general diagram of the apparatus. The quartz monochromator comes from an SF-4 spectrophotometer, from which the lampholder and cell compartment have been removed. The normal exit slit of the spectrophotometer is used as the entrance slit. In place of the lamp is an FEU-18 photomultiplier in a light-tight casing. The hollow-cathode lamps are supplied with stabilized alternating current at 600 V and 50 c/s; they are connected in series with a set of resistors of between 2 and 10 kΩ. The alternating component of the photocurrent is recorded with an LV9-2 valve millivoltmeter, with a most sensitive full-scale reading of 10 mV. To reduce the photomultiplier shot effect, the measuring instrument is shunted with a capacitance between 50 and 200 μF, so that the time constant of the instrument is increased to 5–6 sec. There is a U-shaped filter, between the ballast resistor and the valve millivoltmeter, to cut out high frequencies (>150 c/s); this filter consists of an inductance L=263 H, and capacitors C_1 and C_2 of 680 pF and 10^3 pF. The monochromator slits are 0·015 – 0·080 mm wide.

Poluektov and Vitkun[34] used a simplified photometer for determining mercury; this photometer was described on p. 79, and consisted of a combination of filters and an FEU-20 photomultiplier. The output from the photomultiplier was connected to an M-95 microammeter, with a full-scale deflection for a current of 1 μA. The photomultiplier voltage was regulated to 1·2–1·4 kV.

Malmstadt and Chambers[33] described a photometer constructed by the compensation principle. The recording part of the device consisted of a sulphur-cadmium photoresistive cell and a galvanometer, sensitive to 200 μA, with three shunt coils. The light source consisted of potassium and sodium vapour lamps, whose light passed through the flame, a diaphragm and an interference filter. The sample and reference solutions were interchanged automatically in 1 sec.

Many research workers use the comparatively simple recording system proposed by Box and Walsh.[8] The receiver is an IP28 photomultiplier for the ultra-violet and visible regions and an IP22 photomultiplier for the near infra-red region. The photomultiplier output travels to an a.c. amplifier (Fig. 2.44). The amplified signal is registered by a microammeter (0–250 μA). The full-scale deflection corresponds to an input signal of $5 \cdot 10^{-9}$ amp. To improve the linearity of the amplifier, the first and third stages are in a loop with overall feedback. A 200 μF capacitor shunts the microammeter to increase its time constant. The circuit provides for a potentiometer recorder with a full-scale sensitivity of 10 mV. With the input open, amplifier noise does not exceed 1 per cent. The light beam is modulated by supplying interrupted current at 100 and 50 c/s to the hollow-cathode lamps, using a rectifier (p. 40).

Some research workers use narrow-band-pass a.c. amplifiers for amplifying modulated signals. In my own early work,[54] I used a line amplifier, with an *RC* filter adjusted to a modulation frequency of 925 c/s. The band half-width was about 30 c/s. The signal was recorded with an EPP-09 potentiometer recorder. Doerffel, Geyer and Müller[55] used a selective amplifier adjusted to 125 c/s. The signal was recorded with a dial-type Zeiss galvanometer. The beam of light was modulated in both the above cases by a rotating sector-disk with a synchronous motor.

Logarithmic amplifiers were first used for recording atomic absorption by the

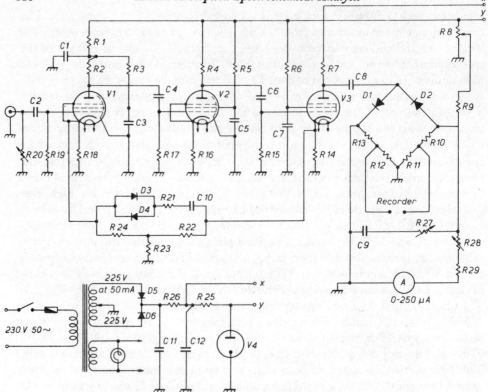

FIG. 2.44. Diagram of a.c. wide-band amplifier.[8] V_1, V_2—Z729; V_3—6AM5; V_4—OA2; D_1, D_2—OA85; D_3, D_4—OA73; R_1—10 kΩ; R_2—470 kΩ; R_3—2·2 MΩ; R_4—56 kΩ; R_5—100 kΩ; R_6—220 kΩ; R_7—27 kΩ; R_8—500 kΩ; R_9—330 kΩ; R_{10}, R_{13}—1500 Ω; R_{11}, R_{12}—15 Ω; R_{14}—510 kΩ; R_{15}—2·2 kΩ; R_{16}—820 Ω; R_{17}—1 MΩ; R_{18}—1 kΩ; R_{19}—10 MΩ; R_{20}—5 MΩ; R_{21}—82 kΩ; R_{22}—15 kΩ; R_{23}—5 kΩ; R_{24}—15 kΩ; R_{25}, R_{26}—2·5 kΩ; R_{27}—2·5 kΩ; R_{28}—1 kΩ; R_{29}—3 kΩ; C_1, C_2, C_7—8 μF; C_2, C_4—0·01 μF; C_5—0·5 μF; C_6, C_{10}—0·1 μF; C_8—2 μF; C_9—200 μF; C_{11}—16 μF; C_{12}—24 μF.

author.[56,57] The logarithmic amplifier used by L'vov[57] operated on the principle of approximating to a logarithmic curve by a series of linear segments; it used several modules, consisting of crystal diodes and resistors, connected in series to define the limits of each segment and included in the grid circuits of a four-stage amplifier.[58] Since, however, the characteristics of semiconductor diodes vary with temperature, the logarithmic response of the amplifier was extremely unstable.

Nikolaev[59] investigated the possibility of using the single-stage logarithmic amplifier described by Dianov-Klokov[60] for making spectrophotometric atomic absorption measurements. The amplifier exploits the logarithmic characteristic of a diode included in the circuit as the photomultiplier ballast. The signal from the photomultiplier passes to a cathode repeater, which operates under 'floating grid' conditions. Although the input unit is screened together with the photomultiplier, this amplifier is sensitive to high-frequency interference.

The author has proved by experience that the circuit shown in Fig. 2.45 has quite good qualities. The amplifier consists of a cathode repeater, a narrow-band stage with a T-shaped *RC* filter, adjusted to a light-beam modulation frequency of 150 c/s, a logarithmic amplifier, a detector and a cathode repeater to match the amplifier output with the low-impedance input of the EPP-09 recorder. The time constant for the complete amplifier is determined by the *RC* circuits at the cathode repeater output.

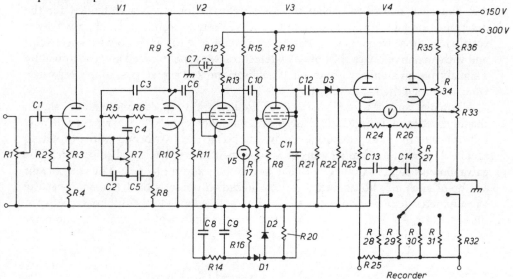

FIG. 2.45. Diagram of the logarithmic amplifier. V_1, V_4—6N3P; V_2—6K1P; V_3—6Zh9P; V_5—SG2S; D_1, D_2, D_3—D2E; R_1—4·7 MΩ; R_2, R_8—1 MΩ; R_3, R_{25}—220Ω; R_4—2 kΩ; R_5, R_6—110 kΩ; R_7—50 kΩ; R_9, R_{29}—20 kΩ; R_{10}, R_{18}—150 Ω; R_{11}, R_{14}—360 kΩ; R_{12}—4·7 kΩ; R_{13}—18 kΩ; R_{15}—5·1 kΩ; R_{16}—200 kΩ; R_{17}—36 kΩ; R_{19}, R_{28}—6·2 kΩ; R_{20}, R_{21}—51 kΩ; R_{22}—510 kΩ; R_{23}, R_{27}—12 kΩ; R_{24}, R_{26}—1·3 kΩ; R_{30}—43 kΩ; R_{31}—100 kΩ; R_{32}—160 kΩ; R_{33}, R_{34}—150 kΩ; R_{35}—220 kΩ; R_{36}—390 kΩ; C_1, C_3, C_6, C_{12}—0·1 μF; C_2, C_5—0·01 μF; C_4—0·02 μF; C_7—10 μF; C_8, C_9—0·2 μF; C_{10}—6800 pF; C_{11}—0·5 μF; C_{13}—4 μF; C_{14}—50 μF.

A modified circuit for the low-frequency amplifier, with automatic amplification regulation (AAR) using the control grid, described by Volkov,[61] is used for converting signals into logarithmic form. Logarithmic amplification is accomplished by automatically regulating the position of the working point on the anode-grid characteristic of the 6K1P valve. The AAR circuit consists of an amplifier stage (6Zh9P valve) and a detector. This amplifier provides a logarithmic output over a 30 dB range of input signals, i.e. it allows absorbances up to 1·5 to be measured. The maximum deviation from the ideal logarithmic amplitude characteristic does not exceed ±0·5 dB.

In order to measure absorbance ($A = \log I_0 - \log I$), directly, the zero signal $\log I_0$ is balanced by means of the second arm of the cathode follower. The value of A therefore corresponds to $-\log I$. The minus sign corresponds to the inverse relationship between A and log I. The signal is recorded by means of a type EPP-09 potentiometer recorder, the pen of which takes 1 sec to traverse the entire scale. The scale can be graduated for absorbances between 0·2 and 1·5 by using additional shunt coils for adjusting the sensitivity. Experience in the use of this circuit has shown that the amplifier characteristic is sufficiently stable.

Double-beam devices have also been used by a number of workers. In their original work, Russell, Shelton and Walsh[5] used a double-beam spectrophotometer that had been described earlier.[62] This spectrophotometer was intended for the determination of ozone. The sample and reference beams were chopped by a rotating sector at frequencies differing by a factor of 2 (93 and 186 c/s respectively) and with a relative phase shift of 90°. Detection was synchronized and controlled by a signal from an auxiliary photocell. The discriminated signals passed to a recorder, which recorded their ratio directly.

Alkemade and Milatz[63] used the double-beam spectrophotometer shown diagrammatically in Fig. 2.46. Sample and reference signals of the same frequency, but opposite in phase, are obtained by chopping the two beams with a rotating sector. The photomultiplier output is amplified and fed to a galvanometer. The galvanometer coil is supplied with a.c. at the chopping frequency. The coil current can be of such amplitude and so synchronized in phase with the signals that the galvanometer deflection to either side of zero corresponds to the difference between the signals: $I_0 - I$ or $I - I_0$. The galvanometer time constant is 1 sec. Galvanometer needle flutter amounts to about 1 per cent of the light signal.

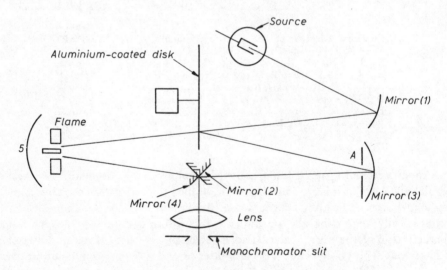

FIG. 2.46. Diagram of the double-beam spectrophotometer used by Alkemade and Milatz.[63]

Herrmann and Lang[64] described a double-beam spectrophotometer in which the signals compared are modulated in opposite phase by means of a rotating sector. The signals are isolated by means of a phase discriminator synchronized with the modulation frequency by means of an additional phototube and a filament bulb. The null indicator is a galvanometer.

Baker and Garton[65] also used a double-beam spectrophotometer with the beams modulated in opposite phase. Fig. 2.47 is a block diagram of the apparatus. The beam from the source (a hollow-cathode lamp) is reflected by the mirror (1) to the aluminium-coated disk, which rotates at 50 rev/s; this disk has six cut-out segments. The beam passing through the segments travels through the flame twice; it is reflected from the small plane mirror (2) and focused on the monochromator slit. The comparison beam, reflected from the surface of the disk, is also focused on the slit by means of mirrors (3) and (4). The beams travel in different planes and are focused on the slit at different heights. An E-492 spectrograph is used as monochromator. The signal from the IP28 photomultiplier, or the EMI-6095 photomultiplier for the red region, is amplified by the narrow-bandpass amplifier and detected by a phase discriminator synchronized with the rotation of the disk. The signals are balanced by means of the attenuator A, which stops down the comparison beam. The scale of the attenuator is calibrated in I/I_0 units.

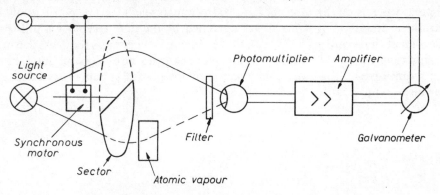

FIG. 2.47. Diagram of the Baker and Garton double-beam spectrophotometer.[65]

Menzies[67] used a single-beam two-channel spectrometer, based on a direct-reading Hilger medium quartz spectrograph. The second channel was for recording the comparison line, which was not absorbed in the flame. The system was tested on the Cu 3248 Å–Cu 2824 Å lines.

Butler, Strasheim, Strelow, Matthews and Feast[66] described a single-beam two-channel spectrophotometer intended for the determination of gold. The second channel in this spectrophotometer was for isolating the line of an internal standard (palladium) introduced into the solutions analysed. The output slit of a standard monochromator was replaced by a double slit with components corresponding to the positions of the Au 2428 Å and Pd 2476 Å lines. Light passed through the two slits reached the same photomultiplier. A diaphragm, vibrated horizontally, passed each of the lines alternately. The signals from the photomultiplier were separated by means of a switch, synchronized with the movement of the diaphragm, and sent to two separate amplifiers. The read-out device (an automatic recorder or galvanometer) registered the difference between the amplified signals. The source was a lamp with a composite hollow cathode, lined with palladium and gold. It should be pointed out that although, as the authors stated, this circuit had good qualities as regards improving the accuracy and precision of analysis in comparison with standard single-channel devices, it would be better to record, not the difference between signals, but their ratio or the difference between their logarithms.

Integrating systems were used by Menzies,[67] Baker,[65] Gidley,[68] and a number of other workers.

Warren[69] used a scanning spectrometer, with which a spectral region (0·6 Å) wider than the line could be recorded. The time required for a single scan was 15 sec. A d.c. amplifier was used for amplifying the photocurrent. The record of the narrow spectral region scanned provided a direct measure of the signal above the background of photomultiplier dark noise and continuous lamp and flame emission.

It should, however, be noted that the scanning method greatly reduces the useful measurement time τ; for the instantaneous peak signal is recorded, and not the mean signal throughout the 15 sec period of the scan. The time required to record the peak depends on the time constant of the apparatus. The time constant of the instrument described by Warren was governed by the time taken for a full-scale deflection of the recorder pen (2 sec). Since the precision of measurements depends on $\sqrt{\tau}$, it can be assumed roughly that the error in measuring by the scanning method is $\sqrt{7\cdot5}$ times greater than when, for instance, an integrating system is used with the same measurement time.

Chapter 3

ATOMIZATION OF SAMPLES
GENERAL PRINCIPLES

INTRODUCTION

The substance being investigated by atomic absorption spectroscopy must be converted into a state in which the elements being determined are in the form of free atoms, capable of absorbing light. Any of the numerous methods of atomization used in emission spectroscopy can be used for this purpose. At a first glance, this purely empirical approach to selecting the method of atomizing is quite justified, since sources of spectrum excitation normally also effect atomization. Paradoxical though it may be, however, the selection of the best method of atomizing is a much more complex problem in atomic absorption analysis than in atomic emission analysis. In atomic absorption analysis, there is a trend towards greatly simplified apparatus and methods, entailing a departure from the principles of preparing standards similar in composition and properties to the sample being analysed.

These principles were introduced into emission analysis to allow for uncontrollable variation in the composition of samples and for variable experimental conditions:

1. Atomization of samples
2. Excitation conditions
3. Recording of signals

In atomic absorption spectroscopy, owing to the inherent advantages considered on p. 33, the last two stages of analysis need not be taken into consideration, but atomization of the sample remains as important as in emission.

Having, therefore, entirely or partially departed from the principle of individual standardization, we are obliged to impose higher requirements as regards the methods of atomization. When we attempt to assess or select a method of atomization suitable for atomic absorption analysis, we must first decide whether it is possible to select atomization conditions that will guarantee a simple relationship, independent of the composition of the sample between the analysis signal and the concentration of an element in the sample. Let us therefore dwell on a few of the general laws relating to the atomization of samples and then try to establish possible methods of recording analysis signals and atomization conditions (independent of the actual mechanism of atomization) which will guarantee this conformity.

METHODS OF RECORDING ANALYSIS SIGNALS AND SAMPLE ATOMIZATION CONDITIONS UNAFFECTED BY THE SAMPLE COMPOSITION

On the basis of the ideas formulated by Mandel'shtam,[14] and later by Raikbaum,[15] we can represent sample atomization as taking place within a cell through which the sample passes. We shall take the cell as being the analysis volume which directly participates in producing the analysis signal observed by the experimenter. For the sake of general discussion, we shall not yet specify which signal, absorption or emission, is being measured.

For simplicity, we shall confine our examination to the passage of the atomic vapour through the analysis volume, eliminating all the intermediate stages of the transport of material from the sample up to the moment of its entering the analysis volume. We shall also ignore any complicating processes that might accompany the transport of substance through the cell, e.g. dissociation of molecules and ionization of atoms, and shall assume that all of the element we are determining is in the atomic state while inside the cell.

We shall use the following symbols: N_0 is the number of atoms of an element in the sample, N the total number of atoms entering the analysis volume at any instant t, τ_1 the overall length of time during which atomization occurs, τ_2 the mean length of time spent by an atom in the analysis volume, i.e. the transit time, and τ_3 the length of time during which the signal is recorded.

The balance between the number of atoms entering the analysis volume $n_1(t)$ and escaping from it $n_2(t)$ (in unit time) is given by

$$\frac{dN}{dt} = n_1(t) - n_2(t) \tag{3.1}$$

The form of functions $n_1(t)$ and $n_2(t)$ depends on the method of atomization used, but in many instances we can approximate the functions as

$$\left. \begin{array}{l} n_1(t) = \dfrac{N_0}{\tau_1} \\[1em] n_2(t) = \dfrac{N}{\tau_2} \end{array} \right\} \tag{3.2}$$

Equations (3.2) are valid for a uniform process of transfer of atomic vapour to the analysis volume and for the diffusion mechanism of atoms escaping from the analysis volume. Thus equation (3.1) can be written in the form

$$\frac{dN}{dt} = \frac{N_0}{\tau_1} - \frac{N}{\tau_2} \tag{3.3}$$

If we resolve the variables and integrate (3.3), we get the following equations for the instantaneous values of N at times when t is both less than and greater than τ_1:

$$N_{t<\tau_1} = N_0 \frac{\tau_2}{\tau_1}(1 - e^{-t/\tau_2}) \tag{3.4}$$

and
$$N_{t>\tau_1} = N_0 \frac{\tau_2}{\tau_1}(1 - e^{-\tau_1/\tau_2})e^{-(t-\tau_1)/\tau_2} \qquad (3.5)$$

These equations portray the kinetics of change in the number of atoms in an analysis volume for the simplified model of vapour transport. Fig. 3.1 contains kinetic curves for different ratios τ_1/τ_2.

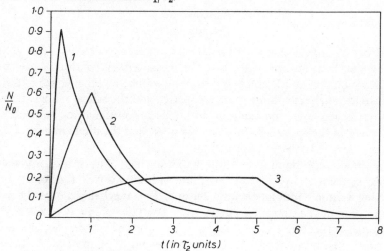

FIG. 3.1. Variation in the number of atoms in an analysis volume for different ratios τ_1/τ_2.
1. $\tau_1/\tau_2 = 0.2$. 2. $\tau_1/\tau_2 = 1$. 3. $\tau_1/\tau_2 = 5$.

We can now move on to the problem in which we are directly interested, namely that of finding sample atomization conditions and methods of recording analysis signals which will ensure that there is a simple relationship between the magnitude of the analysis signal and the concentration of an element in a sample.

Consideration of equation (3.4) and the curves in Fig. 3.1 shows that, as the value of t increases (i.e. with high ratios τ_1/τ_2), the magnitude of N tends towards some equilibrium value N_{equil}, governed by the equation

$$N_{equil} = \frac{N_0}{\tau_1}\tau_2 = n_1 \tau_2 \qquad (3.6)$$

where n_1 in contrast to $n_1(t)$ has a constant rate of atomization.

By measuring the value of N_{equil} with τ_2 constant, we can draw a simple conclusion regarding the magnitude of n_1. If we now assume that the rate at which atoms enter the analysis volume is proportional to the concentration C of atoms in a sample and the rate w at which the sample enters the analysis volume, i.e.

$$n_1 = wC \qquad (3.7)$$

by substituting equation (3.7) in equation (3.6) we get:

$$N_{equil} = w\tau_2 C \qquad (3.8)$$

With a constant product $w\tau_2$, the value of N_{equil} must bear a simple relationship to the concentration of atoms in the sample. In practice, the rate at which a sample enters the analysis volume and the length of time spent by atoms in the volume are in most cases independent of one another. In order, therefore, to fulfil the condition

$$w\tau_2 = \text{constant} \tag{3.9}$$

both the parameters must also remain unchanged, i.e.

$$w = \text{constant}$$
$$\tau_2 = \text{constant} \tag{3.10}$$

Measuring N_{equil} thus enables us to determine the concentration C of atoms in a sample by simple means, provided that the measurement conditions are such that the rate at which the sample enters the analysis volume and the length of time spent by atoms in the analysis volume remain constant and the rate at which atoms of the element concerned enter the analysis volume is proportional to the concentration of the element in the specimen. It is also assumed that the total amount of a sample greatly exceeds $w\tau_2$, so that N approximates N_{equil}.

Here is the place to point out a quite important fact. The equilibrium method of measuring concentrations imposes definite limitations on the mechanism for transporting a sample into an analysis volume. This mechanism must be such that the composition of the substance entering the analysis volume is equivalent to that of the sample. If this is not so, and larger proportions of certain elements are introduced, there must be an apparent change in the composition of the initial sample, i.e. a change in the concentration C of the element being determined; the magnitude of N_{equil} will change.

There are different methods of introducing substances into analysis volumes, mechanical, electrical (cathodic sputtering) and thermal. Of these methods, only the first is unaffected by the individual chemical components of samples and therefore ensures that the composition of the sample is precisely the same as that of the substance entering the volume. The cathodic sputtering process is rather sensitive with respect to the different components of a substance being atomized, though in some cases the differences can be ignored. The sample will enter the analysis volume least uniformly if a thermal mechanism is used for its transportation. In this case, the rate at which the individual components of a sample are atomized depends not only on the composition of the sample, but also on the vapour pressures of its components; there can therefore be no conformity between the compositions of the vapour and the sample.

Let us now consider possible methods of recording analysis signals. It follows from Fig. 3.1 that, at the moment when a sample has all been atomized ($t=\tau_1$), N reaches its peak value:

$$N_{peak} = N_0 \frac{\tau_2}{\tau_1}(1 - e^{-\tau_1/\tau_2}) \tag{3.11}$$

If in addition $\tau_1/\tau_2 \ll 1$,

$$1 - e^{-\tau_1/\tau_2} \simeq \frac{\tau_1}{\tau_2} \tag{3.12}$$

and so

$$N_{peak} = N_0 \tag{3.13}$$

Atomization of Samples

If a sample is completely atomized, therefore, in a length of time much shorter than the time spent by the atoms in the cell, the peak value N_{peak} must correspond to the total number of atoms in the sample. It is a remarkable fact that the variations in the values of τ_1 and τ_2 which may take place during measurements have practically no effect on relationship (3.13), provided that the condition $\tau_1/\tau_2 \ll 1$ is fulfilled. For instance, with $\tau_1/\tau_2 = 0.1$, $N_{peak} = 0.95 N_0$, while $\tau_1/\tau_2 = 0.2$, $N_{peak} = 0.90 N_0$, i.e. doubling the length of time to atomize a sample only alters the peak value N_{peak} by 5 per cent. With a ratio $\tau_1/\tau_2 < 0.1$, the effects of variations in the parameters τ_1 and τ_2 must be even less.

Attention should be drawn to the fact that, when this method of measuring an analysis signal is used, direct information is obtained regarding the absolute number N_0 of atoms of an element being determined in a sample, as opposed for instance to the method of measuring the equilibrium value of the concentration, which only enables the concentration C of atoms in samples to be determined with an accuracy of up to $w\tau_2$. This is extremely important for certain research based on measuring the signals from atomic vapours with known populations of atoms (for instance when determining the absolute values of oscillator strengths).

Of course the peak method of measuring absorption can be used not only for cases where $\tau_1 \ll \tau_2$. If $\tau_1 \geqslant \tau_2$, the value N_{peak} depends on τ_1 and τ_2, so that a simple relationship between N_{peak} and N_0 is possible only if

$$\frac{\tau_1}{\tau_2} = \text{constant}$$

Another possible method of measuring analysis signals bearing a simple relationship to the number of atoms in a sample is that of measuring the integral value Q_N of concentration during the entire period for which atoms are atomized and remain in the analysis volume:

$$Q_N = \int_0^{\tau_3} N\, dt \tag{3.14}$$

On the basis of equations (3.4) and (3.5),

$$Q_N = \int_0^{\tau_1} N_0 \frac{\tau_2}{\tau_1}(1 - e^{-t/\tau_2})\, dt + \int_{\tau_1}^{\tau_3} N_0 \frac{\tau_2}{\tau_1}(1 - e^{-\tau_1/\tau_2}) e^{-t-\tau_1/\tau_2}\, dt \tag{3.15}$$

After carrying out the necessary transformations, we get:

$$Q_N = N_0 \tau_2 [1 - \tau_2/\tau_1 (1 - e^{-\tau_1/\tau_2}) e^{-(\tau_3 - \tau_1)/\tau_2}] \tag{3.16}$$

When we consider equation (3.16), we find that the second term within the square brackets tends towards 0 as τ_3 is increased. Even with $\tau_2/\tau_1 \gg 1$, the value of Q_N does not differ from $\int_0^\infty N\, dt$ by more than 2 per cent, provided that

$$\tau_3 \geqslant \tau_1 + 4\tau_2 \tag{3.17}$$

Under condition (3.17), with an accuracy of up to 2 per cent we can therefore write

$$Q_N = N_0 \tau_2 \tag{3.18}$$

It is extremely significant that the value of Q_N is entirely independent of τ_1, the sample atomization time, i.e. no variations in the kinetics of entry by a sample into an analysis volume, caused by changes in the atomization conditions or in the properties of the sample, can affect the magnitude of the analysis signal (the areas beneath the curves in Fig. 3.1 are identical). Here the only requirement regarding the atomization conditions is the necessity for maintaining experimental conditions with which the length of time spent by the atoms in the analysis volume τ_2 remains unchanged.

Attention should now be drawn to the following. In using the term 'analysis signal' we have nowhere up to now clarified to what actual quantity the term relates. In general, when equilibrium and peak values of concentrations are measured, the signal could be any function $f(N)$ bearing a simple relationship to the number of atoms N, for instance or the absorbance or transmittance (I/I_0) when recording absorption. When integral concentrations are measured, the analysis signal $f(N)$ must be proportional to the concentration N of atoms, since when we determine the value of $\int_0^{\tau_3} f(N)dt$ in an experiment, we must then convert to the value of $\int_0^{\tau_3} Ndt$, and this is only possible if

$$f(N) \propto N$$

In absorption measurements, therefore, the analysis signal, expressed in absorbance (A), should be integrated, while in emission measurements the intensity (I) should be integrated; these quantities are associated with the concentration of atoms by the relationships

and
$$\left. \begin{array}{c} A = k_A N \\ \\ I = k_I N \end{array} \right\} \tag{3.19}$$

Equation (3.14) can then be written in the form

$$Q_A = \int_0^{\tau_3} A\,dt = A_0 \tau_2 = k_A \tau_2 N_0 \tag{3.20}$$

for atomic absorption measurements, and

$$Q_I = \int_0^{\tau_3} I\,dt = I_0 \tau_2 = k_I \tau_2 N_0 \tag{3.21}$$

for emission measurements.

The essentials of the method represented by equation (3.21) were formulated in Slavin's classic work[16] as long ago as 1938. It was asserted in this work, on the basis of qualitative considerations, that

$$Q_I = k N_0 \tag{3.22}$$

The nature of the relationship of the proportionality factor k to the experimental conditions, and in particular to the length of time spent by atoms in an analysis volume, was not however discussed. If we compare equations (3.21) and (3.22), we find that

$$k = k_I \tau_2 \tag{3.23}$$

Quite correctly, the method of recording proposed by Slavin was called the total energy method, since when emission is measured the energy emitted by the source of light is accumulated. In our more general case of considering any analysis signal (either emission or absorption), it is more correct to call this an integration method, since when absorption is recorded a signal proportional to a dimensionless quantity ($A = \log I_0/I$) is accumulated.

By analogy, we shall call the first two of the methods considered above the equilibrium and peak methods.

When we compare the methods which have been considered of measuring analysis signals, we can draw the conclusion that the integration method must have the widest field of application; the conditions for its application are: total atomization of the element determined and constancy of the length of time spent by the atoms in the analysis volume; these conditions are natural and quite easily fulfilled. The equilibrium and peak methods impose much more rigid conditions on the methods of atomization, and the fields of their application are therefore much narrower.

It is interesting to compare these methods as regards sensitivity of determination. We shall confine ourselves to considering absolute sensitivity. The problems associated with assessing the relative sensitivity of the different methods will be discussed on pp. 237–45.

The ratio of the measured magnitude of an analysis signal (in N unit) to the amount N_0 of an element consumed can be used as the criterion of absolute sensitivity.

1. With the equilibrium method of measuring, this relationship is

$$\frac{N_\text{equil}}{N_0} = \frac{n_1 \tau_2}{n_1 \tau_1} = \frac{\tau_2}{\tau_1} \tag{3.24}$$

It follows that the more rapidly a sample is introduced into the analysis volume, and the longer the atoms dwell there, the higher will be the sensitivity of the equilibrium method. It should not, however, be forgotten that one of the conditions for using the equilibrium method must be:

$$\frac{\tau_2}{\tau_1} \leqslant 1 \tag{3.25}$$

so that τ_2 can only be increased, or τ_1 reduced, in (3.24) to values governed by (3.25).

2. In the peak method of measurement, provided $\tau_1/\tau_2 \leqslant 1$

$$\frac{N_\text{peak}}{N_0} = \frac{N_0}{N_0} = 1 \tag{3.26}$$

The peak method is τ_1/τ_2 times more sensitive than the equilibrium method (τ_1 and τ_2 here relate to the equilibrium method).

3. In the integration method, the following relationship is the criterion of absolute sensitivity:

$$\frac{Q_N}{N_0} = \frac{N_0 \tau_2}{N_0} = \tau_2 \qquad (3.27)$$

The sensitivity is proportional to the length of time spent by atoms in the analysis volume. As in the case of the equilibrium methods, therefore, methods of atomization providing the maximum values of τ_2 must be the most sensitive.

To end our discussion and comparison of the general particulars of the different methods of measuring analysis signals, we shall also point out the special features of the recording systems used in either case. Signals are most easily measured in the equilibrium method, since measurement then amounts to measuring a constant signal. This can be done by taking readings from a read-out meter, by collecting a charge corresponding to the value of N_{equil} in a capacitor during a specific length of time, or by making a continuous recording of the signal with automatic chart recorder.

In the integration method, the analysis signal can be measured by making a continuous recording of I (for emission) or A (for absorption), and then determining the areas beneath the $I=f(t)$ or $A=f(t)$ curves, or alternatively by accumulating a current proportional to I or A in a capacitor.

Finally, in the peak method, analysis signals can only be measured by the continuous recording of a signal associated with N, i.e. the selection of recording methods is extremely limited. The recording of peak signals is made even more difficult because the time constant of the recording circuit must be sufficiently small to enable N_{peak} to be recorded without any great distortion.

METHODS USED IN ATOMIC ABSORPTION ANALYSIS FOR ATOMIZING SAMPLES

In proceeding to consider methods of atomizing samples which have either been used or merely tested in atomic absorption analysis, we shall pay most attention to the conditions needed to ensure that the methods of measuring analysis signals considered above can effectively be utilized.

The flame

Flames were used by Alkemade and Milatz[1] and by Walsh and his colleagues[2] as the first means of atomizing samples for atomic absorption spectrochemical analysis. Usually the equilibrium zone of the flame above the inner cone is used as the analysis volume. The time τ_2 spent by atoms in the analysis volume is governed by the rate of flow of gas, and remains constant provided that the composition and amount of gases consumed also remain constant. The mean value of $\tau_2 \simeq 10^{-4}$ sec.[17]

Any of the recording procedures can be used for measuring analysis signals. When the equilibrium method is used, the sample is introduced into the analysis volume by compressed air, or occasionally ultrasonic, nebulization of a solution of the substance to be analysed. The constancy of the rate at which a sample is introduced into the flame (w) is governed by the stability of the characteristics of the mechanical systems concerned, also by the constancy of the physical and chemical particulars of the samples atomized.

Naturally, everything which has been stated above regarding the rate at which samples are introduced is accurately true up to the stage at which the mist becomes vapour.

We shall consider the process of atomization of the mist in greater detail later (p. 175). Here, to simplify the model, we shall assume that all the mist (or at least the element being determined) is atomized. This is a justifiable assumption for most of the elements atomized in high temperature flames.

The equilibrium method of measuring atomic absorption in flames is the method most used in analysis practice, and is up to now the only method of atomization used in mass-produced devices. The integral procedure can also, however, be used for recording atomic absorption in flames. In this case, samples can be introduced into the flame either by the normal method (mechanical nebulization of a solution), or by any other method, such as introducing a dry residue into a flame on a wire.

The integration method of measuring must have the advantage over the generally adopted equilibrium method that the unstable characteristics of nebulization systems and variations in the properties of solutions have no effect on the results of analyses. When a sample is atomized by the thermal method from a wire, the reservation made above regarding the stage of atomization of mist no longer applies. When the integration method of measuring is used, the only requirement is that the sample should be completely atomized. In this connection, when solutions are atomized, it is necessary to avoid any unaccountable losses of solution in the cloud chamber, for instance by using direct injection burners for this.

Up to now, the integration method of measurement has been used only in emission flame photometry practice.[18,19] Preliminary research into this method as applied to absorption measurements in flames was conducted by L'vov and Plyushch (see p. 189).

Kahn et al.[20] have employed the peak method of measuring absorption. They used a flame and evaporated the sample from a tantalum boat. Obviously in their experiments $\tau_1 > \tau_2$.

Before considering the other methods of atomization which are used in atomic absorption, it should be pointed out that the conclusions reached in this section apply equally to systems such as the d.c. plasma jet and induction-coupled plasma which are similar to flames.[21] For this reason, we shall not consider these methods of atomization separately. The only difference between the methods, from the point of view of the atomization conditions considered here, concerns the length of time spent by atoms in the analysis volume. For instance, the value of τ_2 in a plasmotron jet is, according to experimental measurements,[17] an order of magnitude lower than in a flame. On the other hand, the length of time spent by atoms in the analysis volume of an induction-coupled plasma is $\simeq 10^{-2}$ sec,[21] i.e. much longer than in a flame. These differences are governed by the optimal rates of flow of the gas bearing the mist for each method.

Other conditions being equal (the rate at which the sample enters the analysis volume and the dimensions of the latter), as has already been pointed out (on p. 121), the magnitude of τ_2 indicates the possibility of attaining higher or lower measurement sensitivities.

Hollow-cathode discharge

Gatehouse and Walsh were the first to use the hollow-cathode discharge as the method of atomization for atomic absorption analysis.[3] The analysis volume is the space extending for the full length of a cylindrical cathode. The length of time spent by atoms within the cathode is governed by the diffusion rate of atoms to the open ends of the cathode, i.e. by the length of the cathode and the pressure, temperature and nature of the foreign gas filling hollow-cathode lamp. With water-cooling, temperatures of about 300°K and argon pressures of about 3 torr, $\tau_2 \simeq 0.1$ sec in a cathode 50 mm long for heavy atoms such as atoms of mercury [equation (6.29) can be used for estimating τ_2]. With hot cathode temperatures ($T \simeq 1000°K$), otherwise under the same conditions, $\tau_2 \simeq 0.01$ sec.

As in the case of flames, the equilibrium and integration methods can be used for recording atomic absorption. When the equilibrium method is used, the rate at which atoms are introduced into the analysis volume achieves constancy owing to the constant cathode sputtering conditions.

The substance to be analysed is either applied uniformly to the inner surface of the cathode or, in the case of metallic samples, the cathode is made entirely of the substance to be analysed. When samples are introduced in the form of compounds, the compounds are decomposed and the elements reduced to metals by preliminary running of the lamp. In spite, however, of the comparatively uniform state of the surface atomized, the amount of cathode sputtering of the element being determined is obviously not independent of the presence of foreign components in the substance.

In this respect, the integration method of measuring is bound to provide much better facilities, since it guarantees that the analysis results are independent of the kinetics of sample vaporization.

Up to now, only the equilibrium method has been used for recording atomic absorption in hollow cathodes. The integration method has proved effective only in emission spectrochemical analysis.[22]

Graphite furnaces

The first type of graphite furnace suggested by King[4] has been extensively used in research spectroscopy for studying and interpreting atomic spectra and measuring oscillator strengths; it consists of a graphite tube, in a vacuum chamber, heated by a.c. to 3000°C. The element being investigated is placed in a special boat inside the furnace. The sample atomized passes through apertures in the tube and is deposited against the cold parts of the chamber. Although, when the furnace has reached a steady temperature, a constant concentration of the element is established within it and maintained for a long time, it is not possible to carry out quantitative analysis by measuring the equilibrium concentration of atoms. The fact is that the rate at which the atoms enter the analysis volume depends in this case, not on the amount of an element in the sample, but on the saturated vapour pressure of the element at the furnace temperature; it remains practically the same whatever the amount of the element in the specimen.

In view of this, L'vov[5] proposed using a pulsed method of atomizing samples in a graphite furnace. The sample, applied to the tip of a carbon electrode, was introduced into the heated furnace through the transverse aperture at the centre of the tube. In order to accelerate sample atomization, the electrode head was preheated by a powerful d.c. arc struck between the electrode introduced into the furnace and an additional electrode positioned below the tube. The arc method of preheating electrodes was later replaced by the more effective, and technically simpler, resistance method of preheating.[13, 28]

The graphite furnace itself has also been much improved. In order to reduce the rate at which atoms leave the cell by diffusion, the graphite furnace was housed in a chamber filled with inert gas at a pressure above atmospheric[23] and graphite tubes impervious to gases were used for making the furnace. These measures lengthened the time atoms spent in the tube to $\tau_2 \simeq 1$ sec.

Thanks to these improvements to the furnace and to the pulsed method of atomizing samples, it proved possible to fulfil the basic condition for using the peak method of measurement ($\tau_1 \ll \tau_2$), and to use this method of atomizing as the basis for spectrochemical analysis and scientific research by atomic absorption.

As opposed to the King furnace, in which the heated tube is the means of atomizing samples, with the method of atomization described above, the furnace merely acts as a cell hindering the loss of the atoms; in other words, it plays the part of a cuvette. Hence, when considering this method of atomization, we shall use the word cuvette from now on.

The peak method of measuring absorption is not, however, the only method of measurement used when samples are atomized in a cuvette, though most of the available results were actually obtained by the peak method. Since the conditions governing the length of time spent by atoms within the cuvette can be kept constant during the experiments, the integration method can equally well be used for measuring absorption.

The arc

Belyaev[24] and his colleagues used the d.c arc from a carbon electrode as the means of atomization for atomic absorption analysis. The discharge cloud was the analysis volume. The time τ_2 spent by atoms in the arc volume is basically governed by diffusion rate, and is $(1-5) \cdot 10^{-3}$ sec, depending on the atomic weight of the element.[17]

Samples were atomized from a channel in the electrode (with a current of 5–10 amp) until the elements being determined had been completely consumed. The atomization time τ_1 was $\simeq 20$ sec for volatile elements such as Zn, Cd, Pb, Pl and Bi, and was about 60 sec for elements of low or average volatility, such as Mo, Cr and Sn.

Belyaev et al.[24] used an integration system, with charge storage by capacitor, for measuring absorption; the integrated signal was expressed, not in absorbance, but in intensity difference units $(I_0 - I)$. This restricted the effectiveness of the integration method.

Nevertheless, even in the preliminary stage of their research, Belyaev et al.[24]

achieved high relative sensitivities for determining the elements listed above in carbon powder (between 10^{-4} and 10^{-6} per cent) and satisfactory reproducibility with a variance of 15 per cent. The effects of the parent substance on the results of analysis were not investigated.

Here we should make the following remark. The integration method of measuring the analysis signal from samples atomized in arcs has been extensively used in emission spectrochemical analysis. This applies particularly to photographic methods of analysis based on the complete atomization of samples. The integration method has the potential advantage that the results of measurements are independent of the kinetics of sample atomization in a vast number of cases, but we have not profited from the advantage, except in certain successful experiments, of which the work described in reference 16 is an example.

When Boumans[25] considered the probable reasons for this in detail, he indicated that the atomization rate may affect the subsequent stages of emission spectrochemical analysis:

1. The process of combustion and the extent to which substances enter the arc discharge plasma.
2. The transfer of a substance through the discharge zone (τ_2).
3. The excitation of spectra (through change in the temperature and the concentration of electrons in the plasma).
4. Self-absorption of spectral lines.
5. The Schwartzschild effect (reciprocity failure).

Since the measurement of atomic absorption eliminates the effects of the last three factors on analysis results, it can be anticipated that the integration method will be more effective in atomic absorption than in emission when the samples are atomized in arcs.

When mechanical methods are used for introducing samples into arcs (as is extensively done in emission analysis, e.g. by pouring or blowing powdered samples or solutions into the arc or introducing solutions by means of a rotating electrode, etc.), the equilibrium method of measuring absorption can also be used.

Pulsed lamps

Pulsed lamps were used by Nelson and Kuebler[6] for recording the atomic absorption spectra of certain elements with average and low volatility. The analysis volume in this case is within a quartz tube housed inside a spiral capacitative discharge tube. The substance being analysed is applied uniformly to a tungsten grid or graphite base layer and inserted in the tube. A powerful source of light with a flash energy of 30 J/cm², the flashes lasting about 3 ms, enables the surface of the wire or base layer to be heated to several thousand degrees Centigrade, and the substance which has been applied is atomized. To lengthen the time spent by atoms from a sample in the analysis volume, the tube is filled with inert gas, which delays the spreading of the atoms and their deposition on the tube walls.

These authors used the peak method of measuring absorption. To do this, a light pulse from another Lyman-type lamp emitting a continuous spectrum was passed through the tube with a time-lag of 1 ms relative to the start of the main pulse. The illuminating pulses lasted 20 μs. The spectra of the flashes were photographed with a grating spectrograph.

The analytical possibilities of the method were not investigated quantitatively. Unfortunately, no data are available regarding the kinetics of atomization and the completeness with which samples are atomized. The effectiveness of the peak method used for recording is therefore not clear.

Lasers

Hagenah, Laqua and Mossotti,[11] and Atwill[26] attempted to use laser energy as a means of atomizing solid samples for atomic absorption measurements. The most detailed experimental research into the analytical particulars of this method was carried out by Mossotti, Laqua and Hagenah.[27]

The technique for atomizing samples by means of pulsed laser beams has been described in detail in a number of earlier published works, such as that by Karyakin et al.[12] A parallel beam of light generated by a laser is focused, by means of a lens, on the surface of the object being analysed. Normally the pulse energy is between a few joules and a few tens of joules. Emission or absorption is observed in the atomic vapour leaving the surface of the substance as a result of the pulsed heating of a small region of surface, with a diameter of $\sim$0·1 mm, to 5000–10 000°C.

Mossotti, Laqua and Hagenah[27] used for atomizing an average-power ruby laser consuming between 0·1 and 10μg of sample per pulse. Oscillograph recordings of the kinetics of changes in absorption proved that, in accordance with the stream transfer of the atomic vapour through the analysis volume, the absorption pulses represent symmetrical peaks with a total duration of about 100 μs. The amount of substance expelled with the atomic vapour depends to a great extent on the type of base substance, the state of the surface, and the pulse energy.

Under these conditions, not one of the methods which have been considered for measuring absorption can be effective. The equilibrium method is not suitable owing to the pulsed manner in which the sample is atomized; the integration and peak methods cannot be used because the sample is not completely atomized.

Mossotti et al.[27] used the peak photoelectric method of recording absorption and, by means of a separate measuring channel, took into account the amount of substance atomized, determining it from the amount of light emitted by the principal component in the sample. As was to be expected, however, reproducibility was very poor (the scatter was 30–60 per cent) and the base substance had a tremendous effect. As an instance, the absolute sensitivity for manganese, copper, silver and chromium varied by three orders of magnitude, between 10^{-9} and 10^{-6} g, depending on the base substance, iron, zinc or lead.

Since a recording system with a high resolving power (circuit time constant 15 μs) was used, the shot noises were 4 per cent and absorptions less than 10 per cent could not be measured.

Having now concluded our discussion of the principal methods of atomization separately, let us try to draw certain general conclusions. During the period of more than fifteen years in which atomic absorption has been developed, quite a few attempts have been made to adapt for its purposes methods of atomizing samples which have been used for practical analytical emission spectroscopy. Nevertheless, the situation regarding this problem is in general still unsatisfactory.

For this reason, a great deal of attention is still being devoted to the use for atomic absorption of fashionable methods of atomizing samples, such as the laser or the plasma jet, and to the discovery of methods of atomization which are new in principle (an example is the method of burning solid sample, in powder form, in admixture with pyrotechnical combustible mixtures[28]). Our discussion has, however, shown that the possibilities of old methods of atomizing which have already been used for atomic absorption (flame and cuvette), are still far from exhausted and may be greatly extended when other methods of measuring absorption are used.

Each of the methods of atomization considered above has been used in combination with only one, and sometimes not the best, method of measurement. The most effective integration method has not yet been introduced at all.

Preliminary assessment has shown that the integration method will enable many analysis problems, associated with the atomization of solid samples in flames, increasing determination sensitivity, eliminating the effects of incomplete atomization of samples in flames, and improving and technically simplifying the method of atomization in a cuvette, to be solved. We should therefore first take advantage of the possibilities afforded by the integration method of measuring absorption, in combination with methods of atomization, flames and cuvettes, which are already known and have been thoroughly investigated.

These tasks, which in my opinion are the ones for solving many of the problems of atomic absorption analysis, do not however mean that other methods of atomizing samples, in particular the arc and the hollow cathode, are not useful and should not be developed and improved. This research is definitely interesting, for instance for investigating the problem, mentioned earlier, of the effects of the kinetics of the atomization of samples in arcs on the results of emission analysis. In addition to this, for a number of reasons on which we shall not dwell, these methods of atomization may be effective for solving certain special analytical tasks. For instance, the method of atomizing samples in hollow cathodes will no doubt be useful for atomic absorption analysis of isotopes and gases.

DISSOCIATION

Let us consider the general laws for the processes which lead to the breakdown of a molecule into atoms as well as the converse, the association of atoms and the formation of molecules which are stable in the gas phase. This information is necessary if we are to understand the phenomena which take place in an atomic vapour.

Since, at high temperature, kinetic factors do not play any important part, we shall confine ourselves to the thermodynamic consideration of systems. For the general form of breakdown reaction,

$$Me_nX_m \rightleftharpoons nMe + mX \qquad (3.28)$$

which takes place in a gas phase at constant pressure, the partial pressures of the components in the reaction are subject, under equilibrium conditions, to the law of mass action:

$$K_p = \frac{P_{Me}^n P_X^m}{P_{Me_n X_m}} \qquad (3.29)$$

where K_p is the equilibrium constant and depends on the temperature and the physical and chemical characteristics of the components, and P represents the partial pressures of the components (in atmospheres).

The ratio of the partial pressure of a free metal to the pressure of a combined metal is:

$$\frac{P_{Me}}{P_{Me_n X_m}} = \left(\frac{K_p}{P_X^m P_{Me_n X_m}^{n-1}}\right)^{1/n} \qquad (3.30)$$

It follows from equation (3.30) that, when the partial pressures P_X and $P_{Me_n X_m}$ are reduced (in the case of $n>1$), the ratio $P_{Me}/P_{Me_n X_m}$ will increase; the higher the values of m and n, the more rapid the increase. The stability of multi-atomic compounds with components with low partial pressures is thus extremely low, and only diatomic molecules are therefore of importance ($m=1$ and $n=1$). In this latter case,

$$\frac{P_{Me}}{P_{MeX}} = \frac{K_p}{P_X} \qquad (3.31)$$

i.e. the ratio of the free and combined atoms of the metal depends solely on the equilibrium constant and the partial pressure of the atom X.

When atomic vapours are being obtained during atomic absorption measurements, usually the partial pressures of the components are so small that the possibility of the formation of multi-atomic molecules can be ignored; only the diatomic molecules need be considered.

The atoms of metals may form molecules with atoms of O, H, N, C, F, Br, Cl, I and with OH radicals. Under the normal conditions in which atomic vapours are obtained, the concentrations of free atoms of C, F, Cl, Br, and I are extremely small, and the formation of the corresponding diatomic molecules can therefore also be ignored (if, of course, an $H_2 + F_2$ flame is used, the formation of MeF type molecules must in the first place be taken into account).

Of the remaining compounds, MeO, MeH, MeN and MeOH, the MeO and MeOH molecules are the most stable. Aluminium is an illustration of this (Table 3.1).

TABLE 3.1 Equilibrium constants for the reactions in the dissociation of aluminium compounds[7]

T (°K)	AlO	AlH	AlN
2000	$2 \cdot 1 \cdot 10^{-7}$	$7 \cdot 1 \cdot 10^{-3}$	$2 \cdot 7 \cdot 10^{-3}$
2500	$8 \cdot 4 \cdot 10^{-5}$	$2 \cdot 5 \cdot 10^{-1}$	$1 \cdot 1 \cdot 10^{-1}$
3000	$4 \cdot 6 \cdot 10^{-3}$	$2 \cdot 7$	$1 \cdot 4$

Under normal conditions for the production of atomic vapour, therefore, molecules of the MeO and MeOH type are the most stable. In the case of most elements, the MeO molecule is the more stable of these. Alkali elements and the third-group elements, gallium and indium, have more stable MeOH molecules. For comparison purposes, Table 3.2 gives the equilibrium constants for the LiOH, LiO and LiH molecules.

TABLE 3.2 Equilibrium constants for the reactions of dissociation of lithium compounds[7]

$T(°K)$	LiOH	LiO	LiH
2000	$3\cdot 3 \cdot 10^{-12}$	$4\cdot 2 \cdot 10^{-4}$	$3\cdot 5 \cdot 10^{-2}$
2500	$1\cdot 5 \cdot 10^{-7}$	$2\cdot 6 \cdot 10^{-2}$	$7\cdot 0 \cdot 10^{-1}$
3000	$2\cdot 0 \cdot 10^{-4}$	$4\cdot 0 \cdot 10^{-1}$	$5\cdot 1$

The equilibrium constants K_p are calculated by the equation

$$\ln K_p = \frac{\Delta\phi}{R} - \frac{\Delta H}{RT} \qquad (3.32)$$

where R is the universal gas constant, equal to $1\cdot 987$ cal/mol°C, $\Delta\phi$ is the variation in the reduced thermodynamic potential (cal/mol°C); ΔH is the change in enthalpy resulting from the chemical reaction or thermal effect of the reaction at 0°K (in cal/mol.); for diatomic molecules the quantity ΔH represents the dissociation energy E_d; T is the absolute temperature.

The reduced thermodynamic potential ϕ for the components in the reaction is calculated by statistical thermodynamics.[7] The thermal effect of the reaction at 0°K is determined experimentally on the basis of calorimetric measurements of the thermal effect ΔH_T of the reaction at a temperature T, and the calculated values for the variation in enthalpy $H_T - H$, or alternatively by experimentally measuring equilibrium constants (spectroscopic investigation of flames, the mass spectrographic analysis of saturated vapours, measurements of the pressure during adiabatic explosion, etc.) for a particular temperature and using equation (3.32). The energy E_d of dissociation of diatomic molecules can also be determined on the basis of research into the optical spectra of molecules.[8]

The values of $\Delta\phi$ for different oxides do not differ by more than a few cal/mol°C. In many cases, the differences between the values of E_d/T amount to several tens of cal/mol°C. The differences between the equilibrium constants for different oxides are therefore basically governed by the dissociation energies E_d.

Table 3.3 gives all the dissociation energies known in 1962 for diatomic oxides of metals, also dissociation energies for the most stable metal hydroxides (Me + OH).

The most stable oxides are those of B, Ba, Ce, Gd, Ge, La, Nb, Nd, Sc, Si, Sn, Ta, Th, Ti, U, V, W, Y and Zr.

The dissociation process is represented by the degree of dissociation α, which corresponds to the number of dissociated molecules with relation to the initial

TABLE 3.3 Dissociation energies of diatomic oxides and hydroxides[9]

Molecule	E_d (kcal/mol)	Molecule	E_d (kcal/mol)	Molecule	E_d (kcal/mol)
AgO	57±10	InO	23±10	SO	124
AlO	115±5	InOH	86±7	SbO	102±20
AsO	113±2	KOH	87·3	ScO	160±30
BO	184±11	LaO	186±5	SeO	81±10
BaO	137±2	LiO	78±5	SiO	192±5
BeO	106±4	LiOH	102	SnO	134±2
BiO	85±20	LuO	122±25	SrO	112±2
CaO	115±4	MgO	100±3	TaO	194±11
CdO	88	MgOH	56±5	TeO	63
CeO	185±5	MnO	96±3	ThO	196
CrO	101±7	MoO	116±15	TiO	156±6
CuO	113±10	NaOH	90±3	UO	179±17
FeO	98±10	NbO	180±35	VO	148±5
GaO	58±12	NdO	165±6	WO	154±10
GaOH	102±5	NiO	97	YO	167±7
GdO	148±45	PO	142	ZnO	92
GeO	159±4	PbO	94±1	ZrO	181±8
IO	44±5	PrO	171±7		

number (the entire amount of the element). The partial pressures of the components Me and MeX in the dissociation reaction

$$\mathrm{MeX} \rightleftharpoons \mathrm{Me} + \mathrm{X} \qquad (3.33)$$

can be expressed in terms of the degree of dissociation α and the total pressure $P_{\mathrm{Me}} + P_{\mathrm{MeX}} = P$, which remains constant during changes in equilibrium in the system:

$$\begin{aligned} P_{\mathrm{Me}} &= \alpha P, \\ P_{\mathrm{MeX}} &= (1 - \alpha)P \end{aligned} \qquad (3.34)$$

The partial pressure of the component X (O or OH) when atomic vapour is being produced is governed by the secondary reactions taking place in the analysis volume (for instance combustion of the fuel gas or oxidation of free carbon in the cuvette), and it remains practically constant, whatever the amount of the component MeX introduced.

If we substitute the partial pressures of the components in equation (3.31), and solve it for α, we get:

$$\alpha = \frac{1}{1 + P_\mathrm{X}/K_p} \qquad (3.35)$$

When equation (3.35) is considered, the following facts can be noted:

1. The degree of dissociation of the MeX compound does not depend on the amount of it which was introduced into the analysis volume. The number of free atoms of the elements is therefore directly proportional to the total number of molecules of the compound introduced into the analysis volume.

2. The degree of dissociation depends equally on the variation in K_p, the equilibrium constant of the reaction, and the change in P_X, the partial pressure; these act

in opposite directions. When attempts are made to increase the degree of dissociation of a compound, the effects of both factors must be taken into consideration. In particular, when the temperature of the medium is raised, in order to increase α it is necessary to take into account the fact that, as K_p increases, the partial pressure P_X may also increase, so that the value of α may not only fail to increase, but may even decrease.

3. The degree of dissociation of compounds tends towards unity ($\alpha \to 1$) on condition that P_X/K_p tends towards zero ($P_X/K_p \to 0$). In practice, however, dissociation can be considered to be complete if $P_X/K_p < 0.01$.

IONIZATION

Under thermodynamic equilibrium conditions, the ionization of free atoms in an atomic vapour can be regarded as a process of the equilibrium dissociation of an atom A into a positive ion A^+ and a free electron

$$A \rightleftharpoons A^+ + e \qquad (3.36)$$

When the process of ionization is considered, there is usually no need to take the time factor into consideration, since the rate at which equilibrium is established is so great that, even in a flame (excluding the reaction zone of the inner cone), there is an equilibrium ionization state.

By analogy with dissociation, the ionization process can be characterized by the equilibrium constant K_i, which is expressed in terms of the partial pressures of the components (in atmospheres) as follows:

$$K_i = \frac{P_e P_{A^+}}{P_A} \qquad (3.37)$$

and the degree of ionization K, which is the proportion of ionized atoms to the total number of atoms. If we represent the total pressure of the atoms, $(P_A) + (P_{A^+})$, as P, the partial pressures can be represented in the following form:

$$\left. \begin{array}{l} P_{A^+} = KP \\ P_A = (1-K)P \\ P_e = KP \end{array} \right\} \qquad (3.38)$$

If we substitute equation (3.38) in equation (3.37), and solve the quadratic equation, we get:

$$K = \frac{K_i}{2P} + \frac{K_i}{2P} \sqrt{\left(1 + \frac{4P}{K_i}\right)} \qquad (3.39)$$

If we put

$$\frac{P}{K_i} = t$$

we finally get:

$$K = -\frac{1}{2t} + \frac{1}{2t}\sqrt{(1+4t)} \qquad (3.40)$$

By means of equation (3.40) we can estimate the degree of ionization of atoms with relation to the ratio of the total vapour pressure of an element in a system to the ionization constant. Table 3.4 gives the results of calculating K using equation (3.40), for several values of t.

TABLE 3.4 Relationship of degree of ionization K to t

	\multicolumn{5}{c}{t}
	10^{-1}	1	10	10^2	10^3
K	0.92	0.42	0.27	0.10	0.03

When Table 3.4 is examined, we find that the degree of ionization can be ignored (with an error of $\leqslant 10$ per cent) only when the vapour pressure of the element in the system is given by

$$P > 10^2 K_i \tag{3.41}$$

The ionization equilibrium constant is calculated, with the medium at any temperature, by the Saha equation:

$$\log K_i = -\frac{V_i}{4 \cdot 573 T} + \frac{5}{2} \log T - 6 \cdot 49 + \log \frac{2g_A^+}{g_A} \tag{3.42}$$

where V_i is the ionization potential of the atoms (in cal/mol), T is the absolute temperature, and g_A and g_A^+ are the statistical weights of an atom and an ion.

It is interesting to estimate the elements for which the partial ionization of atoms should be taken into account. We shall assume that the temperature of the medium is 3000°K, and that the total pressure of the neutral and ionized atoms is 10^{-8} atm. These values correspond to the marginal experimental conditions with which ionization is bound to be maximum. It was indicated earlier that ionization can be ignored in cases in which $K_i < 10^{-2} P$. For the selected maximum value $P = 10^{-8}$ atm, therefore, the value of K_i at 3000°K must be less than 10^{-10} atm.

The following elements are among those for which the ionization constant K_i at 3000°K is higher than 10^{-10}: alkali elements, alkaline earth elements (except for Be), rare earth elements, group III elements (Al, Ga, In, and Tl), and Ti, V, Cr, Y and Mo. The degree of ionization of these elements at 3000°K and a vapour pressure of 10^{-8} atm is greater than 0.1.

When degrees of ionization were being estimated it was assumed that $P_e = P_{A^+}$, i.e. that the concentration of electrons in a system is governed solely by the ionization of the one element being determined. In fact, a considerable number of foreign atoms are usually present in an atomic vapour, either as contaminations or as atoms of the principal substance being analysed. The ionization of gases, water vapour and products of flame combustion can be ignored, since the ionization potentials of these components are higher than 10–12 eV, and the degree of ionization is accordingly very low for them.

Sodium is a constant impurity in air and gases obtained from air: oxygen, nitrogen, and inert gases; under normal conditions of pressure and temperature, the mean concentration of sodium in the atmosphere is $5 \cdot 10^{10}$ atoms/cm^3.[10] At a temperature of 2400–3200°K an electron concentration of $(3-2) \cdot 10^9$ electrons/cm^3 corresponds to this number of atoms (as the temperature is increased within the limits indicated above, the concentration of electrons decreases slightly, since the increase in the degree of ionization is more than compensated by the thermal expansion of the gas).

The minimum concentration of electrons recorded in oxidizing and stoichiometric flames is governed by this 'background' concentration of electrons. In reducing hydrocarbon flames, the minimum concentration of electrons is an order of magnitude greater, being up to $3 \cdot 6 \cdot 10^{10}$ electrons/cm^3. This special feature is due to the thermal ionization of particles of carbon formed in a flame when there is insufficient oxygen present; the dimensions of these carbon particles are 100–1000Å. The effective ionization potential of particles, measured in reference 10 from the concentration of electrons in an oxy-acetylene flame, is $\simeq 8 \cdot 5$ eV—the mean of the ionization potential of atoms of carbon (11·265 eV) and the work function for solid carbon (4·35 eV).

The 'background' concentration of electrons reduces the degree to which the atoms are ionized in comparison with the calculated degree (assuming $P_e = P_A+$). As an instance, with a concentration of $1 \cdot 10^{-7}$ atm for the atoms of lithium in a flame, and a background concentration of electrons amounting to $1 \cdot 10^{-7}$ atm ($T = 3000°K$), the degree of ionization of lithium must be 0·49, not 0·68.

An excess of foreign elements (particularly easily ionized elements) may have an even greater effect on reducing the degree of ionization of an element being investigated. As an instance, in the presence of 10^{-4} atm of atoms of calcium, the electron concentration at 3000°K must be $2 \cdot 10^{-6}$ atm. Under these conditions, the degree of ionization of atoms of lithium, with a vapour pressure amounting to 10^{-7} atm, is only 0·07.

Increasing the electron concentration by introducing an excess of an easily ionized component is a method used in practice (for instance when determining alkali elements by flame photometry) for suppressing the ionization of atoms of elements being determined.

Chapter 4

THE FLAME

The flame is the medium most widely used for atomizing a sample in atomic absorption spectroscopy. It plays the same part as in flame emission spectroscopy, with the sole difference that in the latter case the flame is also the means of exciting atoms. Naturally, therefore, the technique for the flame atomization of samples in atomic absorption spectrochemical analysis is in many respects the same as the technique for emission flame photometry.

A number of works have been published on problems relating to flame photometry, setting forth the principal factors associated with the use, properties, and special features of different flames; these published works have been sufficiently well covered in the reviews by Mavrodineanu,[1] Margoshes,[2] and Dean.[3]

Of the monographs published in recent years, we should particularly mention the book by Poluektov *Methods of analysis by flame photometry* (1959), the book *Flame photometry* (1957) by Burriel-Marti and Ramirez-Muñoz, which was translated into Russian, also the monographs by Dean (1960), Herrmann and Alkemade (1960) and Schuhknecht (1961).[4,5,6]

The monographs *The spectroscopy of flames* by Gaydon (1957) and *Flames, their structure, radiation and temperature* by Gaydon and Wolfhard (1953), which were translated into Russian, described the properties of different flames, including their spectral characteristics. There is no need, therefore, to dwell in any detail on descriptions of different types of flame, designs of burner, or working techniques. We shall concentrate our main attention on the special features of the use of flames in atomic absorption spectrochemical analysis.

NEBULIZERS

As in flame photometry, solutions are nebulized in order to produce a mist of a sample by means of nebulizing systems consisting of the nebulizer itself and a spray chamber. In other cases the direct atomization of solutions in flames is used.

Compressed air nebulizers consist of two capillary tubes positioned either at right angles to one another (angled nebulizers), or concentrically (concentric nebulizers). The sample solution is fed or sucked into one of the capillary tubes, and the nebulizing gas (nitrous oxide, oxygen, or air) is fed into the other. The current of compressed gas covers the surface layer of liquid and nebulizes it into small droplets. The nebulizing gas can be fed into either the central or the outer nozzle of a concentric nebulizer. The Lündegardh nebulizer works on the principle of nebulizing solutions with a central jet of gas, while in Beckman burners the jet of gas is fed through the outer nozzle.

Concentric nebulizers provide a more highly dispersed mist; this can, for instance, be seen from Filcek's experiments.[7] This author proved that the effects of phosphorus on determining calcium, due to the incomplete atomization of refractory calcium phosphates in flames, depend on the dispersion of the solution introduced into the flame. The effects were respectively 92, 62, 15 and 0·5 per cent for the four types of nebulizers of which diagrams are given in Fig. 4.1.

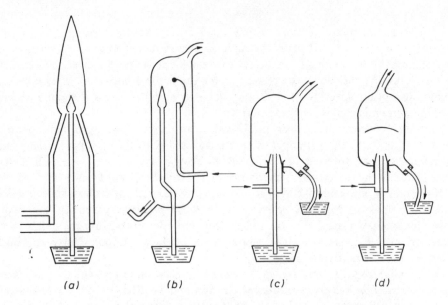

FIG. 4.1. Types of nebulizer investigated by Filcek.[7] (a) Direct injection; (b) Premix chamber with an angled nebulizer; (c) Premix chamber with a concentric nebulizer; (d) As (c) but with a baffle.

Davies, Venn and Willis described a pneumatic nebulizer intended for nebulizing solutions.[70] The position of the end of the capillary tube (a hypodermic needle) relative to the tip (a Venturi tube) can be regulated smoothly by means of a micrometer screw; the rate at which solutions are nebulized can be altered over a wide range: between 1 and 8 ml/min. The nebulizer is so designed that the capillary tube and tip can rapidly be changed.

In the spray chamber, the mist is homogenized by the precipitation of large droplets on the walls. The precipitated droplets are removed from the chamber through a drain hole, and are then either discarded or added to the solution being analysed. In some cases, in order to reduce condensation, the walls of the chamber are preheated, or there is intensive circulation of air along the walls.

The sensitivity and accuracy of atomic absorption measurements depend largely (just as in emission flame photometry) on the design of the nebulizing systems. For instance, according to Herrmann and Lang,[8] with the same amount of solution and other conditions equal, the sensitivity of measurements varied in a proportion of 20:1 for the eight nebulizers investigated.

In order effectively to utilize a solution, nebulization must be as efficient as possible and there must be no condensation losses of the solution in the chamber. The effects of the different factors (surface tension, viscosity and density of solutions, the velocity of the jet of nebulizing gas, and the ratio between the volumes of gas and liquid) on the dimensions of mist droplets were investigated by Nukiyama and Tanasawa.[9] These authors proved that changing the velocity of the gas from 50 to 350 m/s reduces the mean diameter of the droplets from 65 μm to 10 μm. Further increasing the velocity of the gas scarcely affects the droplet dimensions. It follows from this that the gas velocity in nebulizers must not be lower than the speed of sound. The dimensions and design of the spray chamber also affect the effectiveness with which solutions are nebulized. According to Herrmann and Lang,[8] large chambers make it possible to improve nebulization to some extent, but this is at the expense of large consumptions of solution and increases in measurement time (before the nebulizing conditions are stabilized). Preheating the chamber or the jet of air likewise does not provide any substantial improvement, since it is accompanied by deterioration in the uniformity of nebulizing of solutions.

Chambers with the air circulating in a vortex around the walls (Fig. 4.2) provide far better results. The chamber consists of a 6 cm diameter tube with two nozzles, for introducing additional air. The efficiency of nebulization in this type of chamber is 21 per cent compared with 2–10 per cent for chambers of the standard design.

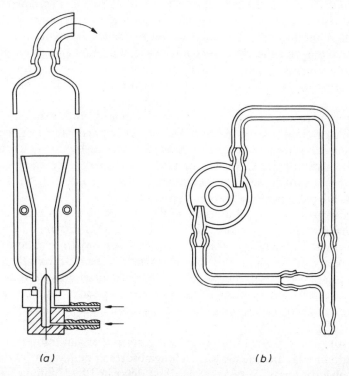

Fig. 4.2. Premix chamber with vortex air circulation;[8] (a) elevation and (b) plan.

Herrmann and Lang[8] pointed out that the improvement in the sensitivity of atomic absorption measurements made when using chambers of this type is greater than the difference between the nebulization efficiencies, and is evidently due to the smaller size of the droplets.

An original design of spray chamber has been developed by the firm of Beckman.[73] The droplets of nebulized solution first enter a chamber heated to 300°C or more, where all the mist is vaporized. The vapour of the solvent, together with the dry aerosol particles, then enters a cooled trap, in which most of the solvent condenses on the walls; the remainder of the vapour, together with the dry aerosol particles, enters the flame. Owing to the concentration of the substance being analysed relative to the solvent, sensitivity increases several times in comparison with nebulizing chambers of standard designs.

BURNERS

The same fuel gases are used for atomic absorption spectrochemical analysis as are used for flame photometry: town gas, propane, butane, hydrogen, acetylene, and cyanogen; the oxidizing agents are air, oxygen, or nitrous oxide.

The fuel gas and oxidizing agent can be fed into the torch nozzle either along separate channels, with the components subsequently mixed by diffusion (diffusion flames), or by first mixing the components in a mixing chamber (premixed flame).

When a mixture of gases is fed, the flame front is above the nozzle of the burner, owing to the rapid flow of gas through the burner slot. The velocity of the current of gas is usually 3–10 times greater than the speed of the flame. The speed of the flame differs by many times for different mixtures. For instance the combustion speed for a mixture of C_2N_2 and air is 20 cm/s, while for mixtures of H_2 and air it is 440 cm/s.

Certain mixtures, particularly when oxygen is present, may detonate. When this occurs the flame front progresses with a speed several orders of magnitude higher than the normal combustion speed; it is, for instance, 2810 m/s for H_2–O_2 mixtures and 2920 m/s for C_2H_2–O_2 mixtures. Naturally in the case of detonating flames the gases cannot be mixed before they are fed into the burner, and the components are fed separately into the burner nozzle.

In accordance with this subdivision of flames there are two principal types of burner. The standard laboratory Méker burner is, for instance, a burner for premixed flames. In burners of this type the fuel gas and air are mixed as the gas leaves a small nozzle within the tube. Air is drawn into the tube from the atmosphere through circular apertures in the gas tube. There is a metal grid above the mouth of the burner, and this prevents the flame from striking back into the tube. In most cases, however, there is a system for the forced supply of air into the mixing chamber.

The advantage of a flame from pre-mixed gases is that slot burners can be used; these provide thin laminar flames with a long absorbing path. A burner of this type for atomic absorption measurements was first described by Clinton.[10] The burner consists of two identical aluminium alloy castings, and is shown diagrammatically in Fig. 4.3. The aluminium is anodized in order to prevent corrosion by acid solutions. The upper parts of the castings form a rectangular slot 12 cm long, 0·7 mm wide and 1 cm deep, thus creating a laminar flow of gas. The parts of the burner are clamped together with screws. The gas enters the burner through a branch pipe with a 15 mm diameter aperture, secured to the burner by means of a flange. The burner can be rotated around its axis, so that the effective length of the absorbing path can be varied between 1 and 12 cm.

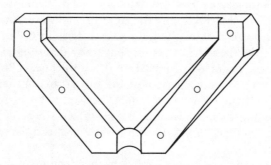

FIG. 4.3. Slot burner.[10]

This burner was used with town gas and a mixture of air and acetylene. The general flow of gas was 9 l/min, 8 l/min of this being the current of air. With different gas flows, the flame either blew out or struck back. The height of the flame above the burner was 4 cm.

Recent research by Amos and Willis has proved that slot burners are effective not only with mixtures of air and acetylene, but also with acetylene mixed with 50 per cent O_2 + 50 per cent N_2, or with nitrous oxide N_2O.[63] The acetylene and nitrous oxide flame has a great advantage over air-acetylene flames, for with the same flame velocity its temperature is 400°C higher. Amos and Willis[63] emphasized that, when a slot burner with a 100 × 0·4 mm slot is used, the C_2H_2–N_2O flame is completely safe, whatever the ratios of the components and whatever the sequence in which the gases are switched off. To prevent the burner from overheating, it must be sufficiently massive. The flame from the mixture of C_2H_2 and (50 per cent O_2 + 50 per cent N_2) also has a higher temperature than the acetylene-air flame, but this flame must nevertheless be used with care.

An elongated burner made of glass was described by Schüler and Jansen.[11] The top was made from a titanium plate, with 1·2 mm diameter gas apertures arranged checkerwise in five rows, covering an area of 130 × 12 mm². The plate is pressed against an asbestos insert on the burner by means of springs. The titanium nebulizer is fitted in the side inlet of the burner. The use of glass and titanium eliminates any corrosion of the substance of which the burner is made when highly acid solutions are used. This type of burner has been used for mixtures of propane–butane and air.

Butler[12] described an elongated burner made of plexiglass with a flat top made of brass; it was intended for an air-acetylene flame. The plate was 6 mm thick and the diameter of the apertures 1 mm. The plate was cooled with running water passing through six transverse channels. The rate of flow of air was 6 l/min.

It should be noted that elongated burners with tops in which there are several rows of apertures are evidently less effective than slot burners, since the true pathlength in the flame, made up of several small separate flames, is much less than the length of the top (particularly towards the bottom region of the flame, above the internal cones, which is used for atomic absorption measurements). This was confirmed by Willis's experiments.[63]

The drawback of the flames from standard single-slot burners is their small thickness. Owing to the considerable angular aperture of the beam of light passing through a flame, a proportion of the light passes through the upper layers of the flame, or even past the flame. As a result, sensitivity deteriorates slightly, and the calibration graphs are distorted. To increase flame thickness, Boling[74] proposed a multi-slot burner for air-acetylene flames; the top of this burner contained three parallel slots 0·45 mm wide and 111 mm long, separated by a gap of 1·57 mm. As would be expected this burner provided a gain in sensitivity of 50–100 per cent, linearization of the calibration graphs and greatly reduced flame noise (by about an order of magnitude).

The Beckman burner, a direct-injection burner, comes under the heading of burners for diffuse flames (Fig. 4.4). Burners of this type can be made of either glass or metal.

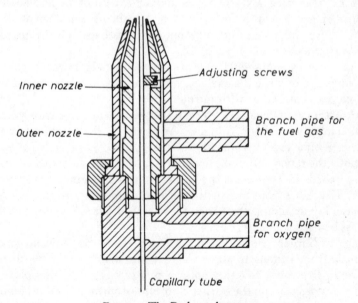

FIG. 4.4. The Beckman burner.

The Beckman burners, and also burners for premixed flames, have the limitation that the rate at which the solution is nebulized into the flame cannot be adjusted independently from the supply of one of the components of the flame (for instance the oxygen). In order to eliminate this limitation, Robinson and Harris,[13] designed a burner with a mechanical system for supplying the solution into the nebulizer (Fig. 4.5) In this burner the entry of the solution into the nebulizer nozzle is regulated by a piston device, in which a Teflon vessel is moved steadily relative to a fixed piston. The maximum rate of feed of solution was 4 ml/min. The solution was nebulized by an axial jet of oxygen at a pressure of up to 10 kg/cm². The main flow of oxygen was regulated independently.

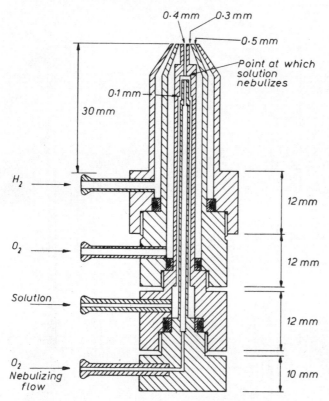

FIG. 4.5. Direct injection burner with forced solution feed.[13]

The advantage of this design is that nebulization is highly effective; very small mist droplets enter the flame, whatever the viscosity of the solution (the results of the experiments made with this burner will be discussed on pp. 143–6).

Diffuse flames have the disadvantage, in comparison with mixed gas flames, of being less stable owing to the turbulent movement of the fuel gases in the burner. Noise is particularly perceptible from highly reducing flames.

The other failing of the Beckman burner is that deposits of solid cause a build-up in the oxygen nozzle. The formation of this deposit alters the speed of the flow of oxygen, and consequently also the flame burning conditions and the effectiveness with which the solution is nebulized.

In order to eliminate these failings of highly reducing oxy-acetylene flames, Fassel and his colleagues[14] developed a special jet for Beckman burners, which acted as a mixer for the gases and the mist. The jet [Fig. 4.6(a)] is a graphite tube with cover in the form of a copper cylinder; this is secured to the body of the burner by means of three pairs of screws 120° apart in the base of the cylinder. The gap between the burner nozzle and the jet must be 0·2–0·1 mm. Three 2 mm diameter apertures are drilled in the copper tube at the level of the gap; these prevent the mixture from exploding in the space between the burner and the jet should the flame enter the tube.

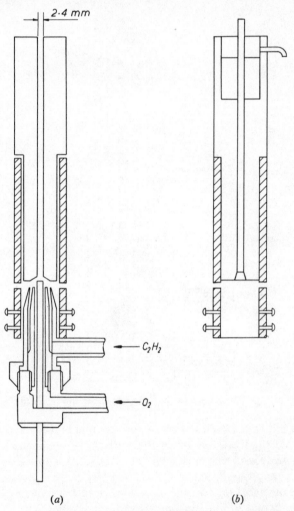

(a) (b)

FIG. 4.6. Mixing tubes for Beckman burners; (a) made of graphite; (b) made of stainless steel and Teflon.[14,64]

When the jet is adjusted relative to the burner, great care must be taken to ensure that the orifices of the jet and the nebulization capillary tube are coaxial. A slight displacement of their axes will disrupt stable operation through the clogging of the jet with the solution being nebulized. The authors of reference 14 suggest that correct adjustment should be controlled by observing the position relative to the copper cylinder of a stream of water atomized through the oxygen channel of the burner. To make this test, the burner is mounted vertically with a nozzle downwards, and the water is fed into the burner through the appropriate branch pipe. Rough adjustment is made with the tube removed, fine adjustment with the graphite tube in position. When the burner is well adjusted the flame does not strike back and the graphite does not become heated.

The same authors[64] later suggested a modification of this burner [Fig. 4.6(b)]. The new feature was the use of a stainless steel tube for the internal channel in the jet, while the remainder was made of Teflon. The upper part of the jet contains a reservoir for excess liquid sprayed from the channel. The advantages of the later design are that liquid is less likely to flow into the channel and that the walls 'remember' the previous specimen less. Its disadvantages are that the upper end of the steel tube has to be cooled with water poured into the reservoir and that there is a risk that the tube will be corroded when corrosive solutions are analysed.

When a highly reducing oxy-acetylene flame is used (C_2H_2 and O_2 flow rates 4·2 and 3·4 l/min respectively), and solutions containing not less than 50 per cent of organic solvents are nebulized, flame fluctuations recorded from changes in background emission are reduced by more than an order of magnitude in comparison with the flame from the same type of Beckman burner with no special jet.

The burner does not operate stably with aqueous solutions. If the salts in the solution exceed 0·5 per cent the jet may be flooded with liquid. Under the conditions indicated above, the rate at which the solution is nebulized is about 1 ml/min.

EFFECTS OF SOLVENTS ON THE NEBULIZATION PROCESS

Substances being analysed are introduced into flames by nebulizing solutions of these substances in water or organic liquids.

Experiments have established that, in many cases, organic solvents are superior to aqueous solutions from the point of view of sensitivity of measurement. In estimating the effects of organic solvents quantitatively, different research workers differ greatly and even contradict one another. For instance, Allan[15] obtained an absolutely identical increase in sensitivity for different elements with one and the same organic solvent. Lockyer and his colleagues[16] found sensitivities increased differently (between 1·5 and 10 times) for different elements. Robinson[17] obtained the same result. According to Warren[18] the use of organic solvents, acetone or methanol, does not increase sensitivity at all. This indicates that the effects of organic solvents are governed by a number of circumstances, not all of which have been taken into consideration by individual researchers.

When we discuss the parts played by solvents, we must differentiate between two groups of factors:
1. The effects of solvents on the nebulizing of solutions.
2. The effects of introducing a solvent on the properties of the flame, i.e. the temperature and composition of the products of combustion.

The second group of factors will be considered shortly (pp. 146–53). Here we shall

Atomic Absorption Spectrochemical Analysis

discuss only the effects of solvents on the nebulizing process. We must at once note that this effect on the results of measurements is evidently of precisely the same nature in both absorption and emission research, for these processes are associated solely with producing a cloud of atoms of a sample in a flame.

The efficiency with which solutions are nebulized depends on the surface tension, viscosity and density of the solution. The mean diameter d of the droplets produced when a solution is nebulized by a jet of gas moving at a velocity u can be expressed by the following empirical equation:[9]

$$d = \frac{585}{u}\frac{\sqrt{\sigma}}{\sqrt{\rho}} + 597\left[\frac{\eta}{\sqrt{(\rho\sigma)}}\right]^{0\cdot45}\left(\frac{1000 R}{V_g}\right)^{1\cdot5} \qquad (4.1)$$

Here u is the velocity of the jet in m/sec, σ is the surface tension of the solution in dyne/cm, η is the coefficient of viscosity of the solution in poise, ρ is the density of the solvent in g/cm^3, and R and V_g are the consumption by volume of solution and gas in unit time. Droplet diameter d is expressed in micrometres (μm).

Let us explain the manner in which the properties of a solvent affect the dimensions of droplets. We shall take a jet velocity of 350 m/s (increasing the velocity above this value scarcely affects droplet dimensions[9]). We shall also assume that the ratio R/V_g is close to a mean value of 0·00033 (e.g. 1 ml of solution to 3 l of oxygen a minute for burners with direct injection, and 2·4 ml of solution to 8 l of air a minute for burners with premix system).

Table 4·1 gives viscosity and surface tension figures for several solvents, together with droplet diameters d calculated from equation (4.1) with the assumptions mentioned above.

TABLE 4.1 Characteristics of the solvents used for nebulizing solutions

Solvent	Formula	ρ (g/cm^3)	$t_{boiling}$ (°C)	$\eta \cdot 1000$ in P (20°C)	σ (dyne/cm)	d (μm)
Acetone	C$_3$H$_6$O	0·79	57	3·3	23·7	13·1
n-Heptane	C$_7$H$_{16}$	0·68	98	4·2	18·4	13·4
Ethyl acetate	C$_4$H$_8$O$_2$	0·92	77	4·5	23·6	12·8
Methyl alcohol	CH$_4$O	0·79	65	5·9	22·6	13·8
4-methyl-2-pentanone	C$_6$H$_{12}$O	0·80	117	5·9	22·7	13·9
Benzene	C$_6$H$_6$	0·88	80	6·5	28·9	14·5
Toluene	C$_7$H$_8$	0·87	111	7·7	28·4	15·0
Amyl acetate	C$_7$H$_{14}$O$_2$	0·88	144	8·9	24·7	14·8
Cyclohexane	C$_6$H$_{12}$	0·78	81	9·3	26·5	15·9
Carbon tetrachloride	CCl$_4$	1·63	77	9·6	27·0	11·9
Ethyl alcohol	C$_2$H$_6$O	0·79	79	12·0	22·8	15·3
Nitrobenzene	C$_6$H$_5$O$_2$N	1·73	211	19·8	42·6	16·4
Isopropyl alcohol	C$_3$H$_8$O	0·79	82	22·5	21·7	18·1
Water	H$_2$O	1·00	100	10·0	73·0	19·0

The following conclusions can be drawn from the results given in Table 4.1:
1. Droplet dimensions are smaller for organic solvents than for water. This is

due to the substantial difference in surface tensions. The measurements made by Dean[4] proved that, when solutions are nebulized in a burner, the diameters of droplets of water and chloroform (the properties of which are close to those of carbon tetrachloride) were 20 μm and 14 μm respectively; these figures agree well with calculated data.

2. The viscosity, which varies by almost 7:1 for different organic solvents, has only a small effect on droplet dimensions. It should, however, be taken into consideration that reduction in the viscosity of a solvent is conducive to increasing R, the consumption of the solution; this is bound to cause a proportional increase in the second term in equation (4.2), i.e. to increase the dimensions of droplets.

3. Although the properties of solvents may affect the dimensions of droplets, and thereby dimensions and vaporization rates of the solid particles remaining when the solvent has dried, their importance should not be overestimated. To judge from the figures in the table, the greatest possible difference in the dimensions of droplets owing to changes in the properties of the solvent cannot exceed 1·5 : 1. The vaporization rate of particles is proportional to the square of the diameter, and so the total vaporization time cannot vary by more than $(1·5)^2$ times. Of all the possible effects of solvents on sensitivity, the effects on droplet dimensions are obviously the least substantial.

Let us now consider the effects of the type of solvent on the rate of consumption of a solution. We know that the amount R of liquid flowing through a capillary tube during a unit of time follows Poiseuille's law:

$$R = \frac{\Delta P \pi r^4}{8l\eta} \tag{4.2}$$

Here r is the radius of the capillary tube in m, l is the length of the capillary tube in m, η is the viscosity of the solution in s . kg/m^2, ΔP is the pressure causing the current of liquid in kg/m^2. The pressure ΔP is the difference between the hydrostatic pressure of the column of liquid in the capillary tube and the pressure caused by surface tension. Pungor and Mahr,[19] however, taking the nebulizing of aqueous solutions of alcohol as an example in which the viscosity may vary by 3:1 proved that during pneumatic nebulization of solutions, the surface of the column of liquid is so distorted by the jet of gas that the surface tension ceases to affect the velocity of the flow in the capillary tube, and R is governed solely by the viscosity, the hydrostatic pressure of the column, and the dimensions of the capillary tube.

Some of the results obtained by Dean,[3] which confirm the effects of viscosity and the fact that surface tension has no influence on the consumption of solutions when using a Lündegardh nebulizer, are given in Table 4.2.

The effects of the properties of solvents on the nebulizing process thus amount firstly to the effects of viscosity on the consumption of the solution, and secondly to the effects of the surface tensions of solvents on the dispersion of the mist. In addition to this, in the case of premix burners, the dispersion, the surface tension and the boiling point of the solvent have a substantial effect on the condensation of mist in the spray chamber.

On the basis of the mechanism, described above, for the effects of organic solvents

TABLE 4.2 Effects of the viscosity of solvents on the consumption of solutions

Solvent	Viscosity (cP)	Surface tension (dyne/cm)	Nebulizing rate (ml/min)
4-methyl-2-pentanone	0·59	22·7	2·00
Amylacetate	0·89	24·7	1·47
Carbon tetrachloride	0·96	27·0	1·09
Water	1·00	73·0	0·93

on the process of nebulization of solutions, let us consider the possible causes of deviations between the results obtained by different research experts. Allan[15] described his experiments with an air-acetylene flame. When water was replaced by any organic solvent, the amount of acetylene supplied was reduced until the luminous flame disappeared, so that the temperature and gaseous composition of the flame were practically the same in the presence of organic solvents. On the other hand, the vaporization of organic metal complexes (Zn, Cu, Mg, Fe, Mn) introduced into the flame was complete. Replacing the solvent therefore only affected the process by which the mist entered the flame. Actually it was proved, by making direct measurements of the amount of mist entering the flame, that the changes in sensitivity which were observed correspond to changes in the amount of mist entering the flame. Under these conditions, i.e. with the solvent not affecting the nebulization process or the dissociation of compounds, it was naturally expected that there would be the same increase in sensitivity for all the elements investigated.

In Lockyer's experiments,[16] variations in the temperature and composition of the gases were not compensated when different solvents were used. It is therefore quite likely that the different increases in sensitivity for different elements when organic solvents were used was associated with variations in the degree of dissociation of compounds. This assumption was made all the more likely because Hinnov and Kohn[20] proved that iron, for which the increase in sensitivity was greatest (it was 10 times better, in comparison with 1·5–3 times for nickel, caesium, silver, calcium and zinc), is not completely dissociated in the air-acetylene flame.

These examples show how carefully we must approach the assessment and comparison of the results obtained by different research workers. On the other hand, they prove the necessity for more detailed information regarding the characteristics of solvents, burners, fuel mixture compositions, and the measurement procedures used in each particular research project.

TEMPERATURE AND GASEOUS COMPOSITION OF THE FLAME

Temperature and gaseous composition are the principal characteristics of flames and govern the dissociation of compounds introduced into them or formed in them. Both characteristics depend on the type of fuel (gas plus oxidizing agent) and on the amount of solvent.

The products of combustion in flames burning at atmospheric pressure are, except in the reaction zone of the inner cone of laminar flames, in a state of thermal equilibrium. The temperature of the flame and the composition of the products of combustion can therefore be calculated from thermochemical data. The basic difficulty in the calculation is that the temperature and gaseous composition of the flame are interconnected; increasing the temperature increases the degree to which the products of combustion are dissociated, and the energy consumed by the dissociation of products of combustion reduces the temperature of the flame. Usually, therefore, the composition and temperature of a flame are calculated by a method of successive approximation.[21]

The results of calculating the temperatures and gaseous compositions of the flames most used in spectrochemical analysis are given in Table 4.3 (no solvent present). Experimentally measured flame temperatures agree with the calculated temperatures.

The figures in Table 4.3 show that replacing oxygen by air lowers the temperature of the flame by about 700°C. The reason is that about 70 per cent of the air in the flame takes no part in the combustion and acts as a form of ballast. The temperature

TABLE 4.3 Temperatures and gaseous compositions of flames[22]

	$H_2 + 0.5\ O_2$	H_2 + air (stoichiometric)	$C_2H_2 + 2.5\ O_2$	C_2H_2 + air (stoichiometric)	$C_2N_2 + O_2$
Enthalpy (kcal/mol)	0.00	0.00	54.6	54.6	73.9
T (°C)	2810	2100	3030	2250	4577
Composition (atm)					
CO_2	—	—	0.14	0.12	0.00
CO	—	—	0.32	0.04	0.66
H_2O	0.57	0.32	0.11	0.07	—
O_2	0.05	0.00	0.12	0.02	0.00
O	0.04	$5 \cdot 10^{-4}$	0.10	$2.2 \cdot 10^{-3}$	$8 \cdot 10^{-3}$
H_2	0.16	0.02	0.03	0.00	—
H	0.08	0.00	0.06	0.00	—
OH	0.10	0.01	0.11	0.01	—
NO	—	—	—	0.01	$3 \cdot 10^{-4}$
N_2	—	0.65	—	0.73	0.32

of the cyanogen flame is 1500°C higher than that of the oxy-acetylene flame, although the fuel gases for these flames have very nearly the same enthalpies. This is because the products of combustion of the cyanogen flame (CO and N_2) scarcely dissociate at the flame temperature, while the product of combustion of the oxy-acetylene flame, i.e. water, is 50 per cent dissociated at 3030°C.

Let us now consider how the behaviour of solvents when aqueous or organic solutions are atomized affects the temperature of the flame. Allowance for the effects of water on the temperature of a reaction amounts to calculating the thermal balance between the reaction heat of ΔH of a pure flame and the heats $Q_p(T)$ and

$Q_{H_2O}(T)$ respectively for heating and dissociating the products of the reaction and the introduced water; the calculation is performed from the equation:

$$\Delta H - Q_p(T) - xQ_{H_2O}(T) = 0 \qquad (4.3)$$

where x is the amount of water relative to the combustion mixture. If we know the functions $Q_p(T)$ and $Q_{H_2O}(T)$, we can find the relationship between T and x. Baker and Vallee[23] used this method of calculating in proposing the following empirical equations for calculating the temperatures of cyanogen and hydrogen flames:

$$T_{(C_2N_2-O_2)} = 4850 \frac{1}{1+0.39 \ln(1+4x)} \,°K \qquad (4.4)$$

$$T_{(H_2-O_2)} = 3000 \frac{1}{1+0.113x+0.046x^2} \,°K \qquad (4.5)$$

The variation in the temperatures of oxy-cyanogen and oxy-hydrogen flames with relation to the ratio of the molar water content to the molar oxygen content of the combustion mixture is plotted in Fig. 4.7. This graph shows that the introduction of water has a particularly great effect on the temperature of the cyanogen flame. If 0.5 of a mol of H_2O is introduced per mol of O_2 (with the C_2N_2 rate of 0.5 ml/min this ratio corresponds to a requirement of 0.2 ml H_2O/min), the temperature of the flame drops by 28.5 per cent (from 4850 to 3400°K), while the temperature of the hydrogen flame drops by 6.7 per cent when water is introduced. The great reduction in the temperature of the cyanogen flame is understandable; for, as we indicated earlier, the high temperature of the pure cyanogen flame is associated with the absence of dissociating products.

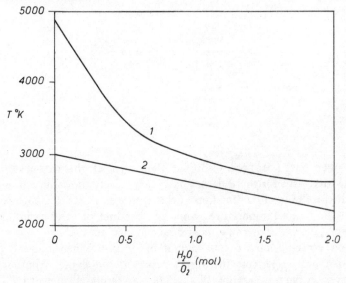

FIG. 4.7. Variations in flame temperature when water is introduced.[23]
1. Oxy-cyanogen flame. 2. Oxy-hydrogen flame.

Foster and Hume[24] measured the temperatures of oxy-hydrogen flames experimentally, with different amounts of introduced water, and confirmed the theoretical conclusions. With 6 ml/min of water and an oxygen flow rate of 3·2 l/min ($x=2\cdot3$) the temperature dropped by 23 per cent, as against a theoretical figure of 33 per cent. The discrepancy between the experimental and calculated figures arose because the flame was lean (3·2 l O_2 + 5·7 l H_2) compared with its stoichiometric composition.

Measurements[25] of the emission of impurity elements in cyanogen and hydrogen flames, relative to the amounts of aqueous solutions of these elements introduced into the flames, further confirmed the theoretical conclusions, in particular the great reduction in the temperature of the cyanogen flame when water is present.

We shall now consider the effects of organic solvents on flame temperature. Organic solvent introduced into a flame should be regarded as an additional source of fuel gas. The heats of combustion of certain compounds as far as their production of CO_2 and H_2O are given in Table 4.4. This table also contains the experimental results of measuring the absorption and emission of the Ni 3415 Å line in an oxy-hydrogen flame, with the solutions introduced at the same rate of 1 ml/min.[17]

TABLE 4.4 Emission and absorption of the Ni 3415 Å line, using different solvents, in an oxy-hydrogen flame[17]

Solvent	Heat of combustion (kcal/ml)	Absorption ($I_0 - I$)	Emission
Water	0	4	4
Carbon tetrachloride	0·39	8	5
Methyl alcohol	4·23	15	29
Ethyl alcohol	5·58	15	31
Acetone	5·83	14·5	33
Benzene	8·80	16	42

The results of both the emission and the absorption measurements reveal a correlation with the heat of combustion of the solvent. In emission measurements the effects of the solvent are better defined than in absorption measurements, since increasing the temperature is conducive not only to the more complete atomizing and dissociation of compounds introduced into the flame, but also to the more effective excitation of atoms. As an instance, in the case of methyl alcohol and benzene, the heats of combustion of which differ in a ratio of 2:1, nickel absorption is the same, while emission varies in a proportion of 1·5:1.

It was pointed out on p. 130 that the partial pressures of the radicals O and OH are the parameters which most influence the dissociation of stable compounds in flames. The partial pressures of these radicals depend in turn on the concentration of water vapour in the flame, and can be calculated from the equilibrium equations

$$H_2O \rightleftharpoons H_2 + O$$

and

$$H_2O \rightleftharpoons H + OH$$

In flames in which combustion of the fuel forms water, the concentration of the radicals O and OH remains extremely high. In hydrogen flames, where water is the only final product of combustion of the fuel, the concentration of the radicals O and OH in the flame remains high, whatever the ratio of oxidizing agent to fuel gas.

The picture is different for hydrocarbon flames. The most stable final product of the combustion of hydrocarbons is CO (the energy of dissociation of CO is 256 kcal/mol, as against 115·6 kcal/mol, for H_2O). If, therefore, there is a shortage of oxygen, by far the greater part of the oxygen is consumed on the oxidation of carbon, not of hydrogen. As an instance, with equal amounts of acetylene and oxygen, combustion corresponds to the reaction

$$C_2H_2 + O_2 \rightarrow 2CO + H_2$$

It is interesting to note that, formally, this reaction corresponds to the combustion conditions for cyanogen,

$$C_2N_2 + O_2 \rightarrow 2CO + N_2$$

The essential difference between these flames is that only a little N_2 dissociates at 4000°K ($K_p = 3\cdot1.10^{-6}$), while almost all the H_2 dissociates at 4000°K ($K_p = 2\cdot5$). A considerable proportion of the energy generated during the combustion of acetylene is therefore consumed, not on heating the gases, but on the dissociation of H_2, and as a result the temperature of the flame is far lower than the temperature of the cyanogen flame.

While the introduction of a solvent into a hydrogen flame can be regarded from the point of view solely of the change in flame temperature, since the concentration of water, the principal product of combustion in the flame, remains practically unchanged, the introduction of any particular solvent into a hydrocarbon flame may alter the composition of the products of combustion qualitatively and must be estimated from the point of view of change in the reducing or oxidizing properties of the flame. It is of great practical interest to consider the effects of solvents on oxy-acetylene flames, which with certain ratios of fuel gas to oxidizing agent are bound to have reducing properties (and high temperatures). In view of this, the author calculated the temperatures and gaseous compositions of flames with different ratios of acetylene and oxygen, and with water or ethyl alcohol used as the solvent. In every case, the amount of oxygen was the same, since the rate of flow of oxygen governs the amount of solvent required. The amount of solvent (H_2O or C_2H_5OH) introduced was 1 ml/min in every case; the amount of oxygen was 2·8 l/min, and the amount of acetylene was varied between 0·7 and 5·5 l/min. The range of ratios of solvent to combustible mixture corresponded to the commonest experimental conditions employed using Beckman burners. The calculations were made with a Ural-2 electronic computer. The initial data were the elementary composition of the fuel and its enthalpy H and are given in Table 4.5.

Consideration of the results obtained confirms our preliminary conclusions regarding the manner in which the properties of flames alter when the hydrocarbon fuel content is increased. As the acetylene content of the flame is increased, its composition steadily alters in the direction of greater reducing properties (i.e. of reduction of the partial pressures of O and OH). When equal carbon and oxygen contents (in gram-atoms) are reached, the composition of the flame undergoes an abrupt change in properties: H_2O, CO_2 and O_2 disappear completely from the composition of the flame, while the concentration of the radicals O and OH decreases by 5–6 orders of magnitude, the concentration of atoms of carbon increases by 4–5 orders of magnitude, and solid carbon (soot) appears, which governs the luminosity of the flame.

As the acetylene content is further increased, the concentration of solid carbon in the flame increases rapidly, but the reducing characteristics of the flame remain practically unchanged. In addition, the temperature of the flame steadily drops. The best conditions for producing a reducing atmosphere are therefore provided by a flame whose elementary composition is in the proportion

$$\frac{C}{O} = 1 \tag{4.6}$$

On the other hand, the results given in Table 4.5 enables us to estimate the effects of solvents on the properties of flames. In the first place, the temperature of a reducing flame is about 200°C higher when ethyl alcohol is used as the solvent than when water is used. As in the example of an oxy-hydrogen flame considered earlier, the change in flame temperature accompanying changing to a different solvent is associated with the variation in the overall enthalpy of the fuel. In the second place, to achieve conditions under which a flame has a reducing effect requires more acetylene in the presence of water than in the presence of ethyl alcohol. This is quite natural, since the molecule of water contains an atom of oxygen but not a single atom of carbon, while the molecule of alcohol contains one atom of oxygen and two of carbon.

From the point of view of the C/O ratio, all solvents in general can be subdivided into three categories: reducing (C/O>1), neutral (C/O=1) and oxidizing (C/O<1). Reducing solvents (most organic solvents, such as C_2H_5OH, C_6H_6, etc.) increase the overall C/O ratio in the flame; neutral solvents (CH_3OH) do not alter it, and oxidizing solvents (HCOOH or H_2O) reduce it.

Let us now give some of the results of our discussion of the flame characteristics which are most important for analysis purposes.

1. For a flame to be of a reducing nature is not compatible with the presence of water in the products of combustion of the fuel. The hydrogen flame, the only product of combustion in which is water, does not therefore produce a reducing medium under any circumstances.

2. For a medium to be of a reducing nature it is necessary that the gram-atom oxygen content of the fuel (combustible gas, oxidizing agent and solvent) should not exceed the total carbon content of that fuel.

3. To ensure reducing properties at a higher temperature, conducive to the vaporizing of low-volatility compounds, mixtures of acetylene with oxygen or nitrous oxide, not with air, should be used.

TABLE 4.5 Temperature and composition of the oxy-acetylene flame with different consumptions of acetylene

Consumptions of components			Elementary composition of the fuel (gram-atoms)			H (kcal/kg)	T (°K)	Partial pressures (atm)									
Solvent (1 ml/min)	O_2(l/min)	C_3H_8(l/min)	H	O	C			CO_2	CO	H_2O	H_2	O_2	OH	H	O	C	C_{solid}
Water	2·8	5·5	53·0	26·8	43·2	839	2874	—	0·374	—	0·342	—	8·4 . 10⁻⁸	0·055	8·2 . 10⁻⁹	1·2 . 10⁻⁵	0·229
	2·8	3·8	47·8	32·5	36·0	575	2876	—	0·526	—	0·355	—	1·1 . 10⁻⁷	0·062	1·2 . 10⁻⁸	1·6 . 10⁻⁵	0·057
	2·8	2·8	43·6	37·3	30·0	351	3049	0·025	0·522	0·093	0·252	0·0004	0·012	0·093	0·0026	1·0 . 10⁻⁹	—
	2·8	1·9	39·0	42·4	23·6	112	3164	0·081	0·406	0·201	0·128	0·014	0·054	0·092	0·023	4·2 . 10⁻¹⁰	—
	2·8	1·4	35·5	46·3	18·6	−71	3132	0·132	0·292	0·257	0·073	0·058	0·082	0·064	0·043	1·1 . 10⁻¹⁰	—
	2·8	0·95	32·0	50·1	13·8	−244	3023	0·177	0·169	0·300	0·038	0·147	0·090	0·033	0·047	1·2 . 10⁻¹¹	—
	2·8	0·7	29·7	52·8	10·6	−364	2883	0·196	0·086	0·328	0·020	0·241	0·077	0·016	0·037	9·6 . 10⁻¹³	—
Ethyl alcohol	2·8	5·5	53·3	23·8	47·1	1148	3013	—	0·310	—	0·309	—	9·5 . 10⁻⁸	0·077	1·4 . 10⁻⁸	4·5 . 10⁻⁵	0·304
	2·8	3·8	48·1	29·1	40·6	944	3052	—	0·429	—	0·307	—	1·3 . 10⁻⁷	0·095	2·4 . 10⁻⁸	7·7 . 10⁻⁵	0·169
	2·8	2·8	43·8	33·4	35·2	771	3084	—	0·552	—	0·306	—	1·7 . 10⁻⁷	0·112	3·6 . 10⁻⁸	1·0 . 10⁻⁴	0·030
	2·8	1·9	30·1	38·2	29·2	583	3249	0·031	0·509	0·091	0·187	0·002	0·027	0·140	0·012	3·0 . 10⁻⁹	—
	2·8	1·4	35·5	41·8	24·6	442	3288	0·066	0·429	0·146	0·119	0·018	0·062	0·123	0·038	1·3 . 10⁻⁹	—
	2·8	0·95	32·0	45·3	20·2	294	3257	0·102	0·340	0·184	0·077	0·057	0·088	0·092	0·061	4·3 . 10⁻¹⁰	—
	2·8	0·7	29·6	47·7	17·2	208	3214	0·126	0·272	0·203	0·056	0·103	0·099	0·070	0·073	1·1 . 10⁻¹⁰	—

The Flame

4. Organic solvents, unlike water, are additional sources of heat generated during combustion. The drop in flame temperature caused by introducing an organic solvent is therefore less than when water is introduced.

5. The use of organic solvents when solutions are atomized in a flame means that, in most instances, the ratio of the carbon to the oxygen in the fuel is increased, while water reduces this ratio.

6. The pure oxy-cyanogen flame has a high temperature because there is practically no dissociation of products of combustion. When solvents (water in particular) are introduced into the flame, its temperature drops abruptly to that of the oxy-acetylene flame.

DETERMINATION OF ELEMENTS WITH REFRACTORY OXIDES

Until 1960–61, the problems of atomizing elements with refractory oxides by means of flames seemed quite insoluble. It was assumed that elements with oxides such as AlO, BeO, MoO, TiO, VO, etc., could not be determined from their atomic spectra in flames. Attempts at solving this problem appeared still more hopeless because, even in oxy-cyanogen flames, the temperature of which is highest, the sensitivity with which these elements could be determined was excessively low. As far back as 1957, however, Knutson[71] had noted the good effect of the fuel-rich air-acetylene flame as regards increasing the emission of atomic lines. This effect did not attract the required attention, since the measurements were made with magnesium, which also dissociates quite well in stoichiometric (oxidizing) flames.

In 1959 Eshelman, Dean, Menis and Rains proved[26] that using an organic solvent for aluminium in an oxy-acetylene flame results in atomic lines a hundred times more intense than when using aqueous solutions. Oxy-acetylene flames with an oxygen/acetylene ratio of 2·0 were tried in the experiments (3·3 l/min of O_2 and 1·65 l/min of C_2H_2). The use of greater amounts of acetylene proved impossible owing to increasing fluctuations in the flame background. The amounts of solution nebulized were 1 ml/min for water and 2 ml/min for 4-methyl-2-pentanone. The gram-atom oxygen/carbon ratio can be calculated from the proportions of fuel gas, oxidizing agent, and solvent. The ratio is 1·74 when water is used and 1·04 when an organic solvent is used. The substitution of organic solvents for water thus made the conditions almost the same as in reducing flames.

The best results were obtained by Fassel and his colleagues;[27,28] they used a highly reducing oxy-acetylene flame, with an oxygen/acetylene ratio of 0·8 for determining elements with refractory oxides. Ethanol was used as the solvent, with an atomizing rate of 1 ml/min. During their first work, Fassel and his colleagues investigated rare earth elements, scandium and yttrium,[27] the atomic spectra of which had never been observed in flames owing to the high dissociation energies of their monoxides (see Table 3.3). Atomic spectra were observed for all these elements in the reducing oxy-acetylene flame, while practically no atomic emission from many of the elements was obtained from the oxy-hydrogen flame. The emission in the oxy-hydrogen flame was about 1000 times weaker for lanthanum, praseodymium, neodymium and erbium. The dissociation energy of the oxides of europium and ytterbium is not great, and the ratios of the emission intensities in the different flames were 3:1 and 5:1 respectively.

The use of ethanol instead of water provided a gain in the emission intensity for all elements (between 3:1 and 5:1 for different lines).

There is no difficulty in interpreting these results on the basis of the arguments set forth in the previous section. As was to be expected, the use of a vigorously reducing oxy-acetylene flame made it possible to dissociate even the most refractory monoxides of these elements, while even with an organic solvent the oxy-hydrogen flame could not provide conditions under which sufficient dissociation took place.

The use of an organic solvent instead of water in a highly reducing flame does not have any great effect on its reducing properties. About the same emission intensity was therefore recorded for all elements, and this intensity was basically associated with the increased flame temperature (increased by 200°C, see Table 4.5). The slightly different amplifications for different lines of the same element were associated with differences in the line excitation energies.

In their work, Fassel and his colleagues[28] investigated low-volatility elements with oxide energies between 6·1 and 6·9 eV. The limits of detection of these elements, corresponding to twice the root mean square error in measuring the background, are given in Table 4.6.

Under normal (turbulent) combustion conditions, flame background fluctuations due to non-uniform mixing of fuel gases and solvent were very great. A Beckman burner with a mixing tube enabled the same authors[14] to reduce the fluctuations by an order of magnitude and accordingly to improve the detection limits.

TABLE 4.6 Detection limits of low-volatility elements in a reducing oxy-acetylene flame from atomic emission spectra[28,14]

Element	Most sensitive line (Å)	Detection limit (ug/ml) Without mixing tube	With mixing tube
V	4379·2	3	—
	4408·2 + 4408·5	—	0·3
Nb	4058·9	12	1·0
Ti	3998·6	5	0·5
Mo	3798·3	0·5	0·03
Re	3460·5	3	0·3
W	4408·8	90	4·0

Research into the possibility of using reducing flames for atomic absorption measurements was at first carried out with mixtures of air and acetylene. Allan[29] proved that, in magnesium determinations, increasing the amount of acetylene from 1·1 to 1·56 l/min, with a constant 8 l/min of air, increases the absorption of the Mg 2852 Å line by 25 per cent. About 0·12 ml of aqueous solution was introduced into the flame a minute. It is easy to calculate that, in Allan's experiments, the O/C ratio varied between 1·5 and 1·07, i.e. the flame had to be reducing.

Later, David[30] established that, if a slightly luminous air-acetylene flame was

The Flame

used, molybdenum could be determined with a sensitivity of $5 \cdot 10^{-3}$ per cent. Gatehouse and Willis[31] investigated the manner in which the sensitivity with which molybdenum was determined varied with different rates of flow of acetylene and a constant 5·5 l/min of air. When the amount of acetylene was increased from 1·4 to 2·0 l/min, five times more absorption of the Mo 3133 Å line was recorded, so that the sensitivity was $5 \cdot 10^{-5}$ per cent. Under the same conditions, the sensitivity for tin (by the Sn 2863 Å line) was $5 \cdot 10^{-4}$ per cent. Attempts at determining aluminium, beryllium, vanadium, niobium and tungsten were, however, unsuccessful. When Allan[32] used a slightly luminous air-acetylene flame he found that the sensitivity for ruthenium and chromium was improved by a factor of 4 to $2·5 \cdot 10^{-5}$ per cent and $5 \cdot 10^{-6}$ per cent respectively. Slot burners 10–12 cm long were used with aqueous solutions in all the research described in these references.

Robinson[17] proved that the cyanogen flame does not solve the problem of determining aluminium, molybdenum, tungsten or other elements with stable oxides, although in this type of flame the V 3184 Å line is absorbed with sensitivity of $3 \cdot 10^{-2}$ per cent. The only possible explanation is that aqueous solutions rob the cyanogen flame of its reducing nature, making it closer in characteristic to the stoichiometric oxy-acetylene flame.

Reducing oxy-acetylene flames were first used for atomic absorption measurements by Fassel and Mossotti[33] and by Slavin and Manning.[34] Using the same atomizing conditions as in emission measurements, Fassel and Mossotti determined the most sensitive lines of vanadium, titanium, niobium, scandium, yttrium and rhenium, and measured the detection limits for these elements (Table 4.7). At the same time, Slavin and Manning used a reducing oxy-acetylene flame for determining aluminium, beryllium, vanadium, titanium and barium (Table 4.7).

It should be noted that, in reference 33, the research was carried out by the photographic recording of emission from a continuous spectrum source (a xenon lamp), the beam of light from which was reflected three times through three successive burners. A single passage of a beam of light from a hollow-cathode lamp through one burner was used by Slavin and Manning[34], and measurements were made with a spectrophotometer. There was a constant discrepancy of 4, therefore, between the results given in references[33,34] as to the sensitivity with which the same elements (vanadium and titanium) were determined.

The right-hand columns of Table 4.7 contain sensitivities achieved by Amos[63] using the flame from a mixture (50 per cent O_2 + 50 per cent N_2)–C_2H_2, and by Willis[40,63] using an N_2O–C_2H_2 mixture flame. Slot burners respectively 3 and 5 cm long, with a premix chamber, were used. The fact that the temperatures of these flames are higher than that of the air-acetylene flame, also the reducing combustion conditions, enabled the authors to obtain appreciable absorption even of low-volatility elements such as tungsten, hafnium and zirconium. The results were better than those obtained by Slavin and Manning[34] as regards the sensitivity with which aluminium, barium, beryllium, titanium and vanadium were determined; this was evidently due to using an elongated slit burner instead of a Beckman burner.

Dowling et al.[35] also were devoted to the determination of aluminium in oxy-

TABLE 4.7 Sensitivity of the atomic absorption determination of elements with stable oxides

Element	Line (Å)	Sensitivities (10^{-4} per cent)			
		Fassel and Mossotti[33]	Slavin and Manning[34]	Amos[63]	Willis[63]
Al	3093			1	1
	3962		6		
B	2497			100	100
	2498				50
Ba	5535		3.5		0.4
Be	2349		0.2	0.03	0.03
Hf	3073			20	
Nb	3580			30	
	4059	250			
Re	3460	25			
Sc	3907	5			
Si	2516			10	5
Ta	2715			33	
Ti	3643			4	3.5
	3653	50	12		
V	3184	25	7	4	1.5
W	2944				12
	2947			17	18
Y	4077	50			
Zr	3601			9	

acetylene and air-hydrogen flames, using organic solvents. These authors established that the oxy-acetylene flame provides nine times better sensitivity than the oxy-hydrogen flame; the parameters of the former were: rate of flow of oxygen 4 l/min, acetylene 2·2 l/min, rate of consumption of 4-methyl-2-pentanone solution 3·8 ml/min. It should be noted that the gram-atom composition of this fuel is: C—29 per cent, O—29 per cent, H—42 per cent, i.e. all the oxygen is consumed in oxidizing carbon. The sensitivity with which aluminium is determined depends critically on the flame region in which absorption is measured. Maximum absorption was recorded 19 mm from the burner nozzle, and absorption being about halved at distances of 15 and 24 mm from the nozzle (Fassel and Mossotti made their measurements 25 mm above the burner nozzle, evidently to benefit from an increase in the reducing properties of the flame). When the flame is moved to one side of the optical axis, two maxima are observed, corresponding to where the beam of light intersects the boundaries of the inner cone of the flame. With the rate at which the solution is nebulized and the flow of oxygen both constant, a reduction in the amount of acetylene fed into the flame also reduces the aluminium absorption. For instance, increasing the oxygen/acetylene ratio from 2·7 to 4·5 about halves the absorption. The sensitivity of aluminium determinations was about $8 \cdot 10^{-4}$ per cent which agrees well with the figure of $6 \cdot 10^{-4}$ per cent given in Table 4.7.

The results obtained in the cited published works are extremely important, since they confirm that elements with refractory oxides can be determined in flames by the atomic absorption spectrochemical method. It is important that these results were achieved by extremely simple technical means: burners with simple designs and commonly used types of flame.

Wendt and Fassel[75] solved the same problem by a method far more complicated technically. They atomized low-volatility elements in an induction-coupled plasma at a temperature of 16 000°K. The discharge took place in a quartz tube which was mounted vertically within the oscillating field of a 5 kW radio-frequency generator (3·4 Mc/s), the magnetic component being axial to the tube. Argon passing through the tube secured laminar flow and isolated the plasma from the tube wall. The plasma flame had a bright opaque central cone 8 mm in diameter and 25 mm long, a concentric zone 16 mm in diameter and 75 mm long, and an outer region extending 150 mm above the central zone. The outer region of the plasma flame above the quartz tube, 80–100 mm from the central cone, was used for measuring absorption. The width of the plasma region through which the beam passed was about 10 mm. Solutions were nebulized at 0·12 ml/min by means of an ultrasonic generator (870 kc/s, 12 W).

The high temperature of the plasma, the low velocity at which the mist passed through it ($\simeq$38 cm/s), and the small dimensions of the mist droplets ($\simeq$5 μm), meant that low-volatility compounds were more effectively atomized than in the flames described earlier. With the beam of light passing three times through the plasma, however, aluminium, niobium, titanium, tungsten, vanadium and yttrium were determined with a sensitivity of the same order as in a flame. This confirms that refractory compounds are nebulized and dissociated almost completely in effectiveness in high-temperature reducing flames.

In addition to the cumbersomeness of the apparatus, the plasma flame has a further shortcoming in that the free atoms are greatly ionized. For instance, there is a difference in sensitivity of 1·5 :1 between atomic absorption determinations of calcium (ionization potential 6·11 V) from the neutral Ca 4227 Å line and the single ionized Ca 3934 Å line.

Plasma nevertheless has important advantages over hydrocarbon reducing flames in giving much less continuous noise and greater transparency. These mean, that in principle, we can expect improved limits of detection through recording lower absorbances than in flames and that the field of atomic absorption analysis can be extended into the vacuum ultra-violet part of the spectrum.

METHODS OF INCREASING THE LENGTH OF THE ABSORBING VOLUME

The effective length of the path followed by light in an atomic vapour can be increased by passing the beam of light several times through the volume (pp. 103–7) or by increasing the length of the volume itself.

The following expedients are used for increasing the length of the absorbing volume when samples are atomized in flames: slot burners, the installation of several burners on the optical axis, and special attachments enabling the gases leaving the flame to be used as the atomic vapour.

When slot burners are used, the cross-sectional area of the outlet aperture from the burner must not exceed the cross-sectional area of concentric burners, since the stability of the flame is critically associated with the velocity at which the gases pass through the burner nozzle. With the same rate of flow of gases, if the length of the burner is increased to 12 cm the width of the slit must be reduced to 0·7 mm.[10] Obviously these dimensions are close to the limit since further reducing the width of the burner slot results in rapid clogging of the slot with particles of salt from the solution nebulized. In addition, reducing the thickness of the flame introduces optical difficulties by restricting the angular aperture of the beam of light from the source and so lessening the luminous flux passing through the flame. Slot burners provide five or six times the sensitivity of simple concentric burners with a flame 2–2·5 cm thick.

The method under consideration is suitable only for flames from premixed gases. In the diffuse flame, the only method of lengthening the absorbing volume is to use several separate burners positioned successively along the path followed by the beam of light. The instrument produced by JACO, for example, includes three Beckman burners, improving the sensitivity three times. It should be noted that the use of several separate burners results in a corresponding increase in the amount of solution required, as opposed to slot burners, with which the amount of solution required depends solely on the nebulizing system and not on the shape of the flame.

A number of research workers have studied the possibility of using the spent gases from flames to increase the sensitivity of atomic absorption measurements.[36–39] For this purpose the beam of light is passed through a long tube, into which the end of the flame is introduced. According to the relative positions of the tube and the burner, there are horizontal, vertical and T-shaped adaptors (Fig. 4.8).

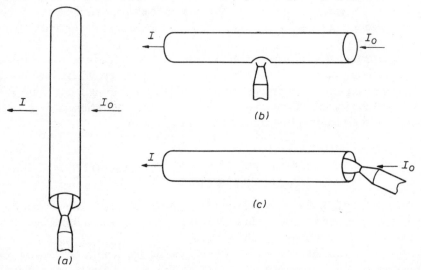

FIG. 4.8. Adaptors for using spent gases from flames: (a) vertical, (b) T-shaped, (c) horizontal.

The Flame

When mercury was being determined,[37] brass and iron tubes with diameters of 1·5–2·0 cm, and between 22 and 68 cm long, were used as T-type adaptors. A propane-butane flame was employed. It was established that the sensitivity increased with the length of the column of gases (the tube). With a tube 69 cm long, using 8–9 ml of solution a minute with a concentration of $1 \cdot 10^{-8}$ g of mercury per ml, the absorbance was about 0·1.

The possibility of using this method for determining other elements has been studied by Zelyukova and Poluektov.[38]

It was established that the use of a 1·5–1·85 cm diameter T-shaped adaptor 30–50 cm long for determining copper, gold, silver, or cadmium provides a considerable gain in sensitivity compared with direct analysis in a propane-butane flame. The gains in sensitivity for these elements were 5·4:1, 8:1, 10·6:1 and 13·7:1 respectively. Zelyukova and Poluektov[38] gave the results of research into the lifetimes of free atoms of metals in the spent gases within a vertical adaptor consisting of a 2·2 cm diameter quartz tube 70 cm long. The lower edge of the tube was positioned 2·5 cm above the burner. The height of the flame within the tube was 10 cm. The results of absorption measurements made at different heights in the tube are given in Fig. 4.9. When this graph is considered, we find that the free atom lifetime is longest in the case of cadmium, mercury, gold and silver, shorter for copper and lead, and shortest for magnesium, thallium and sodium. For this reason, adaptors may not provide any perceptible increase in the sensitivity for magnesium, thallium and sodium. These results are confirmed by Robinson,[36] who pointed out that this type of adaptor is not suitable for determining alkali metals.

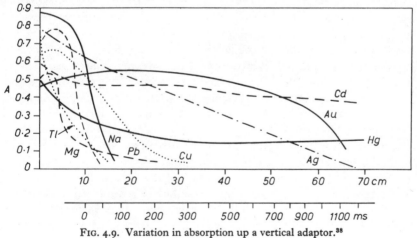

FIG. 4.9. Variation in absorption up a vertical adaptor.[38]

A more effective method of utilizing the spent gases was proposed by Fuwa and Vallee.[39] They used a quartz tube 91 cm long and with an internal diameter of 1 cm. This tube was mounted horizontally within a boron silicate tube 99 cm long. The 5 mm gap between the tubes was filled with powdered magnesium oxide. The tubes were connected to an exhaust system. The solution was directly nebulized into an air-hydrogen flame by means of the Beckman burner positioned at an angle of 15° to the optical axis. The upper part of the burner was chamfered to avoid obstruction to the parallel beam of light passing through the tube. The sensitivity with which cadmium and zinc were determined, in comparison with the normal vertical flame from a Beckman burner, was three orders of magnitude better; for cobalt, nickel, and copper, it was two orders better, and for magnesium it was an order better. It is interesting that the T-shaped adaptor used by Fuwa and Vallee in their experiments provided practically no increase in sensitivity.

Fuwa and Vallee using a horizontal adaptor investigated the relationship between the sensitivity of zinc determinations and the length and diameter of the tube, the substance of which it is made, the interlayer of MgO powder, and partial closing of the discharge aperture at the tube end. The results of these experiments were as follows:

1. Reducing the diameter of the tube increases the sensitivity and with particular abruptness between diameters of 1·65 and 1·0 cm. If the diameter is reduced to less than 1 cm, sensitivity does not increase any more.

2. The sensitivity increases linearly with tube length up to lengths of about 70 cm, beyond which the improvement is less pronounced.

3. The gain in sensitivity is less from tubes, all of the same shape, made of asbestos, graphite or zirconium than from a quartz tube.

4. A layer of MgO powder increases sensitivity by 16 per cent.

5. Partially closing the discharge aperture from the tube increases sensitivity by 30 per cent.

Fuwa and Vallee considered that results (1), (3) and (5) are due to internal reflections of the beam of light from the tube walls; in their opinion these reflections increase the effective length of the absorbing volume. In our opinion, these explanations are not acceptable; for internal reflections are only of infinitesimal importance in tubes if we consider the low coefficient of reflection from the surface and the almost parallel path followed by rays within the adaptor. The explanation of the part played by tube diameter is even more far-fetched. The authors consider that point (5) is due to the increase in pressure in the tube, but it is quite obvious that there is no question of any increase in pressure within a tube both ends of which are open.

In our opinion, results (1), (3) and (4) for increase in sensitivity are associated with changes in the temperature distribution along the adaptor. Actually as the gas passes through the adaptor, its temperature and that of the tube walls decrease, and this causes atomic vapour to condense, and consequently reduces the concentration of atoms along the tube. According to the measurements made in the work described in reference 38 the temperature of the gases at the entry aperture into the vertical adaptor was 1800°C, while at the discharge aperture (70 cm away) it was only 500°C. An even steeper temperature gradient is to be expected between the ends of 90 cm long tubes. This very fact is the reason why the relationship of the increase of sensitivity to tube length is not linear.

Reduction in tube diameter and the use of a layer of powdered substance naturally result in the temperature of the walls being higher and more uniform at different points along them. Less atomic vapour therefore condenses on the walls and the sensitivity improves. The effects of the tube material are evidently also associated with temperature changes in the column of atomic vapour.

The reason why sensitivity is slightly increased when the discharge aperture from the tube is restricted is that the convectional expulsion of atomic vapour at the tube edges is restricted.

From the point of view of the mechanism I have proposed to explain the results, the greatest possible increase in sensitivity is gained from horizontal adaptors specially heated to their maximum temperatures by means of independent heating apparatus.

A horizontal adaptor of the type suggested by Fuwa and Vallee was quite effectively used by Koirtyohann and Feldman[72] for increasing the sensitivity of bismuth, cadmium, copper, mercury, magnesium, manganese, nickel, lead, antimony, strontium, tellurium, thallium, and zinc determinations. They conducted their experiments with 1 cm diameter quartz tubes between 20 and 80 cm long, with an outer layer of asbestos for thermal insulation. For the most part, oxy-hydrogen flames were used for atomizing the solutions, while, as the authors explained, the oxy-acetylene flame has certain advantages as regards reducing the effects of sample composition on analytical results.

The experiments established that tubes of the maximum length are only effective for lead, zinc, cadmium and mercury, the vapour pressures of which are highest among the elements listed above. Accordingly, the greatest gain in sensitivity was achieved for these elements (a gain of 10–25 times). For magnesium, manganese, and nickel the optimum adaptor length is 20–25 cm, and this does not provide more than a 10:1 improvement in sensitivity. The lowest increase in sensitivity applied to strontium, which has a comparatively stable oxide and is therefore prone to rapid conversion to the molecular state.

When solutions containing large amounts of NaCl, $Al(NO_3)_3$ or H_2SO_4 are analysed, a non-selective absorption signal is recorded because of the absorption and scattering of light by condensed particles (or droplets) of the low-volatility compound in the adaptor.

LIMIT OF DETECTION AND SENSITIVITY

In view of the fact that published works have given contradictory interpretations of the concepts of detection limits and sensitivity which are used for assessing the effectiveness of methods of analysing substances, we shall now include a few general comments concerning the further use of these terms.

By the limit of detection we mean the minimum amount of an element required for registering an analysis signal which confirms that that element is present in a sample. On the basis of the statistical interpretation of this concept given in Kaiser's published works,[78–9] the ultimate value is usually taken as being that amount of an element the probability of determining which is sufficiently close to unity.* We know from the law of the distribution of random errors (normal distribution) that,

* See also Kaiser, H., and Menzies A. C. *The Limit of Detection of a Complete Analytical Procedure* (London, Adam Hilger Ltd, 1968).

if the standard deviation in determining any measured quantity is equal to σ, it is safe to assume that 68·3 per cent of the results obtained will not differ from the true value by an amount exceeding 1σ, 95·5 per cent of the results will be within 2σ of the true value, and 99·7 per cent of the results will be within 3σ. If, therefore, σ corresponds to the standard deviation in a blank measurement, it can be stated with a confidence of 68·3 per cent that an analysis measurement equal to 1σ is caused by the presence in the sample of the element being determined. In the case of signals equal to 2σ and 3σ the confidence rises to 95·5 per cent and 99·7 per cent respectively.

In general, therefore, the limit M_{min} can be expressed by

$$M_{min} = k\sigma_M \quad (4.7)$$

where k is a coefficient representing the confidence level of the measurements. To some extent, k is arbitrary; it may vary between 1 and 3 according to choice and circumstance. Most analysts consider that $k=1$ is too low for reliable measurements. As regards selecting either $k=2$ or $k=3$, opinions differ: Kaiser[78] and Zaidel'[80] use $k=3$, while Slavin,[82] Fassell,[33] Winefordner[46] and many others prefer $k=2$. We also shall use $k=2$.

The analysis signal i measured in an experiment is associated with the amount M of an element which a sample contains by the general relationship

$$i = f(M) \quad (4.8)$$

If we expand $f(M)$ in a power series of σ_M, and confine ourselves to the first term of the expansion, we get

$$\sigma_i = \frac{df}{dM}\sigma_M = \frac{di}{dM}\sigma_M \quad (4.9)$$

If we substitute (4.9) in (4.7) we get

$$M_{min} = k\frac{dM}{di}\sigma_i \quad (4.10)$$

Let us now consider the quantities di/dM and dM/di in equations (4.9) and (4.10). The quantity di/dM represents the rate at which the measured analysis signal i changes when the amount of an element in a sample changes, or in other words the slope of the calibration curve, $i=f(M)$. In metrology this quantity is called the sensitivity. We shall retain this interpretation of sensitivity for spectrochemical analysis.

The ratio dM/di then corresponds to the inverse sensitivity. It is useful to draw attention to the formal analogy between these concepts and such concepts in geometrical optics as dispersion and inverse dispersion. It is clearly preferable to assess an analytical method in terms, not of the sensitivity di/dM, but of the inverse sensitivity dM/di, by analogy with the dispersion in spectroscopic devices, which is normally expressed not as the linear dispersion $dl/d\lambda$, but as the inverse linear dispersion $d\lambda/dl$. For simplicity, although improperly, the word inverse is often omitted in both cases, and high sensitivities or dispersions are represented by low numerical values of dM/di and $d\lambda/dl$.

The Flame

Kaiser[81] suggested that the effectiveness of a method of analysis providing a low (or high) limit of detection with high (or low) capacity for identification should be represented, on the basis of this formal analogy, according to the concept of high (or low) resolving power used for spectroscopic devices with low (or high) resolution limits.

According to (4.10), therefore, the limit of detection depends on a coefficient governing the confidence level of measurements, the inverse sensitivity, and the error in measuring the analysis signal.

After these general comments, let us discuss existing concepts in atomic absorption measurements. The analysis signal is expressed in absorbance A. For low values of A, we have

$$\frac{di}{dM} = \frac{dA}{dM} = \frac{A}{M} \tag{4.11}$$

If we substitute (4.11) in (4.10), we get

$$M_{min} = k\frac{M}{A}\sigma_A \tag{4.12}$$

Since the determination of the absorbance A invariably involves measuring the signal intensities I_0 and I, it is advisable to express the error σ_A in terms of the fluctuation in signal intensity σ_I. When low values of absorbance are measured, according to (2.27),

$$\sigma_A \simeq 0\cdot 87\frac{\sigma_I}{I} \tag{4.13}$$

It follows from equation (4.13) that if the coefficient of variation in the intensity of the analysis signal is equal to 1 per cent then $\sigma_A = 0\cdot0087$, while if $\sigma_I/I = 0\cdot 5$ per cent, then $\sigma_A = 0\cdot 0043$.

If we substitute the value of σ_A from (4.13) in equation (4.12), and assume, as we decided earlier, that $k = 2$, we get

$$M_{min} = 1\cdot 74\frac{M}{A}\cdot\frac{\sigma_I}{I} \tag{4.14}$$

The limit M_{min} is expressed either in absolute units by weight (grams, milligrams —10^{-3} g, micrograms—10^{-6} g, nanograms—10^{-9} g, and picograms—10^{-12} g) or in units relative to the weight or volume of the sample (percentage by weight, parts per million (p.p.m.), micrograms per ml of solution, etc.).

Absolute limits of detection usually relate to the determination of pure elements (in the absence of large amounts of foreign substances). When samples are introduced into flames in the form of solutions, limits are expressed in relative units with relation to solutions of pure compounds of the elements concerned. For flame methods of analysis, absolute limits of detection can be calculated from known values of the relative limits of detection C_{min} (for instance, in $\mu g/ml$), amounts R of solution used per unit of time and minimum analysis-signal measurement times τ_{min}:

$$M_{min} = C_{min} \cdot R \cdot \tau_{min}$$

The time τ_{min} depends on the time taken to establish an equilibrium value for the concentration of the element concerned in the analysis volume (see p. 122), but, when flames burning mixed gases are used, it is governed in practice by the time taken for equilibrium conditions of mist entering the flame to be reached. According to Slavin[82] and Kahn,[83] using a burner typical of those produced for atomic absorption measurements by Perkin–Elmer, with a premix chamber, τ_{min} is about 1 sec.

It should be taken into consideration that the time constant τ_c of the recording device must not exceed $0.25\,\tau_{min}$, since if the reverse is the case there will be a steady reduction in the peak value of the analysis signal (p. 119). In view of this, to assess absolute limits of detection it is necessary to use relative limits C_{min} measured, not for any value of τ_c, but for $\tau_c \leqslant 0.25\,\tau_{min}$.

In order to assess the sensitivity with which any particular element is determined, relative to its amount in the solid sample that is dissolved for the analysis, it is necessary to know the maximum permissible amount of sample in the nebulized solution. Because solids are deposited in the burner jets, and since this prevents normal flame conditions, the concentrations of solids are usually restricted to 1–5 per cent. In certain cases, the concentration may be even lower, if the solubility of the solid is limited. Finally, we should take into account the possible effect (usually detrimental) on the sensitivity (dA/dM) if there is a great excess of foreign substances (see pp. 174–89).

It has already been pointed out that usually the sensitivity of a method is denoted by a quantity which is really the inverse of the sensitivity, dM/dA or M/A. With the equilibrium method of measuring absorption in flames, inverse sensitivities are expressed in terms of concentrations of elements in solution (C) causing specific absorptions (A) when they are atomized in a flame. For traditional reasons, $A = 0.0043$ is normally used as the unit of absorbance; this value of A corresponds to 1 per cent absorption of the beam of light. The units most used for expressing C/A are $(\mu g/ml)/1$ per cent abs, or more simply p.p.m./1 per cent abs.

It should be recognized that, although it would no doubt be more correct to express inverse sensitivities in terms of concentrations related to $A = 1$, the method used in practice is more convenient and realistic, firstly because it corresponds better to the methods used (it is rare in practice to make analytical measurements with levels of absorption lower than 1 per cent), and secondly because, with an absorbance $A = 1.0$, the relation between A and M is usually no longer linear, so that the real value of dC/dA in the $A = 1.0$ region is much lower than in the working range of absorbance (<0.5).

When different methods of analysis are compared they are primarily assessed as to their capacity for identifying elements under certain conditions which are relatively common to a large group of elements. Table 4.8 gives the limits of detection achieved for sixty-eight elements by the equilibrium method of measuring atomic absorption and flame photometry. The figures are for data published up to early 1968.

The lowest emission detection limits were obtained by Pickett and Koirtyohann[84] in a nitrous oxide and acetylene flame, with a 50 mm × 0·5 mm slot burner positioned along the optical axis of the spectrometer. By comparison with a burner positioned perpendicular to the optical axis, the coaxial position provided a gain in brilliance of 100–200 per cent, but owing to self-absorption there was considerable curvature of the $I=f(C)$ calibration curve in a 10:1 range of variation in concentration. The rate of flow of N_2O was 9·4 l/min, and the rate of flow of C_2H_2 was varied between 3·4 and 5·3 l/min for each element in order to arrive at the optimum relationship between line intensity and noise. Emission was recorded with a JACO No. 82–000 grating monochromator (inverse dispersion, 16 Å/mm) with a 1P28 photomultiplier. The amplifier time constant was 3 sec, and the signals were recorded on a chart recorder. A full-size image of a 5 mm high region of the flame, 3 mm above the burner top, was projected on the monochromator slit. The rate of consumption of solution was 3 ml/min. Aqueous solutions were used.

For the remaining thirty-eight of the total number of sixty-eight elements, the figures are those given by Fassel and Golightly.[85] These emission measurements were made with a Beckman oxy-acetylene burner fitted with a mixing chamber to produce a laminar flame (p. 141). Emission was again recorded with a JACO No. 82–000 grating monochromator, a spectral slit width of 0·1 Å, and a recording circuit time constant of 1 sec. The spectra were scanned at a constant rate of 10 Å/min. In the region below 2700 Å, a Beckman DU prismatic monochromator with a larger aperture ratio was used. The rate of consumption of solution was 0·7–1 ml/min. Solutions of the chlorides and perchlorates of elements in a mixture of water and ethanol were used. The flame was made the optimum for each element by adjusting the rate of flow of the gases: oxygen between 3·4 and 3·6 l/min, acetylene 2·1–3·6 l/min.

The limits of detection with which the elements were determined by atomic absorption were obtained by Slavin's team using a Perkin–Elmer model 303 double-beam spectrophotometer.[86] Aqueous solutions of elements were used for atomization. Rates of consumption were 3–4 ml/min. Owing to the steady changes in the signal in the double-beam system, the time constant of the recording circuit could be increased to 60–120 sec. An air-acetylene flame 100 mm long was used for determining most of the elements, while an acetylene-nitrous oxide flame 75 mm long was used for elements with low volatilities.

The limits of detection given in Table 4.8 correspond in every case to emission and absorption signals which are double the noise levels. It can be seen from above that conditions for emission and atomic absorption measurements differed very widely. It is scarcely advisable, therefore, to use the results given in Table 4.8 for any quantitative assessment. Nevertheless, we can confidently select a group of eighteen elements from the list in Table 4.8—Au, As, Be, Bi, Cd, Co, Fe, Hg, Ir, Mg, Pb, Pt, Rh, Sb, Se, Si, Te and Zn—the limits of detection for which are much lower by atomic absorption (more than an order of magnitude lower) than by emission. The limits for the remaining fifty elements are almost the same by

TABLE 4.8 Limits of detection (μg/ml) in atomic absorption and flame photometry

Element	Line (Å)	Flame photometry	Atomic absorption	Element	Line (Å)	Flame photometry	Atomic absorption
Ag	3280·68	0·02*	0·005	Nb	4058·94	1*	3*
Al	3092·71		0·1*	Nd	4634·24		2*
	3961·53	0·01*			4883·81	1	
As	1936·96		0·2	Ni	2320·03		0·005
	2349·84	50			3414·77	0·03*	
Au	2427·95		0·02	Os	2909·06		1*
	2675·95	0·5*			4420·47	10	
B	2496·78	30	6*	Pb	2169·99		0·01
	2497·73				4057·82	0·2*	
Ba	5535·48	0·002*	0·05*	Pd	2476·42		0·02
Be	2348·61	1	0·002*		3634·70	0·05*	
Bi	2230·61	40	0·02	Pr	4939·74	2	
Ca	4226·73	0·0001*	0·002		4951·36		6*
Cd	2288·02		0·005	Pt	2659·45	2*	0·1
	3261·06	2*		Rb	7800·23	0·002	0·005
Ce	5200·12		150*†	Re	3460·46	0·2*	1·5*
	5699·23	10		Rh	3434·89		0·03
Co	2407·25		0·005		3962·36	0·3	
	3453·51	0·05*		Ru	3498·94		0·3
Cr	3578·69		0·005		3728·03	0·3	
	4254·33	0·005*		Sb	2175·81		0·1
Cs	8521·10	0·008	0·05		2598·05	20	
Cu	3247·54		0·005	Sc	3911·81		0·1*
	3273·96	0·01*			4020·40	0·03*	
Dy	4211·72	0·1	0·2*	Se	1960·26		0·5
Er	4007·97	0·3	0·1*	Si	2516·11	5	0·1*
Eu	4594·03	0·003	0·08*	Sm	4296·74		2*
Fe	2483·27		0·005		5175·42	0·8	
	3719·94	0·05*		Sn	2246·05		0·06
Ga	2874·24		0·07		2839·99	0·3*	
	4172·06	0·01*		Sr	4607·33	0·0002*	0·01
Gd	3684·13		4·0*	Ta	2714·67		5*
	4519·66	2			4812·75	18	
Ge	2651·18	0·5*	1*	Tb	4326·47	1	2*
Hf	3072·88		15*	Te	2142·75		0·3
	3682·24	75			2383·25	200	
Hg	2536·52	40	0·2	Th	5760·55	150	
Ho	4103·84	0·1		Ti	3642·68		0·1*
	4163·03		0·1*		3998·64	0·2*	
In	3039·36		0·05	Tl	2767·87		0·025
	4511·31	0·005*			5350·46	0·02*	
Ir	2640·72		2	Tm	4094·19		0·1*
	3800·12	100			4105·84	0·2	1*
K	7664·91	0·003	0·005	U	3584·88		12*
La	3927·56		2*		5915·40	10	
	5791·34	1		V	3183·98		0·02*
Li	6708·84	0·000003	0·005		4379·24	0·01*	
Lu	3312·11	0·2	3*	W	4008·75	0·5*	3*
Mg	2852·13	0·005*	0·0003	Y	4077·38	0·3	0·3*
Mn	2794·82		0·002	Yb	3987·98	0·05	0·04*
	4030·76	0·005*					
Mo	3132·59		0·1	Zn	2138·56	50	0·002
	3798·25	0·03					
Na	5889·95	0·0001	0·002	Zr	3601·19	3	5*

* C_2H_2–N_2O flame † P. E. Thomas, *Resonance Lines*, 1969, 1, 6.

either method (within an order of magnitude) and seldom differ by substantially more than an order of magnitude in favour of the emission method (Ba, Li, Sr).

Although the conditions under which the emission and absorption measurements were made were selected rather arbitrarily, and although therefore there is no doubt that the ascertained relationship between the detection limits will be changed greatly in the future, the atomic absorption method will obviously remain superior to the emission method for determining the group of elements named in the previous paragraph, since this superiority is associated with the particular features of the methods. The temperature of flames (about 3000°K) is not high enough to excite strong (with high values of f) lines shorter than 2500 Å. For instance, the population of the excited level for the Zn 2139 Å line at 3000°K is 10^6 times less than for the Na 5890 Å line (see Table 1.1). With approximately identical measurement conditions, therefore, the limit of detection for zinc is bound to be 10^6 times greater than for sodium. We assume, in both cases, that the noise level depends solely on the multiplier dark current. The figures given in Table 4.8 confirm this difference.

The table draws our attention to the fact that both methods have low capacities for detecting such elements as boron, cerium, gadolinium, hafnium, iridium, lanthanum, niobium, neodymium, praseodymium, samarium, tantalum, terbium, thorium, uranium, tungsten and zirconium. The reasons are many: the low vapour pressures of the elements and their comparatively high atomic weights, the low resonance-line oscillator strengths and the multiplet ground levels of the atoms, the stability of the diatomic oxides, and the low ionization potentials, of certain elements. Besides, the poor quality (low brilliance and spectral complexity) of the light sources used for the research limits the atomic absorption determination of these elements.

If we ignore the last reason and confine ourselves to the factors which govern sensitivity, the most important factors in my opinion are the oscillator strengths f, the ground-level populations from which light is absorbed N, and the atomic weights of the elements A_w. Obviously the stability of gaseous oxides when high-temperature reducing flames are used is least important. To confirm these assumptions, we shall compare the sensitivities of some of the elements listed above with those of elements for which there is no doubt that the oxides are fully dissociated in flames.

The constants, the results of calculations, and the results of experimental measurements of these elements under the same experimental conditions necessary for assessing sensitivities (in relative units) are given in Table 4.9. The experiments were conducted with acetylene-nitrous oxide flames using slot burners 50 mm long[87] and 75 mm long.[41] In practice, however, as Amos and Willis[87] established, this difference is unimportant.

The degrees of ionization of aluminium, samarium, and uranium were calculated for a temperature of 3200°K using Saha's equation (3.42) and equation (3.39). The total vapour pressure of an element in a flame was assessed on the basis of concentrations of elements in solutions exceeding the figures given in the table by 10–20 times.

TABLE 4.9 Calculated and measured values of inverse sensitivity for elements with different oxide dissociation energies

Element	Oxide dissociation energies (eV)	Atomic weight (A_w)	Ionization potential (V)	Degree of ionization (K)	Analysis line (Å)	Relative population of ground level (N)	Oscillator strength (f)	Calculated values of dC/dA (relative units)	Measured values of dC/dA p.p.m. 1% abs.	Measured values of dC/dA (relative units)
Mg	4.3	24.3	7.6	0.0	2852.1	1.0	1.2	1.0	0.019[87]	1.0
Al	5.0	26.9	6.0	0.27	3092.7	0.656	0.23	12	1[87]	53
Sm	5.9	150.4	5.7	0.46	4296.7	0.105	0.44	0.3 · 10³	21[87]	1.1 · 10³
W	6.7	183.9	8.0	0.0	4008.8	0.273	0.088	0.4 · 10³	17[87]	0.9 · 10³
Nd	7.2	144.3	6.3	0.0*	4634.2	0.405	0.07	0.25 · 10³	10[41]	0.5 · 10³
U	7.8	238.1	4	0.95	3584.9	0.498	0.16	3 · 10³	120[87]	6.3 · 10³
La	8.1	138.9	5.6	0.0*	5501.3	0.436	0.025	0.6 · 10³	34[41]	1.8 · 10³

* Ionization was suppressed by introducing an excess of potassium [41]

Inverse sensitivities were calculated using the following equation:

$$\frac{(dC/dA)_{Al}}{(dC/dA)_{Mg}} = \frac{A_{Al}f_{Mg}(1-K)_{Mg}N_{Mg}}{A_{Mg}f_{Al}(1-K)_{Al}N_{Al}}$$

To make the meaning more apparent, the equation is written for Al–Mg vapour. The values of dC/dA for all the elements are related to dC/dA for magnesium.

When the calculated and experimental data given in Table 4.9 are compared, we find that the difference in sensitivity recorded in practice agrees satisfactorily with the anticipated difference associated with the physical constants of the elements. At the same time, no correlation exists between certain discrepancies in the calculated and experimental data and the oxide dissociation energies.

The low sensitivity, and therefore also the low limit of detection with which most of the elements enumerated above are determined, are therefore quite conformable to rule. Perhaps boron is the only element the low sensitivity of the determination of which by atomic absorption may cause surprise, since its physical properties (low atomic weight, comparatively high oscillator strength, and high ground-level population) would be expected to provide at least a hundred times greater sensitivity.

Let us now consider certain problems associated with calculating sensitivity and limits of detection. In a series of fundamental research projects, Winefordner and his colleagues[43,46] proved that the present state of the theory makes it possible, at least in principle, to calculate limits of detection in either flame photometry or atomic absorption methods. These calculations, however, necessitate accurate quantitative knowledge, which is generally not available, of instrumental parameters. Instances are the transmittance and effective aperture of the monochromator, the sensitivity of the photomultiplier, the light source noise, the strength of the analysis line from the light source, etc. Consequently, approximations of detection limits from the equations do not accord well with experiment.[46]

Theoretical calculations of sensitivity are easier and more reliable, since all the dubious and to some extent arbitrary instrumental parameters can be ignored and we can expect theoretical results to agree far better with experiment. This in no way belittles the practical importance of the results obtained in the references given above, since these results can effectively be used for selecting the optimum experimental conditions, for estimating the influence of one or more factors on the method and comparing (in relative units) the performance of different types of apparatus or even of different methods (later we shall compare the atomic absorption and atomic fluorescence methods of determination). We shall therefore try to by-pass the difficulties given above and confine ourselves to calculating sensitivity. We shall first establish the relationship between the concentration of an element in a solution and the number of its atoms in the flame. This number can be expressed by the equation:

$$N = \frac{6 \cdot 02 \cdot 10^{23} CR}{A_w V_g} \alpha \epsilon \quad (4.15)$$

Here C is the concentration of an element in the solution (in g/ml); R is the volume of solution consumed per unit of time (ml/min); ϵ is a coefficient for the efficiency with which the solution is utilized; α is the degree of dissociation; A_w is the atomic weight of the element; V_g is the volume of the gas diluting the vapour of the element at flame temperature and can be expressed by:

$$V_g = \frac{vT_{fl}\eta}{300} \qquad (4.16)$$

where v is the total velocity of the gases (taking the vapours of solvent into consideration) flowing through the burner (in cm³/min) at room temperature ($\simeq 300°K$), T_{fl} is the temperature of the flame, and η is a coefficient for variation in the number of mols of gas during combustion.*

Allowing for flame length l, this gives us:

$$Nl = \frac{6 \cdot 02 \cdot 10^{23} \cdot 300 CR\alpha\epsilon l}{vA_w T_{fl}\eta} \qquad (4.17)$$

On the other hand, if certain simplifications are made, the product Nl can be related to the absorbance A. From equation (1.4),

$$A = 0 \cdot 434 k_\nu l$$

where k_ν is the coefficient of absorption, which can be expressed in terms of k_0, the coefficient of absorption at the centre of the line profile, taking into account the results given in Table 1.6. Assuming $a = 1 \cdot 0$ we get:

$$A = 0 \cdot 356 \cdot 0 \cdot 434 \cdot 2\sqrt{\frac{\ln 2}{\pi}} \cdot \frac{e^2}{mc} \cdot \frac{f}{\Delta\nu_D} Nl \qquad (4.18)$$

[see equation (1.12)]. If the value of Nl from equation (4.17) is substituted in equation (4.18), and the latter equation rewritten relative to the C/A ratio, we finally get:

$$\frac{C}{A} = \frac{1 \cdot 44 \cdot 10^{-24} \cdot \Delta\nu_D v A_w T_{fl}\eta}{Rf\alpha l \epsilon} \qquad (4.19)$$

As an example, let us estimate the sensitivity of cadmium determination from the Cd 2288 Å line in an air-acetylene flame. We shall use parameters corresponding to Allan's experiments.[15,32] If these values, given in Table 4.10, are substituted in equation (4.19), we get $C = 2 \cdot 6 \cdot 10^{-8}$ g/ml. According to Allan, the experimentally measured concentration was $2 \cdot 5 \cdot 10^{-8}$ g/ml. The calculated and experimental results agree excellently, but calculated results are not always so accurate. In particular, there may be no justification for adopting the value of the ratio k_ν/k_0 for $a = 1 \cdot 0$. Nevertheless, the calculation is unlikely to be in error by a factor of more than 2.

We shall use equation (4.19) again for estimating the sensitivity achieved for cadmium in Fuwa and Vallee's experiments,[39] in which the spent gases from an oxy-hydrogen flame acted as the atomic vapour. The principal parameters necessary for the calculation are given in Table 4.11 when they differ from those already given in Table 4.10.

* The coefficient η is about $1 \cdot 0$ for air flames, $\simeq 0 \cdot 9$ for oxy-hydrogen flames, and between $1 \cdot 2$ (oxidizing) and $1 \cdot 9$ (reducing) for oxy-acetylene flames.

TABLE 4.10 Parameters for calculating the sensitivity with which cadmium is determined in a flame

$A = 0.0043$ $T_{fl} = 2500°K$ $A_w = 112.4$
$R = 2$ ml/min $f = 1.20$ $\alpha = 1.0$
$v = 10^4$ cm^3/min $\eta = 1.0$ $\epsilon = 0.1$
$\Delta\nu_D = 4.4 \cdot 10^9$ sec^{-1} $l = 12$ cm

TABLE 4.11 Parameters for calculating the sensitivity with which cadmium is determined in the spent gases from a flame

$T_{fl} = 2800°K$ (allowing for the reduction in T when H_2O is introduced)
$\Delta\nu_D = 2.8 \cdot 10^9$ sec^{-1} (with $T = 1000°K$)
$R = 1.5$ ml/min $\epsilon = 1.0$
$l = 91$ cm $\eta = 0.9$

It should be noted that, in this particular case, the temperature of the flame, which governs the concentration of atoms in the spent gas, is different from the temperature of the absorption volume, which governs the Doppler absorption line broadening. We can take the temperature of the spent gases as being approximately 1000°K.

Using equation (4.19) we get $C = 3 \cdot 10^{-10}$ g/ml. According to Fuwa and Vallee, $C_{exp} = 4 \cdot 10^{-10}$ g/ml. The results agree very well and give further confirmation of our arguments.

Calculations of this sort can, of course, be used only for the few most favourable cases, when the mist is fully atomized, the molecules all dissociated, and the resonance lines of the source are sufficiently narrow for the signal to be considered monochromatic.

Nevertheless, the theoretical representation of sensitivity as a function of experimental parameters is extremely convenient for selecting the conditions for maximum sensitivity. It follows in particular from equation (4.19) that reductions in the rate of flow of the gases and the temperature of the flame are conducive to increase in sensitivity. This has been pointed out in a number of experimental works, for instance those by Allan,[15] Gatehouse and Willis,[31] and Herrmann and Lang.[45]

To conclude this section, we shall dwell on the comparative assessment of the detection limits of the atomic absorption and atomic fluorescence methods of analysis. Atomic fluorescence, as a method of spectrochemical analysis, was studied and used by Winefordner and his colleagues,[65-7] but had been discussed earlier by Alkemade.[68] It is essentially an emission method. Unlike flame photometry, however, it is based on radiative rather than thermal excitation.

Atomic fluorescence analysis requires much the same apparatus as atomic absorption analysis, i.e. a powerful source of resonance emission for the optical excitation of atoms of the element being determined, a flame playing the part of an analysis cell, and a spectrometer with photoelectric recording. The spectrometer must be set up at right angles to the direction of the beam of light through the flame; it is used to measure the fluorescence of atoms in the flame. The exciting luminous flux is modulated and the measured signal then isolated by electronic means in order to separate fluorescence from the foreign radiation from the flame.

The energy emitted in the form of resonance fluorescence E_F is proportional to the absorbed radiation E_A, i.e.

$$E_F = \phi E_A \qquad (4.20)$$

where ϕ, the quantum efficiency of the fluorescence, allows for consumption of a proportion of the absorbed energy on non-emission processes (transfer of energy from excited atoms during collisions with foreign particles known as 'quenching') and for the transfer of excited atoms to other levels with emission at frequencies differing from the principal (exciting) frequency.

In order to compare the potentials of the methods, let us make certain assumptions:

1. We shall assume that, in both methods, the apparatus enables us to achieve the theoretical limits of detection set by statistical fluctuations in the detector photocurrent.

2. We shall assume that, in both methods, the photocurrent is governed solely by the useful signal, i.e. that the photomultiplier dark current and other radiations superimposed on the signal are negligible. It should be noted that, while this condition is commonly justifiable in practice for absorption measurements, in the case of fluorescence it is applicable, and even then only partly so, solely for fluorescence in the far ultra-violet region, where thermal excitation and the flame background can be ignored.

3. We shall assume that the brightness of the light source and the quality of the measuring system (the aperture ratio of the monochromator, the sensitivity of the detector, the amplifier bandpass, the signal measurement time, etc.) are the same in both methods.

In both methods, a signal equal to double the noise level will be taken to set the limit of detection. Using equation (2.34) and assumptions (1) and (2), we can then express the minimum recorded photocurrent i_ϕ for atomic fluorescence as follows:

$$i_\phi = 3 \cdot 3 \cdot 10^{-18} \Delta f \qquad (4.21)$$

Taking the same assumptions (1) and (2) into account, the minimum determinable absorbance in the absorption method can be represented, according to equations (4.14) and (2.34), in the following form:

$$A = 1 \cdot 74 \frac{\Delta i_\alpha^0}{i_\alpha^0} = \frac{1 \cdot 74}{1 \cdot 1 \cdot 10^9 \sqrt{(i_\alpha^0/\Delta f)}} \qquad (4.22)$$

Here i_α^0 is the detector photocurrent corresponding to the unabsorbed signal from the source of light.

In addition to this, in accordance with assumption (3) we can replace the energies E_F and E_A in equation (4.20) by the detector photocurrents i_ϕ and $i_\alpha^0 - i_\alpha$ proportional to them. We must allow for the fact that, in atomic fluorescence, we measure only the proportion ω of the total energy E_F emitted within the solid angle determined by the dimensions of the condenser. We therefore get:

$$i_\phi = \omega\phi(i^0_\alpha - i_\alpha) \qquad (4.23)$$

If we divide both sides of equation (4.23) by i^0_α and use the following relationship, which is true for low absorbance

$$\frac{i^0_\alpha - i_\alpha}{i^0_\alpha} \simeq 2\cdot 3 \log \frac{i^0_\alpha}{i_x} = 2\cdot 3 A$$

we get

$$\frac{i_\phi}{i^0_\alpha} = 2\cdot 3\omega\phi A \qquad (4.24)$$

If the three equations (4.21), (4.22) and (4.24) are solved simultaneously, we can determine the photocurrent i^0_α which is bound to give both methods the same detection limit. We find that:

$$i^0_\alpha = 0\cdot 82 \cdot 10^{-18} \frac{\Delta f}{(\omega\phi)^2} \qquad (4.25)$$

To estimate the magnitude of i^0_α, we shall use the values of $\omega\phi$ found experimentally by Winefordner and Staab.[66] The mean value of $\omega\phi$ for cadmium, zinc, and mercury was about $7 \cdot 10^{-5}$. The half-width of the transmission band Δf can be assumed to be 1 c/s. If these values are substituted in equation (4.25) we get:

$$i^0_\alpha \simeq 10^{-10} \text{ amp}$$

It follows from equations (4.22) and (4.24) that the detection limit of the atomic absorption method is inversely proportional to $\sqrt{i^0_\alpha}$ and the detection limit of the atomic fluorescence method inversely proportional to i^0_α. When $i^0_\alpha > 10^{-10}$ amp, therefore, the fluorescence method is bound to give the lower limits, but when $i^0_\alpha < 10^{-10}$ amp, the absorption method is superior.

In practice, even when the most intense known source of line spectra is used (for instance, a spherical radio-frequency lamp), the detector photocurrent i^0_α does not exceed 10^{-10} amp. In the case of hollow-cathode lamps, the maximum photocurrents are one or two orders of magnitude lower. With the emission intensities achieved by present-day light sources, the absorption procedure is in principle certain to provide the lower limits of detection.

Technically, it is quite a complicated matter to take advantage of this superior quality with high luminous fluxes, since, as an instance, with $i^0_\alpha = 10^{-11}$ amp and $\Delta f = 1$ c/s, the absorbance corresponding to the theoretical limit of determination is 0·0005. Nevertheless, when double-beam systems are used, this problem is quite easily solved (see pp. 95–103 and 107–14).

It has already been pointed out, on the other hand, that when fluorescence is measured condition (2) is valid for only a limited number of elements whose resonance lines lie in the extreme ultra-violet region of the spectrum and whose compounds are atomized and dissociated in flames with a low background level (for instance in air oxidizing flames or oxy-hydrogen flames). Therefore, atomic fluorescence must in principle achieve poorer limits of detection than the atomic absorption method and can determine fewer elements.* Without claiming, however, that the atomic fluorescence method can be used for determining any element, we can admit it as an extremely useful supplement to flame photometry, compensating for the limitations of flames in the thermal excitation of ultra-violet spectra.

EFFECTS OF FOREIGN ELEMENTS ON THE RESULTS OF MEASUREMENTS

The purpose of this section is to consider the effects of foreign elements on the results of atomic absorption measurements made in flames. Putting the problem in this manner does not mean that foreign elements will always and inevitably cause the results of determination to be distorted. Often, indeed, such an effect does not exist. On the other hand, detailed consideration of factors associated with composition effects on analytical results enables us to examine the process of atomization in flames more deeply and to put the selection of methods to eliminate or allow for the effect on a sure footing.

Foreign elements in a solution being analysed may affect the following stages of the atomization process:

1. Nebulization of solutions.
2. Atomization of the mist in the flame.
3. Ionization.

Obviously the dissociation of stable molecules is least affected by foreign elements, since the temperature of a flame and the composition of its gases (in particular, the concentration of the O radical) do not change greatly with the introduction of additional substances into the flame. Actually, when highly concentrated solutions containing up to 5 per cent of the substance, are atomized in a flame, the partial pressure of the vapour of the principal element in the sample (for instance calcium when $CaCO_3$ is being analysed) is $\simeq 10^{-3}$ atm when the solution is sprayed direct, and $\simeq 10^{-4}$ atm when preliminary nebulization is carried out. It is obvious that the presence of such small amounts of an element will not appreciably affect the equilibrium of the reactions

$$H_2O \rightleftharpoons H_2 + O$$

and

$$CO \rightleftharpoons C + O \quad \text{(for reducing flames)}$$

which govern the concentration of O radicals in flames, since the total pressure of the products of these reactions is two or three orders of magnitude greater.

Let us now consider the effects of composition on the different stages of the

---

* Here we have not dealt fully with other drawbacks of the atomic fluorescence method, such as the non-linear concentration relationship $i_\phi = f(C)$, and the additional effect exerted, by comparison with atomic absorption, by the medium on quenching fluorescence, and consequently on the analysis results.

atomization process listed above. The presence of foreign substances in a solution being analysed changes the specific gravity, viscosity, and surface tension of the solution. Table 4.12 gives the viscosities and surface tensions of 5 per cent solutions of some of the inorganic substances most frequently encountered.

TABLE 4.12 Viscosity (at 20°C) and surface tension (at 18°C) of 5 per cent solutions of certain inorganic substances[47]

Substance	η (cP)	σ (dyn/cm)
H_2O	1·00	73·05
NaCl	1·07	73·95
$MgSO_4$	1·28	73·75
$CaCl_2$	1·10	73·7
$Al_2(SO_4)_3$	1·05	74·5
$FeSO_4$	1·07	73·62
$ZnSO_4$	1·10	73·25

It follows from the table that the surface tensions of solutions do not change by more than 1–2 per cent when up to 5 per cent of the inorganic substances is present. If a solution contains up to 20 per cent of acids (HCl, H_2SO_4 or HNO_3), the surface tension does not change by more than 3 per cent.[47] In so far as changes in the surface tension of a solution are concerned, the effect of composition on nebulization can therefore be ignored in practice.

In certain cases, changes in viscosity approach 30 per cent. Increase in the viscosity of a solution reduces in proportion the rate at which the liquid flows through the capillary tube in the nebulizer, and this affects the amount of solution which enters the flame. The effects of viscosity on the dimensions of the droplets of the nebulized solution are hardly important (see p. 145).

Finally, the effects on nebulization of variations in specific gravity between 1 and 1·05 g/cm³ can also, in most cases, be ignored.

The effects of foreign substances on the process by which solutions enter the flame thus principally lie in changes in the amount of liquid nebulized resulting from variations in its viscosity. The most radical means of eliminating this effect lies in replacing nebulizers with a suction capillary by nebulizers with a forced-feed system, for instance a piston device (p. 141).

Let us now consider the effects of foreign elements on the atomization of mists in flames. The process by which a substance which is solute in droplets of a solution is converted into the gaseous state can be subdivided into the following stages:

1. Evaporation of the solvent from droplets of nebulized solution.
2. Melting of the solid residue and formation of molten droplets.
3. Vaporization of the molten droplets.

The first two stages of the process take place very rapidly, since the boiling points of solvents and the melting points of most salts are much lower than the temperature of the flame. The longest stage is that of the vaporization of the molten droplets of solute. The droplets are atomized by heat transfer from the surrounding medium. Heat is absorbed in raising the droplet temperature and vaporizing the droplets; as the droplet temperature rises the rate of vaporization increases until an equilibrium temperature is obtained. This temperature (analogous to the wet-bulb temperature in a hygrometer) corresponds to a steady state in which the heat absorbed by the droplets is used entirely in vaporization processes (losses of heat by radiation can be ignored). Lykov[48] has shown that the following equation for the temperature T_d at the vaporization surface can be obtained from the equations for heat transfer and mass transfer during the vaporization of a liquid from droplets:

$$T_d = T_{fl} - \frac{sMD}{\lambda R T_{fl}}(P_d - P_{fl}) \qquad (4.26)$$

Here T_{fl} is the absolute temperature of the flame, s is the specific heat of vaporization of the liquid in kcal/kg, M is the molecular weight of the substance (solute) in kg/mol, D is the coefficient of diffusion of the gaseous substance at the temperature T_{fl} in m²/h, λ is the thermal conductivity of the gas in kcal/m.h.°C, **R** is the gas constant equal to 0·062 torr.m³/mol.°C, P_d is the saturated vapour pressure of the substance above the droplet surface in torrs, P_{fl} is the vapour pressure of the substance in the flame in torr.

The vapour pressure of the substance in the flame is much lower than the saturated vapour pressure above the droplet surface ($P_{fl} \ll P_d$). Equation (4.26) can therefore be rewritten as follows:

$$\Delta T = T_{fl} - T_d = \frac{sMD}{\lambda R T_{fl}} P_d \qquad (4.27)$$

The difference between the temperatures of the droplet and the flame is thus basically a function of P_d, the vapour pressure above the droplets. To take an

TABLE 4.13 Mean values of the parameters for calculating the temperature of droplets in a flame at a temperature of 3000°K

$s = 5 \cdot 10^2$ kcal/kg
$M = 0\cdot 1$ kg/mol
$D = 3\cdot 2$ m²/h
$\lambda = 2\cdot 4 \cdot 10^{-1}$ kcal/m . h . °C

instance, with the mean parameter values given in Table 4.13 and a flame temperature $T_{fl} = 3000°$K, the difference between the temperatures of the flame and the droplet will be:

$$\Delta T \simeq 3\cdot 6 P_d \qquad (4.28)$$

Two conclusions, which are very important with a view to understanding the vaporization process, arise out of this approximation:

1. The vaporization of droplets in flames takes place at temperatures much lower than the boiling point of the molten substance concerned. Otherwise (with $P_d = 760$ torr), the difference between the temperatures must be about 2600°K, i.e. the boiling point of the molten substance would have to be lower than 400°K.

2. The droplets are only heated to the temperature of the flame in the case of substances with low volatilities. For instance, for the temperature at the droplet

surface (T_d) to differ by 3°C from the temperature of the flame (T_{fl}), the saturated vapour pressure of the solute at that temperature must be lower than 1 torr.

The theoretical estimation of the time taken to vaporize droplets relative to the time they spend in the flame is very important. To make this estimate, we can use the following equation,[48] expressing the relationship of the rate $dm/d\tau$ at which a substance is vaporized to the amount of heat acquired by the droplet:

$$\frac{dm}{d\tau} = \frac{\alpha \pi d^2}{s} \Delta T \qquad (4.29)$$

Here πd^2 is the surface of a droplet with a diameter d (in m), and α is the coefficient of heat transfer, which can be written as follows for the vaporization of droplets with small dimensions:

$$\alpha = \frac{2\lambda}{d} \qquad (4.30)$$

The mass of substance in a droplet can be expressed as

$$m = \frac{\pi d^3}{6} \rho \qquad (4.31)$$

where ρ is the density of the liquid, in kg/m³. From equations (4.29), (4.30), and (4.31), we get

$$d\tau = \frac{s\rho}{4\lambda \Delta T} d \cdot dd \qquad (4.32)$$

If we integrate equation (4.32) assuming that ΔT remains constant while a droplet is being vaporized, we get the following expression for τ (in hours), the droplet vaporization time:

$$\tau = \frac{s\rho d^2}{8\lambda \Delta T} \qquad (4.33)$$

If we substitute the value of ΔT from equation (4.27), we finally get:

$$\tau = \frac{\rho d^2 \mathbf{R} T_{fl}}{8MDP_d} \text{ (in hours)} \qquad (4.34)$$

The time taken for droplets to be vaporized is thus proportional to the surface area of the droplets and in inverse proportion to the vapour pressure of the substance above the droplet surface. It is interesting to note that equation (4.34) does not include s, the specific heat of vaporization of the liquid, or λ, the thermal conductivity of the gas.

As an example, let us estimate the time taken for droplets of Al_2O_3 to vaporize in a reducing air-acetylene flame. The maximum temperature of this flame (above the inner cone) is 2125°C.[22] If 0·5 ml of H_2O is introduced into the flame a minute, this is certain to reduce its temperature to 2075°C. The saturated vapour pressure of Al_2O_3 at 2075°C is $\simeq 3 \cdot 10^{-3}$ torr,[49] so that the difference between the temperatures of the droplets and the flame can be ignored.

The dimensions of droplets of molten Al_2O_3 are easily determined assuming the diameters of the droplets of mist to be 20 μm (see p. 145), and assuming the concentration of aluminium in the solution to be equal, for instance, to 0·01 per cent. Taking the specific gravity of Al_2O_3 as about 4 g/cm³, and the concentration of aluminium, in Al_2O_3, as about 0·02 per cent, we get a value of $d \simeq 1$ μm for the diameter of the droplets of Al_2O_3. Table 4.14 gives the principal parameters for calculating the time taken to vaporize Al_2O_3 under the condition selected.

TABLE 4.14 Parameters for calculating the time taken to vaporize droplets of a 0·01 per cent solution of aluminium in a reducing air-acetylene flame

$\rho = 4 \cdot 10^3$ kg/m³ $D = 2\cdot 2$ m²/h (for a mixture of N_2 and CO_2 [50])
$d = 10^{-6}$ m $P_d = 3 \cdot 10^{-3}$ torr [49]
$T_{fl} = 2350°$K $M = 0\cdot 1$ kg/mol

According to equation (4.34), we get: $\tau = 4 \cdot 10^{-1}$ sec.

The time which the substance spent in the flame zone was measured experimentally in the work of Malykh and Sedykh.[51] With an air-acetylene mixture flow rate of about 5 l/min, the length of time spent by atoms in the emitting zone was about $3 \cdot 10^{-3}$ sec. The measurements were made with lithium, sodium and thallium, which emit all the way along the flame. When elements with oxides which are difficult to dissociate (of the AlO type) are determined, the narrow flame zone above the inner cone, which has reducing properties, is effective for measurement purposes. The flame region between its base and this zone is not more than 0·1 times the total length of the flame, and the time spent by droplets in it is not therefore longer than $3 \cdot 10^{-4}$ sec.

The time taken to vaporize all the Al_2O_3 from solutions containing 0·01 per cent of aluminium is thus three orders of magnitude longer than the time spent by the droplets within the vaporization zone. This explains why attempts at determining aluminium by the atomic absorption method in an air-acetylene flame have not been successful.

In addition to this, the incomplete vaporization of Al_2O_3 may evidently also have an effect when measurements are made with a high temperature oxy-acetylene flame. We will take as an example the work described by Veits and Gurvich.[52,53] These authors measured the equilibrium constant of the reaction of dissociation of AlO assuming that the salts introduced into an oxy-acetylene flame at a temperature of $\simeq 3100°$K were fully vaporized (according to these authors the maximum calculated flame temperature was 3150°K). The aluminium was introduced into the flame in the form of aqueous solutions of the salts $KAl(SO_4)_2$ and $AlCl_3$.

For the calculation, we shall use the figures obtained by Ingram and his colleagues[49] for the vapour pressure above aluminium oxide.* If we extrapolate the vapour pressures measured at temperature between 2100 and 2600°K, we find that the saturated vapour pressure, at 3100°K, is $P = P_{Al} + P_O + P_{AlO} + P_{Al_2O} = 10$ torr. The dimensions of the Al_2O_3 droplets obtained by the decomposition of alum with

---

* These results are more reliable than data given in previous published works, and this was noted, for instance, by Glushko.[53]

a concentration of 0·1182 mol of the salt in the solution, under the previous assumptions regarding the dimensions of the droplets of mist (20 μm), are about 2·3 μm. The principal parameters for calculating the time taken by droplets of Al_2O_3 to vaporize under the conditions given by Veits and Gurvich[52] are tabulated in Table 4.15.

TABLE 4.15 Parameters for calculating τ under the conditions by Veits and Gurvich[52]

$\rho = 4 \cdot 10^3$ kg/m^3 $M = 0·1$ kg/mol
$d = 2·3 \cdot 10^{-6}$ m $D = 3·3$ m^2/h [50]
$T_{fl} = 3100°$K $P_d = 10$ torr

Equation (4.34) gives us $\tau = 5.6 \cdot 10^{-4}$ sec.

The vaporization time is thus of the same order as the time spent by droplets within the vaporization zone in the flame (the flame region lying 1–5 mm above the reaction zone was reflected full size onto the slit of the spectrograph used for recording flame emissions). Under these conditions it was not possible to be certain that all the Al_2O_3 was vaporized. The incomplete vaporization of Al_2O_3 when aluminium was introduced in the form of an alum solution could result in an exaggerated AlO content being calculated on the assumption that all the mist was vaporized in the flame, and as a result of this the equilibrium constant for the reaction AlO→Al+O must have been too low.

The equilibrium constant measured when aluminium was introduced in the form of alum actually proved to be only half the constant measured for $AlCl_3$ fully vaporized in a flame.[53] It is quite justifiable to relate the discrepancy in the measured values of the constants, noted but not explained by Veits and Gurvich[52] with relation to various salts, to the incomplete vaporization of Al_2O_3 in a flame.

The calculations of droplet vaporization times given here are of an approximate nature, since the principal equation (4.34) was deduced on the basis of a highly simplified basic mechanism. In particular, no allowance was made for the polydispersion of the mist, which owing to the different velocities of the droplets in the current of gases results in the collision of droplets and their merging together. When the equation was deduced, no allowance was made for radiation losses of heat by droplets, the effects of droplet surface curvature on the vapour pressure of the substance, etc. Finally, certain parameters necessary for calculating the vaporization time, in particular the coefficients of diffusion of the molecules of the substance being vaporized in the gaseous medium in the flame are absent. When calculations are made, therefore, it is necessary to use the results of measurements for media with similar compositions.

There are, nevertheless, no grounds for considering that the method of calculating vaporization time described here is unsuitable for drawing semi-quantitative conclusions.

Let us estimate the possibility of the complete vaporization of concentrated solutions in oxy-acetylene flames with temperatures of about 3000°K. As before we shall assume that the time spent by the mist in the vaporization zone is about $3 \cdot 10^{-4}$ sec. Complete vaporization of the droplets cannot therefore take longer than $3 \cdot 10^{-4}$ sec $(8 \cdot 3 \cdot 10^{-8}$ hour).

Equation (4.34) gives us

$$P_d = \frac{\rho d^2 \mathbf{R} T_{fl}}{8MD\tau} \qquad (4.35)$$

The dimensions of the droplets of the molten substance are about $4 \cdot 7$ μm when solutions the concentration of the principal substance in which is 5 per cent, and its specific gravity 4 g/cm³, are vaporized, if the diameter of the droplets in the mist is 20 μm. We shall take the diffusion factor as being equal to the diffusion factor for a mixture of N_2 and CO_2.[50] At 3000°K, $D = 3 \cdot 2$ m²/h. Equation (4.35) gives us: $P_d = 80$ torr.

Droplets of the mist are therefore fully vaporized in a flame with a temperature of 3000°K if the vapour pressure of the substance above the surface of the droplets is not less than 0·1 atm. It should be borne in mind that the temperature of the droplets themselves at a pressure of 80 torr must, according to equation (4.28), be 300°C lower than the temperature of the flame. The vapour pressure of the substance above the droplet surface therefore corresponds to the saturated vapour pressure for this substance at a temperature of about 2700°K. For 5 per cent solutions to be fully vaporized in a flame with a temperature of 3000°K, therefore, the vapour pressure of the solute at 2700°K must be more than 0·1 atmosphere.

This condition is satisfied by the chlorides of most metals. On the other hand, many oxides of metals, Al_2O_3, BeO, CaO, La_2O_3, MgO, MnO, SrO, ThO_2, ZrO_2, and HfO_2, which form molten droplets when solutions of sulphates or nitrates of these metals are introduced into flames, have insufficiently high vapour pressures for complete vaporization to take place. Owing to the incomplete vaporization of substances introduced into the flame, the results of determining the concentrations of particular elements contained in the substance are lower than the true figures. This type of effect exerted by foreign elements on the results of analysis in flames (in both emission and absorption measurements) is the type most frequently encountered.

Let us consider how the sensitivity with which any element is measured must vary according to the concentration of a foreign substance, of which the solution contains an excess, and which is less volatile than the element being determined.

According to (4.29), the rate at which the substance in a droplet is vaporized is proportional to the droplet surface area, i.e. d^2 and the coefficient of heat transfer α, which is in turn inversely proportional to d. As a result, the rate at which the substance of which a droplet is made is vaporized is proportional to d. This form of relationship is, however, only true for the vaporization of the solute in a droplet when vaporization takes place under equilibrium conditions.

Droplets are actually in an overheated state with regard to the low-volatility impurities, and these impurities are therefore vaporized under non-equilibrium conditions by 'excess' (with relation to the impurity) droplet thermal energy. The amount of energy expended on vaporizing an impurity is infinitesimally small by comparison with the total droplet thermal energy, and it does not therefore, in practice, affect the thermal balance for a droplet. It can thus be assumed that the amount of an element (ΔM_i), forming an impurity in a less volatile substance, vaporized in a flame is proportional solely to the vaporization surface area, i.e.

$$\Delta M_i \propto d^2 \qquad (4.36)$$

Since the diameter of a droplet is associated with its mass M by the relationship

$$d \propto M^{1/3}$$

we get

$$\Delta M_i \propto M^{2/3}$$

On the other hand, the total amount M_i of the impurity in the droplet must be proportional to M; hence

$$\frac{\Delta M_i}{M_i} \propto \frac{M^{2/3}}{M} = M^{-1/3} \qquad (4.37)$$

When the amount of the solute in a solution is altered, i.e. the mass M of the droplets changes, the relative amount of the impurity vaporized, $\Delta M_i/M_i$, and consequently also the concentration C in the flame, must alter according to

$$\frac{C_1}{C_2} = \frac{\Delta(M_i)_1/(M_i)_1}{\Delta(M_i)_2/(M_i)_2} = \left(\frac{M_2}{M_1}\right)^{1/3}$$

If, therefore, the mist is not completely vaporized, the concentration of an element being determined in a flame must decrease in proportion to the cube root of the relative increase in the concentration of the solute in the solution.

If the concentration of the solute increases by three orders of magnitude, sensitivity is bound to decrease by one order of magnitude. If the concentration of the principal substance is increased ten times, the sensitivity will be reduced $\sqrt[3]{10}$ times, i.e. 2·15 times.

This relationship of sensitivity to the concentration of a solute not completely vaporized in a flame was confirmed in experiments conducted by Rubeška, Moldan and Valny.[54] These authors investigated the manner in which the sensitivity of the atomic absorption determination of sodium in solution in hydrochloric acid varies when increasing amounts of calcium (up to 0·6 per cent) are present. The concentration of chlorine in the solution and the rate of nebulization were maintained constant in all the experiments.

We can see from Fig. 4.10, which gives the results of the experimental research, that the decrease in sensitivity in an air–coal-gas flame is proportional to $(C_2/C_1)^{1/3}$. Actually, when the amount of calcium was tripled, the sensitivity dropped by 31·6 per cent as against a calculated figure of 30·6 per cent. The sodium determination sensitivity in the hotter air-acetylene flame does not depend on the calcium content, since the mist is completely vaporized. The results given by Elwell and Gidley in their book[55] show that large amounts of aluminium have a similar type of effect on the results of determining magnesium. Increasing the aluminium content of a

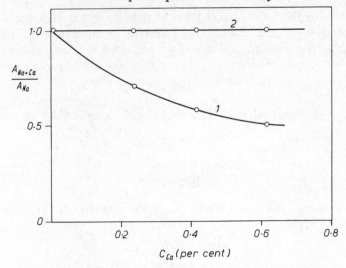

FIG. 4.10. Effects of an excess of calcium on the determination of sodium in different flames. 1. Coal-gas flame; 2. Air-acetylene flame.[54]

solution from 0·2 to 0·5 g per 100 ml reduces the sensitivity with which 5 µg of magnesium per ml of solution is measured by 28·2 per cent; this agrees well with the calculated reduction in sensitivity of $\sqrt[3]{2\cdot5} = 1\cdot36$ times, or 26·6 per cent.

The type of effect exerted by the solute on the measurement results which has just been discussed will only be true when large amounts of a foreign substance are present.

Amounts of foreign substances commensurate with the amount of the element being determined have a different kind of effect. When foreign substances less

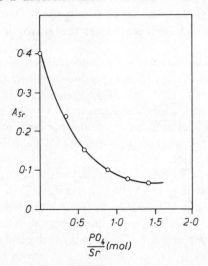

FIG. 4.11. Variation in the sensitivity with which strontium is determined with phosphorus present.[56]

volatile than the element being determined, or substances forming low-volatility compounds with the element being determined, are added to a solution of the element, the amount of the element being determined which enters the flame from the mist decreases abruptly. The result of adding amounts of foreign substances commensurate with the amount of the element being determined reduces the sensitivity of determination in many cases by several tens of times. The abrupt change in the rate at which the element being determined is vaporized is due in particular cases to change in the type of compound or the basic substance in which the element being determined is distributed.

The effects of silicon and phosphorus on the sensitivity with which alkaline earth elements are determined, effects well known in absorption flame photometry, are an example of the effects of foreign substances on the form of the molecular compound in which elements being determined are vaporized. Fig. 4.11 shows the manner in which the sensitivity of the atomic absorption determination of strontium in an air–coal-gas flame varies when H_3PO_4 is added to the solution being analysed,[56] and Fig. 4.12 shows how the sensitivity of the atomic absorption determination of magnesium varies in the same type of flame when silicon is added.[57]

These graphs show clearly that, when equimolar amounts of PO_4^{3-} of SiO_3^{2-} are added to the elements being determined, a situation is reached in which further increasing the amounts of PO_4^{3-} and SiO_3^{2-} has no effect on the sensitivity of determination. This confirms the mechanism, based on the formation of low-volatility compounds, which is proposed for the effect. In these cases, obviously, heat resistant compounds of $Sr_3(PO_4)_2$ and $MgSiO_3$ type are formed.

The effects of aluminium on the sensitivity with which different elements are determined are associated with the fact that, when aluminium salts are dissociated in a flame, the low-volatility oxide Al_2O_3, in which the elements being determined are distributed, is formed. Fig. 4.12 contains the results of measuring the sensitivity

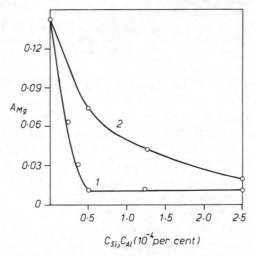

FIG. 4.12. Variation in the sensitivity with which magnesium is determined, with silicon (1) and aluminium (2).[57]

with which magnesium is determined in the presence of $Al(NO_3)_3$, according to data given by Andrew and Nichols.[57] The sensitivity with which magnesium is determined in the presence of aluminium varies more smoothly than in the case of silicon, and over a wider range of variation in the concentration of the foreign element (up to $5 \cdot 10^{-4}$ per cent for aluminium, and up to $5 \cdot 10^{-5}$ per cent for silicon). This difference is real, since the presence of aluminium does not result in the formation of stable molecular compounds (as is often wrongly suggested), but in the distribution of the magnesium in the low-volatility Al_2O_3 base. Naturally the Al_2O_3 base only entirely governs the manner in which the mist is vaporized if there is a sufficient excess of Al_2O_3 over MgO.

When the easily volatilized salt $AlCl_3$ is added to a hydrochloric acid solution being analysed, aluminium has little effect on the sensitivity with which magnesium is determined by atomic absorption. According to Allan,[29] also to Leithe and Hofer,[58] the addition of an amount of aluminium ten times the amount of magnesium reduces the sensitivity of determination by 20 per cent (Fig. 4.13).

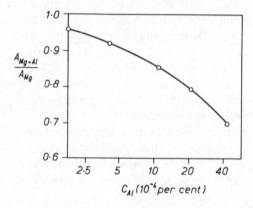 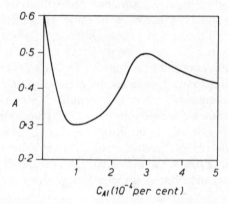

FIG. 4.13. Variation in the sensitivity with which magnesium is determined when $AlCl_3$ is added.[29]

FIG. 4.14. Variation in the sensitivity with which magnesium is determined accompanying the addition of $AlCl_3$ to a neutral solution.[59]

In view of these results it is interesting to refer to the data obtained by Menzies for the effects of $AlCl_3$ on the sensitivity with which magnesium is determined. When the salt $AlCl_3$ was added to an aqueous solution of the salt $MgSO_4 \cdot 7H_2O$, the curve for variation in sensitivity (Fig. 4.14) had an acute minimum, as opposed to the steady variation according to Allan[29] and to Leithe and Hofer.[58] In my opinion this is due to the hydrolysis of the salt $AlCl_3$ in the neutral solutions investigated. The result of hydrolysis is that aluminium hydroxide $Al(OH)_3$ is formed, and this transforms into the low-volatility oxide Al_2O_3 when droplets of the solution are vaporized in a flame. The magnesium is uniformly distributed between the low-volatility oxide Al_2O_3 and the volatile salt $AlCl_3$. We know that the degree of hydrolysis increases as the solution is diluted, and with small concentrations of $AlCl_3$, therefore, hydrolysis is most vigorous. As the concentration of $AlCl_3$ in the solution is increased, the degree of hydrolysis is bound to decrease. Accordingly the amount of magnesium distributed in the volatile $AlCl_3$ increases, and so does the sensitivity of determination.

Allan and Leithe and Hofer used hydrochloric acid solutions of magnesium in their experiments, and there was therefore no hydrolysis of $AlCl_3$.

The proposed mechanism, based on the formation of the low-volatility Al_2O_3 base, in which the elements being determined are distributed and which is only partially vaporized in the flame, proposed here is thus confirmed by the results of numerous experiments.*

The reduction in the sensitivity with which elements are determined in the presence of aluminium salts is therefore, among others, an index to the extent to which aluminium itself is vaporized in a flame. The presence of 20–500 times more aluminium than there is magnesium, which corresponds to a 10^{-3}–10^{-1} per cent concentration of aluminium in the solution, reduces the sensitivity with which magnesium is determined by 20–30 times.[57,59,60] In view of this, it can be assumed that only a few per cent of all the aluminium introduced into the flame are vaporized in this type of flame. This conclusion is in agreement with the results of our calculation of the time taken to vaporize Al_2O_3 in an air–acetylene flame during the vaporization of solutions containing 10^{-2} per cent of aluminium.

The method in which the degree to which low-volatility substances are vaporized is estimated from the reduction in the sensitivity with which some elements are determined by the atomic absorption method can obviously also be used for other compounds.

Let us now consider methods by which we can eliminate the effects of foreign elements on the results of atomic absorption measurements in flames. It is quite obvious that the most radical means of eliminating effects associated with the vaporization of mist in a flame consists of ensuring that all the substance introduced into the flame is vaporized. The most rational methods of achieving this are: increasing the temperature of the flame or the dispersion of the mist.

The effectiveness of these methods has been demonstrated in a number of research projects. For instance, Leithe and Hofer[58] proved that, if an air–acetylene flame is used instead of a coal-gas flame (the temperature of which is lower), foreign elements have less effect on the results of determining magnesium.

The advisability of increasing the temperature of the flame was confirmed by Amos and Willis,[63,40] who used the example of determining magnesium in the presence of aluminium and of determining calcium in the presence of phosphorus in a flame consisting of acetylene and nitrous oxide ($T \simeq 2900°C$). It was established that the presence of 1000 times more aluminium than magnesium does not affect the determination of magnesium, while 200 times more phosphorus than calcium only reduces the results of determining calcium slightly. Nevertheless, the presence of low-volatility metals such as titanium does affect the results of determining other elements.

In his experiments, Filcek proved that increasing the dispersion of the mist

* A similar mechanism of foreign substances influencing the results of emission measurements in flames was suggested independently by Alkemade.[42]

reduces the effects of phosphorus on the results of determining calcium (see p. 136 for greater detail[7]).

Robinson[69] proposed a clever method of increasing the dispersion of the mist introduced into a flame. Solvents with very different boiling points are used for nebulization. Since the more volatile component of the mixture boils vigorously on entering the flame, the droplets of solution are broken down into tiny droplets, and these govern the final dispersion of the mist.

In addition to direct methods based on ensuring that all the substance introduced is completely vaporized, there are also methods conducive to vaporizing a particular element independent of the vaporization of all the mist. These methods consist of adding substances to the solution being analysed, which, by competing with the element being determined or the interfering anion, either prevent low-volatility compounds of the element being determined from forming or protect the element being determined by forming a stable complex with them. These methods can be subdivided into three main groups:

1. The addition of an excess of the competing cation (spectroscopic buffer).
2. The addition of an excess of the competing anion.
3. The addition of a reagent which forms a complex (releasing agent).

It must be noted that, in order to suppress the effects of an interfering cation or anion effectively, there must be a considerable excess of the competing ion. Fig. 4.15 shows the effect of an excess of the competing cation Sr^{2+}, which suppresses the effect of $2.5 \cdot 10^{-3}$ per cent phosphorus on the results of determining $1 \cdot 10^{-3}$ per cent of calcium.[61] This graph shows that the effects of the phosphorus are fully suppressed if there is a 10:1 molar excess of strontium over the PO_4^{3-} (an Sr/P weight ratio of 30:1). In addition to strontium, lanthanum was used for the competing ions for shielding magnesium and calcium from the effects of phosphorus; magnesium and sulphuric acid were used for protecting calcium against phosphorus,[62] and calcium was used for protecting magnesium against aluminium.[58]

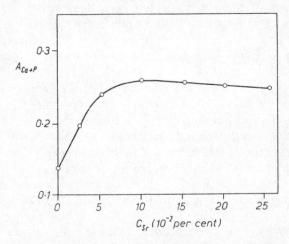

FIG. 4.15. Effects of the competing cation Sr^{2+} in shielding calcium against phosphorus.[61]

The sodium salt of ethylenediamine tetra-acetic acid (EDTA) and 8-hydroxyquinoline were used as complex-forming reagents. Fig. 4.16 shows the effects of adding different amounts of 1 per cent hydroxyquinoline on suppressing the effects of $2 \cdot 10^{-2}$ per cent of aluminium when determining $1\cdot4 \cdot 10^{-3}$ per cent of magnesium; Wallace's figures have been used.[60] About 33 ml of solution is calculated as being necessary to combine all the magnesium into the complex. The graph shows that this precise amount was required to suppress the effects of aluminium on the sensitivity with which magnesium is determined completely.

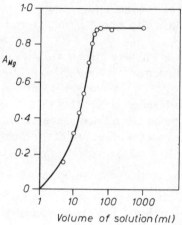

FIG. 4.16. Suppression of the effects of aluminium on magnesium using hydroxyquinoline.[60]

In addition to the direct and indirect methods listed here for ensuring that an element being determined is completely vaporized in a flame, the following methods of allowing for the effects of foreign elements on the results of atomic absorption measurements are also used in analysis practice:

1. The addition to the solution of an excess of the element exerting the effect, in order that the variations in the content of this element in different samples shall not affect the results of determination. This method can only be used if there is a sufficient margin of sensitivity.

2. The preparation of standards with compositions exactly the same as those of the samples determined, or the use of the additions method.

3. The use of an internal standard. As in emission spectrochemical analysis, this method consists in using for comparison the resonance line of an element of similar volatility. A specific amount of this element is added to the samples to be analysed and to standards. In this case, any uncontrolled changes in the degree to which the element being determined is vaporized in a flame will also have the same effects on the comparison element. The ratio of the intensities of the resonance lines after the beam of light has passed through the flame is therefore only sensitive to changes in the concentration of the element being determined. Naturally a double-channel spectrophotometer and a multi-element source of light are necessary for this method to be used.

Finally, in many cases the elements exerting effects are initially isolated by chemical methods.

Methods of eliminating and allowing for the effects of foreign substances on the results of atomic absorption measurements, effects associated with the formation of low-volatility compounds, can be classified under the main headings given in Table 4.16.

In concluding this section, we should note the effect of foreign elements on the degree to which the elements being determined are ionized in a flame. This effect is that, as the concentration of the foreign and easily ionized components increases, the sensitivity of measurement also increases (in both emission and absorption flame photometry methods).

This effect has been studied in very great detail with regard to the mutual effects of alkali and alkaline-earth elements during flame-photometry determination. The theoretical interpretation of the phenomenon is quite obvious: the easily ionized elements increase the concentration of electrons in the flame, and thus suppress the ionization of the impurity being determined. Quantitative conclusions regarding the extent of this effect can be drawn on the basis of the theoretical arguments set forth on pp. 132–4.

TABLE 4.16 Classification of methods of eliminating and allowing for the effects of foreign elements on analysis results

Isolation of interfering element	Without isolation of interfering element		
	Partial vaporization of the element being determined (methods of allowing for effects)	Total vaporization of the element being determined (methods of eliminating effects)	
		Total vaporization of the mist (direct methods)	Fractional vaporization of the mist (indirect methods)
Preliminary chemical isolation of the interfering element (extraction, chromatography, etc.)	Additives method	High temperature flame	Addition of competing cation
	Internal standard method	Increase in dispersion of mist	Addition of competing anion
	Addition of excess of the interfering element		Addition of complex-forming reagent

The methods of eliminating the effects of foreign elements on the degree of ionization are: the use of cooler flames or the cooler part of the flame for analysis, and the introduction of an excess of easily ionized components into the flame. It should be taken into consideration that the use of cool flames may make the analysis less sensitive as a result of incomplete vaporization and a reduction in the degree to which compounds are dissociated. The second method is therefore better.

The introduction of an excess of an easily ionized component (potassium in the form of KCl) was successfully used by Willis[63] for suppressing the ionization of alkali earth elements (Ca, Sr, Ba) during their atomic absorption determination in a high-temperature flame in which a mixture of acetylene and nitrous oxide was burned.

THE INTEGRATION METHOD OF MEASURING ABSORPTION IN FLAMES

It was noted on p. 123 that the integration method of measuring can be used, as well as the equilibrium method, for recording absorption in flames. After a preliminary discussion of these methods, the theory was stated that in certain respects (sensitivity and freedom from the effects of specimen composition on analysis results), the integration method must have considerable advantages over the normal equilibrium method.

We shall now give the results of certain experiments performed by the author in conjunction with Plyushch for the purpose of confirming that the integration method has these advantages. The experiments were performed using a spectrophotometer made of standard units, with a single-beam system. Fig. 2.45 is a diagram of the logarithmic amplifier for this spectrophotometer. The signals were recorded with a chart recorder. The time constant was increased to 1 sec.

As regards the technique for atomizing samples in flames, these experiments did not differ from the beautifully performed work done by Ramsay and his colleagues[76,77]. These scientists were the first to use the integration method of measurement for recording emission in a flame. The only slight differences were that a mixture of coal-gas and air was used for the flame in emission measurements, while in our experiments we used an air-acetylene flame. We employed a Méker burner with a cylindrical top, with nineteen apertures of diameter 0·8 mm uniformly spaced over its area. The consumptions of acetylene and air in the experiments were 0·9 and 6·4 l/min respectively. The diameter of the flame was about 15 mm. The

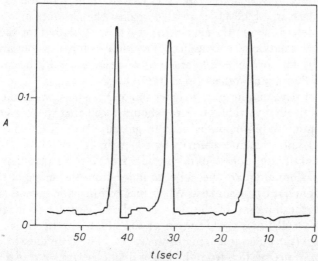

FIG. 4.17. Recordings of the atomic absorption of $1 \cdot 10^{-9}$ g of magnesium in a flame. The samples were introduced on wires.

samples were introduced into the flame by means of 0·1–0·5 mm diameter tungsten wires about 20 mm long, the ends of which were bent in the form of 1 mm diameter loops. The other ends of the wires were fixed in special holders.

The correct amounts of sample were applied to the wires in the form of solutions by means of a micro-pipette; with this pipette it was possible accurately to measure out volumes of 0·5–2 μl with an error of $<$0·025 μl. The droplets of solution applied to the loops were dried with a heating lamp.

The holders and wires were mounted in a support alongside the flame; rotating a holder in the horizontal plane rapidly introduced a loop bearing a specimen into the central zone of the flame (1–2 mm above the inner flame cone apex). The beam of light was focused on the centre of the flame, 5 mm above the position of the loop.

Magnesium was used for the experiments, since the results of measuring magnesium in air–acetylene flames are subject to considerable effects from foreign elements, aluminium in particular. Fig. 4.17 shows how the signal changes when $1 \cdot 10^{-9}$ g of Mg is introduced into a flame. The recording shows that the specimen was very rapidly atomized, in less than 1 sec. Pulse time is governed not by the atomization time, but basically by the time constant of the amplifier (τ_c).

The pulse area corresponding to the integral of absorption Q_A does not depend on τ_c, and τ_c can therefore be selected on the basis of practical considerations: limitations of the total pulse recording time to a reasonable period on the one hand, and elimination of possible distortions of the leading front of the pulse by the recording equipment (through the limited speed at which the pen moves) on the other. Apparently $\tau_c \simeq 1$ sec is close to the optimum value.

Q_A was measured by assessing the regions on the tape corresponding to the pulses, and converting the result into values of Q_A (in units of absorbance times seconds) in accordance with the absorbance calibration of the scale on the instrument and the rate at which the tape was moved. Fig. 4.18 shows the form of a calibration graph in which Q_A is plotted against the absolute amount M of the element being determined in the specimen; it has been plotted for the Mg 2852 Å line, using a pure solution of magnesium. The same graph also contains the results of measuring Q_A for corresponding amounts of magnesium in the presence of an 100:1 excess of aluminium in the form of Al(NO$_3$)$_3$.

The results of determining pure magnesium and magnesium in the presence of aluminium practically coincide. Here we should recall that the presence of 100:1 amounts of aluminium in a solution when magnesium is determined by the normal method reduces measurement sensitivity by 20–30 times.

The kinetics of the atomization of magnesium from 0·1–0·2 mm diameter wires in the presence of aluminium are the same, to judge from the shape of the pulses, as when pure magnesium is atomized. When magnesium is atomized from 0·5 mm diameter wires the duration of the pulses in the absorption of magnesium with aluminium is several times the duration of the pulses during the atomization of pure magnesium (the pulse area remains the same). The difference in the nature of the atomization of magnesium from the larger diameter wires is due to the fact that the loop to which the specimen is applied is more slowly heated; this is even visible to the naked eye.

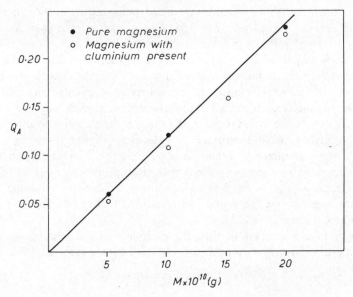

Fig. 4.18. Calibration graph for the integration method of measuring the absorption of the Mg 2852 Å line in a flame.

The reproducibility of the results of the measurements made in these experiments was limited by errors involved in measuring Q_A (cutting out pieces of paper), the small differences in the positions of the wires, when introduced bearing specimens, relative to the flame zone through which the beam of light passed, and errors in the size of the droplets. The overall variation factor was about 5 per cent.

Naturally the method of measuring used here (cutting out areas) is only permissible for preliminary experiments. Absorption can much more conveniently and accurately be integrated by means of special integrating circuits. One such possible circuit has been described by Ramsay et al.[77] It should be recalled that, before integration, the electrical signal must first be converted into logarithmic form, in order that the accumulated value shall be proportional to the absorbance (see p. 120).

The absolute sensitivity with which magnesium is determined, related to 1 per cent absorption in 1 sec, i.e. $Q_A = 0.0043$ sec, is $3.9 \cdot 10^{-11}$ g. Let us compare this with the absolute sensitivity of determination of magnesium in an air-acetylene flame using the equilibrium method of measuring absorption. According to Slavin, Sprague and Manning,[82] when measurements are made with a model 303 Perkin-Elmer double-beam spectrophotometer, 1 per cent absorption corresponds to a concentration of magnesium of 0.015 µg/ml. The smallest amount of solution used for measurement is 0.05 ml (the amount consumed[83] in 1 sec). The absolute sensitivity is therefore $7.5 \cdot 10^{-10}$ g, and this is twenty times worse than when Mg is determined by the method described earlier. This difference is principally due to the inefficiency of utilization of the sample: only a few per cent of the total amount of solution utilized by the nebulizer is introduced into the flame.

High absolute sensitivities were also recorded, using the integration method, when determining other elements: Mn 2795 Å—$1 \cdot 10^{-9}$ g, Cu 3248 Å—$3 \cdot 10^{-9}$ g, Ag 3281 Å—$4 \cdot 10^{-10}$ g, Pb 2833 Å—$1 \cdot 10^{-9}$ g.

As was to be expected from preliminary comparison of the equilibrium and integration methods of measuring absorption, therefore, the latter method provides a higher absolute sensitivity and, most important, eliminates the effects of the compositions of samples, which may contain components with low volatility.

One would expect the integration method also to be superior to the equilibrium method as regards relative sensitivity of measurement. For instance, if a sample weighing 10 mg is atomized in a flame it is possible to achieve a relative sensitivity for magnesium of $4 \cdot 10^{-7}$ per cent, while with the equilibrium method of measurement the relative sensitivity for magnesium, using a solution containing 1 per cent of the solute, must be $1 \cdot 5 \cdot 10^{-4}$ per cent (provided that the presence of the solute does not reduce the sensitivity).

In order to take advantage of this, the technique for atomizing comparatively large amounts of solid samples must be improved. This is not an insoluble problem.*

* See, for example, reference 44

Chapter 5

THE GRAPHITE CUVETTE

APPARATUS

The particulars of the method of atomizing samples in a graphite cuvette were given on p. 124, when different methods of producing atomic vapours were being discussed. Since the technique for analysis in a graphite cuvette is less well known to analysts than the procedure for analysis by flame photometry, we shall now dwell on certain special features of the apparatus and the measurement procedure in greater detail than we did in the case of flame photometry.

Plate 2 shows the general appearance of the spectrophotometer designed and constructed in the author's laboratory for atomic absorption measurements using a graphite cuvette. This spectrophotometer is a single structure on one pedestal, with a rack for instruments secured to four cylindrical columns.

Optical system (Fig. 5.1)

The spectrophotometer is based on a prism monochromator with a single-beam system. In order to allow automatically for interference from molecular absorption and scattering (see p. 235), beams of light from line and continuous sources are recorded simultaneously.[1]

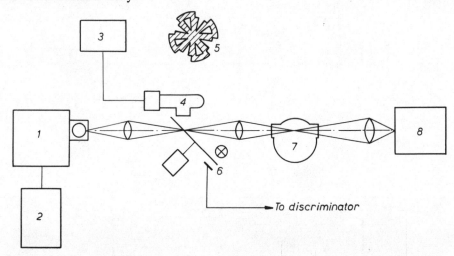

FIG. 5.1. Optical system of the spectrophotometer. 1. High-frequency generator. 2 and 3. Power supply. 4. Deuterium lamp. 5. Chopper. 6. Photoresistor illuminated by a bulb. 7. Chamber. 8. Monochromator.

The beam from the line source (a spherical radio-frequency lamp or a hollow-cathode lamp) and from the continuous spectrum source (a hydrogen lamp) pass alternately along the same optical path to the entrance slit of the monochromator. The beams from both sources are combined and modulated π out of phase, at a frequency of 300 c/s, by means of a rotating mirror disk with four aperture segments. This 10 cm diameter disk is made of stainless steel and coated with aluminium. In order to pass the maximum beams of light through the graphite cuvette, the image of the line source is focused in the plane of the disk by means of an additional lens, and owing to the small size of the lamp envelope the continuous spectrum source is positioned very close to the disk. A two-lens system takes the superimposed beams of light to the monochromator through the cuvette, one lens focusing the images of the sources at the centre of the cuvette, the other focusing them on the entrance slit of the monochromator.

The entire optical system is mounted on a rail 900 mm long (Plate 2).

Recording system

The exit slit of the monochromator isolates the resonance line of the element being determined (in the line spectrum) and the region of the continuous spectrum at the location of the resonance line (in the hydrogen lamp spectrum). The corresponding light pulses reach the photomultiplier (Fig. 5.2) alternately, where they are transformed into voltage pulses and pass to the wide-band amplifier, which has an amplification factor of 10^3 with a range of up to 100 V. The amplified pulses pass through a diode circuit, which sets the zero signal level, and enter a logarithmic circuit with a range of 40 dB (2 decades).

The log pulses pass to a mechanical relay, which sends the signals from the two

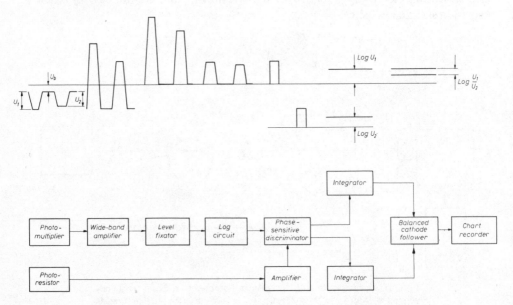

FIG. 5.2. Block diagram of the recording part of spectrophotometer.

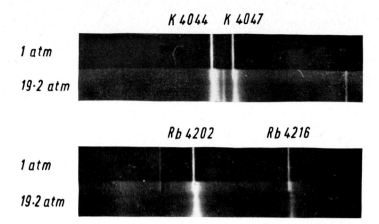

PLATE 1. Shift of potassium and rubidium doublets in an argon atmosphere (see p. 13).

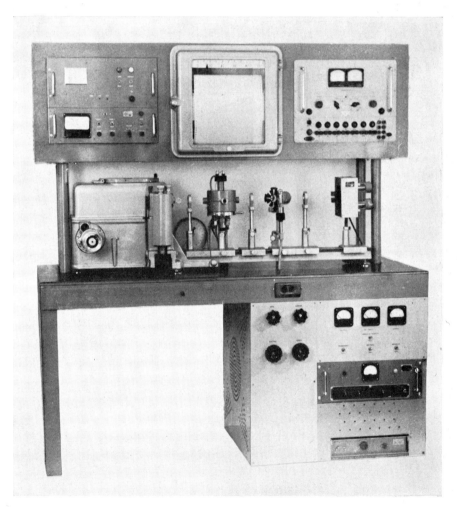

PLATE 2. Atomic absorption spectrophotometer for analysis with a graphite cuvette (see p. 194).

PLATE 3. The chamber which houses the cuvette, open and closed (see p. 195).

sources along different channels. The relay is synchronized with the modulation frequency by pulses from a photoresistor actuated by light from a separate bulb.

The separated signals pass through identical RC-circuits (with the peak method of measurement, the time constant is 0·024 sec); their difference is found in the balancing stage and recorded automatically. If the sensitivity of the device is changed by switching shunt coils, the calibration of the recorder scale can be altered (by 10 mV) in absorbance from 2·0 to 0·25 for the complete scale.

A meter is connected to the RC-circuits through a cathode follower, and the level of each of the signals is controlled by means of this meter. If one of the signals being compared is missing, the other signal is compensated by means of a d.c. reference voltage.

The chamber

A special high-pressure chamber with power supply, temperature control, and water cooling systems is used for atomizing samples in the graphite cuvette. Several types of chamber have now been designed, differing in dimensions, method of sealing, design of the pedestals between which the cuvette is mounted, number of electrodes, etc.[25,30,31,36] One of the latest types of chamber, specially designed by Lebedev and the author of this book for analytical purposes, is described below.

Plate 3 shows the general appearance of the chamber, open and closed. The principal parts are the base, attached to the monochromator rail on a standard rider, the rim, and the cover, which tilts backwards on brackets; all are made of stainless steel. The rim seals the chamber when the pressure of the foreign gas is greater than atmospheric (using a bayonet seal). With the chamber closed, the three projecting segments on the outer surface of the lower part of the cover are beneath the projections in the rim. The chamber is sealed by means of a rubber washer, which, when there is an excess pressure in the chamber, is forced out of its slot near the rim and pressed against the inner surface of the cover and the base. The chamber is sealed by a pressure excess of more than 0·1 atm within the chamber.

On the bottom of the chamber there are two columns, insulated from the base, which are for fixing the cuvette in a horizontal position and supplying current to it; there is also a turntable on which the electrodes are mounted. Running water passes through the columns for cooling. The turntable has five brackets holding the electrodes, and a rod passing through a Teflon powder packing in the base of the chamber. The electrodes are held in sockets in the brackets by means of nuts.

The columns and electrodes are so positioned, relative to one another, that when any of the electrodes enters the cuvette the neighbouring electrodes are protected by coolers from being heated directly by the hot walls of the cuvette.

The chamber cover is a double-walled cylinder cooled by running water. There are three bosses welded to the cylindrical wall of the cover to take windows; two of these are for passing the beam of light through the cuvette, one for visual inspection. Quartz windows 5 mm thick are mounted in the bosses. Rubber packings are used for sealing the windows.

A boss for the telescope of the radiation pyrometer controlling the temperature within the cuvette is welded to the cover cap. The image of the cuvette is focused in the plane of the thermal elements by means of a lens, also positioned within the boss. The radiation pyrometer, connected to a millivoltmeter, is calibrated, by means of an optical pyrometer, for the temperature of the inner cuvette wall (1400–3000°C).

The chamber has an internal volume of 0·7 l. Together with the rider and the radiation pyrometer, it weighs about 5 kg. It is mounted direct on the spectrophotometer rail (Plate 2).

The cuvette and electrodes

The shape and dimensions of cuvettes and electrodes depend on the purpose of the measurements and may vary widely. Fig. 5.3 shows the shape and dimensions of a cuvette and electrodes used for analytical measurements in the chamber described above. They are made of standard 6 mm diameter graphite rods of the type used in emission spectroscopic analysis. The length of the cell is 40 mm, its internal diameter 2·5 mm. The transverse aperture in the wall of the cell, and the ends of the electrodes, are tapered at 30° or 60°.

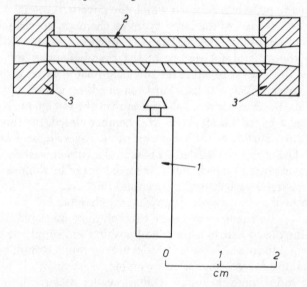

FIG. 5.3. The atomization system. 1. Electrode with the sample. 2. Cuvette. 3. Graphite washers placed inside coolers.

The cuvette is positioned between the two intermediate graphite washers which, in turn, are within the cooled columns of the chamber and tightened by means of steel sleeves threaded within the columns.

Power supply

The graphite cuvette and the electrodes and sample are heated by means of the power supply illustrated in Fig. 5.4.

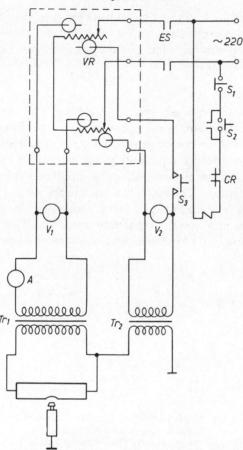

Fig. 5.4. The power supply to the chamber. Tr_1. 4 kW step-down transformer (220 V/10 V). Tr_2. 1 kW step-down transformer (220 V/15 V). VR. Voltage regulator. ES. Electromagnetic switch. S_1 and S_2. Switches for cuvette heating. S_3. Button switch for the electrode heating. CR. Contact relay for control of water cooling system. V_1 and V_2. Voltmeters (250 V). A. Ammeter (30 amp).

The cuvette is heated by a.c. from the transformer Tr_1 (220/10 V, 4 kW), and the electrode introduced into the cuvette is heated by a.c. from the transformer Tr_2 (220/15 V, 1 kW). The temperature to which the cuvette is heated and the electrode heating power are regulated by altering the voltages in the primary circuits of the transformers with the voltage regulator VR.

Current is fed to the chamber columns by means of flexible copper busbars with a cross-sectional area of 100 mm². Heating of the cuvette is switched on by means of a magnetic starter and the switch S_2, and it is switched off by the switch S_1; the electrode heating is switched on with the switch S_3. The electrode turntable is earthed by means of a sliding contact between the rod and the earthed base of the chamber. The power supply units are housed in the table pedestal (Plate 2). On the front panel of the device there are controls for regulating the voltages in the cuvette and electrode circuits, switches for switching the current on and off, and devices for controlling the electrical parameters (voltages and currents) in the transformer primary circuits.

Gas distribution system

Atmospheric oxygen is removed from the chamber. The chamber is then filled with gas to a pressure above atmospheric, and its pressure subsequently reduced, by means of the gas distribution system shown in Fig. 5.5. In addition to the chamber, this system includes two valves, a pressure gauge reading up to 10 kg/cm², and an adjustable valve for automatically equalizing the excess pressure in the system when gas is heated in the chamber. The individual units in the gas distribution system are joined by means of aluminium piping. The valve controls are on the front panel of the console alongside the voltage regulation controls (Plate 2). The pressure gauge is mounted in front of the operator, on the spectrophotometer table.

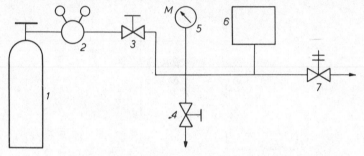

FIG. 5.5. The gas supply system. 1. Cylinder with nitrogen or argon. 2. Reducing valve. 3 and 4. Needle valves. 5. Manometer (10 kg/cm²). 6. Chamber. 7. Regulating valve for releasing automatically excess gas pressure during the heating of the cuvette.

The system is intended for use with foreign gas pressures up to 10 atm.

Argon or nitrogen can be used as the foreign gas with which to fill the chamber.

MEASUREMENT PROCEDURE

Purifying the electrodes and the cuvette

Before measurements are made, any traces of oil left during manufacture, or of the element being determined, must be removed from the cuvette and electrodes. For this purpose, they are baked at temperatures equal to or higher than those at which later analyses will be performed.

The cuvette can be taken as being sufficiently clean if, while it is switched on and being heated to the required temperature, the signal from it does not alter appreciably. The same applies to the electrodes.

When a cuvette and electrodes are being baked (this is also true of later measurements), the state of the contacts between the cuvette and the graphite washers is important. If the contacts are bad, the washers are greatly heated, and impurities are

Amounts of samples

Liquid samples with volumes of 0·5–5 µl are applied to the electrode tips by means of a special pipette and micro-screw. The pipette can easily be made from the 0·2–0·3 mm diameter capillary tube of a laboratory thermometer, to which a polyethylene tube with an elongated end is fitted. With pipettes of this type, micro-litre volumes of liquid can be applied with an error of less than 0·025 µl.

The electrodes are first mounted on a stand; without touching the electrode heads with the fingers, the end of one of them is gently rubbed against the ends of the remainder. This removes the film of graphite formed when the electrodes are heated in the cuvette, which prevents the ends of the electrodes from being wetted by droplets of aqueous solutions. The electrode heads are then briefly dipped in an 0·1 per cent solution of polystyrene in benzene, which on drying forms a water-repellent film which prevents the solution from seeping into the graphite.

The droplets of solutions applied to the electrodes are dried under a heating lamp or in a drying cabinet.

Weighed batches of solid samples are introduced into channels drilled in the electrode heads. The amount of sample introduced is determined by weighing the electrode, before and after it is filled, with a micro-balance.

In some cases, other methods of applying weighed amounts of samples may be used. For instance, when micro-impurities in hairs are being determined, measured lengths of hair may be used.

It will be proved below that, if a double-beam spectrophotometer is used, there is no need to use accurately weighed samples (liquid or powder) in an internal-standard method.

Measurement procedure

The electrodes, bearing the samples, are fitted in the sockets in the brackets, and the nuts tightened. Without closing the chamber cover, the positioning of the electrodes and cuvette is checked: the head of each of the electrodes must enter the aperture in the cuvette accurately, without any displacement.

The cover is then lowered, and inert gas is passed through the chamber for 10 sec at ∼0·5 l/s. The process of purifying the gas in a system with a volume V_0 follows the relationship

$$\frac{V}{V_0} = \ln \frac{1 - C_1}{1 - C_2}$$

where C_1 and C_2 are the initial and final concentrations of an undesirable impurity (in relative units), and V is the amount of gas, free from the undesirable impurity, passed through the system. As an instance, if the amount of oxygen, the atmospheric content of which is 21 per cent ($C_1 = 0·21$), is to be reduced in the chamber to 0·1 per cent, it is sufficient to pass through the chamber a volume of gas 5·3 times greater than the chamber in volume. When this has been done the chamber seal is closed by rotating the rim through 30°, and gas is passed in until the required pressure is reached.

The system for heating the cuvette is switched on. The cuvette reaches its maximum temperature in 10–20 sec. When the required temperature has been reached the measurement procedure begins. Watching the electrode through the front window of the chamber, which is fitted with a dark filter, the head of the electrode is introduced into the aperture in the cuvette, and the switch for heating the electrode is depressed immediately the electrode has made contact with the cuvette. The electrode takes 1–2 sec to heat up. With the peak method of measuring, the voltage in the heating up circuit can be so set that, during the period for which heating up continues, the pen recorder can record an absorption peak maximum. When the integration method of measuring is used the requirements as regards the rate at which the element being determined is atomized, and consequently also the requirements as to the manner in which the electrode is heated up, are less rigid.

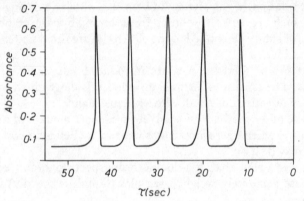

FIG. 5.6. Recorder traces of atomic absorption from 8.10^{-12}g magnesium in a 4·5 mm diameter cuvette at 1800°C, with a pressure of 2 atm of argon.

When absorption has decreased, the electrode is lowered into its initial position and the next electrode brought into position by rotating the turntable. Fig. 5.6 shows a recorded series of measurements.

When all the samples have been atomized, the cuvette-heating circuit is switched off, the excess pressure in the chamber is equalized by opening the outlet valve, and the results are processed.

Processing the results

In accordance with the conclusions drawn in Chapter 3, p. 118, the pulse amplitude is used as the measure of concentration in the peak method of measurement, and the area below the curve of absorption *versus* time during atomization in the integration method.

In the peak method, pulse amplitudes are converted into absorbances A_{peak} in accordance with the calibration of the recording device. In the integration method, values of Q_A are found by weighing the areas of paper below the absorption curves on the recorder charts; the weights M_P (in grams) are converted into values of Q_A after allowing for the speed at which the recorder chart moved and according to the calibration of the recorder in absorbance. Q_A is found in units of absorbance × seconds.

The amount M of an element corresponding to the measured values of A_{peak} or Q_A are found from the calibration graphs for $A_{peak}=f(M)$ or $Q_A=f(M)$. If the values of A_{peak} or Q_A are sufficiently small, M can be calculated from the conversion factors K_{peak} or K_{int}:

$$A_{peak} = K_{peak} M$$
$$Q_A = K_{int} M$$

THE ATOMIC VAPOUR IN A CUVETTE

Let us consider the atomization of samples and the presence of atomic vapour in the cuvette in greater detail. We explained earlier (p. 121) that these factors govern the effectiveness with which absorption is measured by the peak and integral methods. In addition, we can draw certain quantitative conclusions regarding the relationship between the effective length of the absorbing volume in the cuvette and the concentration of an element in a sample.

We shall first discuss the ultimate possibilities of this method of atomization with relation to the rate at which samples are atomized, and shall then discuss the process by which atomic vapour leaves the cuvette and measures to restrict the amount of vapour lost while samples are being analysed.

According to the kinetic theory of gases the following expression gives us the rate at which a sample is atomized in a vacuum with an open surface:

$$G = P \sqrt{\left(\frac{M}{2\pi RT}\right)} \tag{5.1}$$

G is the mass which a sample heated to a temperature T loses per unit of time from unit surface area, M is the molecular weight, P is the saturated vapour pressure at the temperature T. Since the quantities in equation (5.1) are dimensional, we get:

$$G = 5 \cdot 8 \cdot 10^{-2} P \sqrt{\frac{M}{T}} \text{ (g/cm}^2 \cdot \text{s)} \tag{5.2}$$

Here P is in torrs.

The atomization rate will be lower in the presence of a gas which does not react chemically with the sample atomized. Langmuir explained this in terms of the existence of a film of gas close to the surface of the sample atomized; he considered that atoms leaving the surface must diffuse through this film. When Fonda[8] atomized tungsten wire at 2870°K, he found that as the pressure of a mixture of 86 per cent Ar and 14 per cent N_2 was increased, the rate at which tungsten was atomized decreased; at a pressure of 1 atm the atomization rate was about one-sixtieth of the rate in a vacuum.

Let us calculate what the vapour pressure of any element being determined must be for it to atomize in less than 0·1 sec. We shall assume the maximum weight of the element to be 10^{-8} g, the surface area of the sample atomized (equal to the end face area of the carbon electrode) to be $\sim$0·01 cm², the atomic weight of the element to be 50, and the temperature to be 3500°K. Using equation (5.2), and assuming the rate of atomization in the presence of a foreign gas to be reduced by 60 times, we get:

$$P \geqslant 0\cdot 1 \text{ torr}$$

For most elements, the saturated vapour pressure at 3500°K is greater than 0·1 torr. The exceptions to this are tantalum, tungsten, rhenium, and a number of other elements: hafnium, niobium, zirconium, thorium which form carbides with extremely low volatility. In the case of most elements, therefore, we can attain conditions under which samples are atomized more rapidly than in 0·1 sec.

For a sample to be atomized rapidly, it is a necessary, but not sufficient, condition that the temperature T_{vap} at which the saturated vapour pressure is greater than 0·1 torr should be reached. We must recognize that the atomization of a sample begins at a temperature lower than T_{vap} and takes place throughout a certain range of temperature close to T_{vap}. To a first approximation, the atomization rate is related to the temperature by an exponential law (this also applies to the vapour pressure). Increasing the temperature by 100–200°C corresponds to increasing the atomization rate by an order of magnitude. It can therefore be stated that the time

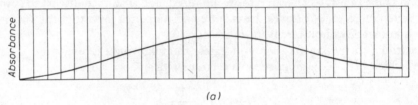

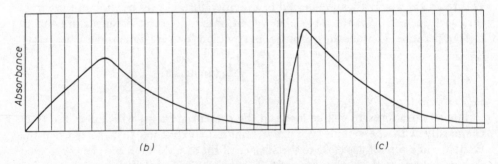

FIG. 5.7. Variation in the absorbance of the Zn 3075·9 Å line with time during the atomization of zinc under different conditions (argon pressure 1 atm).[5] Interval of time scale 0·2 sec. (a) Electrode heated by contact with the cuvette at $T = 1740°$K. (b) The same as (a) but with $T = 2100°$K. (c) The electrode heated simultaneously by the cuvette at $T = 1740°$K and an arc with current of 50 amp.

taken to atomize a sample is equal to the time necessary to heat the electrode by 100–200°C close to the temperature T_{vap}.

It is important to emphasize that, although the maximum temperature T_{vap} necessary for the rapid atomization of a sample depends on the pressure of the foreign gas, the atomization time when the electrode is heated at a constant rate does not necessarily depend on the change in the pressure of the foreign gas.

If the electrode is heated by a separate heater, it takes 2–3 sec to rise from room temperature to its maximum temperature, while, if it is heated solely by conduction from the cuvette, it takes about 10 sec. We can, therefore, assume that the temperature takes roughly 0·1–0·2 sec to rise by 100–200°C with separate heating and 1 sec without it.

Nikolaev[5–7] described how he made direct measurements of the time taken to atomize samples from the tip of an electrode using an oscillographic attachment to a spectrophotometer. Fig. 5.7 contains examples of oscillograms showing how the absorbance of the Zn 3075·9 Å line varies while 1.10^{-7} g of zinc is atomized under various conditions. The arc method of heating,[2,30] which is less effective than heating by resistance, was used for the separate heating of the electrodes. Nevertheless, the results are similar in both cases. It follows from examination of the oscillograms that, without additional heating of the electrode, the times taken to atomize the zinc were 2·5 sec and 1 sec (depending on the temperature of the cuvette), while when arc heating was used the time was 0·3 sec, i.e. it was in conformity with the estimated figures given earlier. Similar oscillograms were also obtained for the atomization of aluminium in a cuvette at a temperature of 2600°K.

The oscillograms in Fig. 5.8, recorded for the atomization of zinc with foreign gas pressures varying by 18:1, make it possible to state that the atomization time does not depend on the pressure of the foreign gas in the chamber. This conclusion

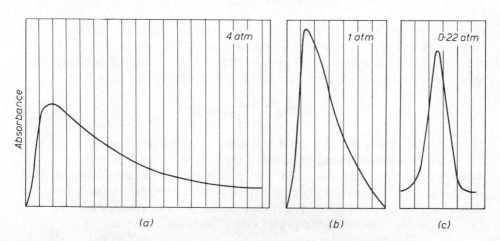

FIG. 5.8. Variation in the absorbance of the Zn 3075·9 Å line during the atomization of zinc, with different argon pressures.[7] Cuvette length 36 mm. $T = 1740°K$. Interval of time scale 0·2 sec.

is in conformity with the atomization mechanism described above, according to which the rate at which an electrode is heated, this being the rate governing the time taken to atomize a sample, does not depend on the pressure.

From all that has been said above, therefore, it follows that, if the electrode carrying a sample is additionally heated, most of the elements can be fully atomized in a few tenths of a second.

We come now to the processes governing the time spent by the atomic vapour in the cuvette.

The loss of atomic vapour from a cuvette is associated with three different processes: the diffusion of atoms through apertures and the porous walls of the cuvette, convection close to the heated surface of the cuvette, and the expulsion of excess atomic vapour of a sample if the volume of the vapour exceeds a certain proportion of the space within the cuvette.

As the maximum amount of the sample for which we can ignore the expulsion of excess atomic vapour, we shall take the amount whose atomic vapour has a volume equal to 10 per cent of the volume of the cuvette. In the case of a 2·5 mm diameter cuvette 40 mm long at a temperature of 2500°K, a foreign gas pressure of 4 atm, and an atomized sample with a molecular weight of 30, the maximum amount is $\sim 5 \cdot 10^{-5}$ g.

If the cuvette is horizontal and its apertures are small compared with its length, the amount of atomic vapour lost by convection currents that carry it to the apertures is small in comparison with the losses by diffusion.

Therefore, if the amount of sample introduced into a cuvette is small and the shape of the cuvette efficient, atomic vapour is lost only by *diffusion* through the apertures and the porous walls of the cuvette.

With the process in equilibrium, the amount of diffusion can be assessed from

$$\frac{dM}{d\tau} = -D \frac{d\rho}{dx} S \qquad (5.3)$$

where $dM/d\tau$ is the rate of diffusion by mass, $d\rho/dx$ is the density gradient in the direction of diffusion, S is the cross-sectional area of the aperture, and D is the diffusion coefficient, which with a pressure P and a temperature T is expressed in terms of the diffusion coefficient D_0 at S.T.P. ($P = 1$ atm, $T_0 = 273°$K) thus

$$D = D_0 \left(\frac{T}{T_0}\right)^n \frac{P_0}{P} \text{ (cm}^2\text{/s)} \qquad (5.4)$$

The value of n varies between 1·5 and 2 for different combinations of gases.

Let us estimate the amount of diffusion through apertures in a tubular cuvette with a length l and a cross-sectional area S. For the sake of simplicity, we shall assume that the density of the atomic vapour in the cuvette falls linearly from the value of ρ at the centre of the cuvette, where the sample is atomized, to zero where it leaves the cuvette. We can then express $d\rho/dx$ in the following form:

$$\frac{d\rho}{dx} = \frac{\rho}{l/2} = \frac{M}{(l/2)(V/2)} = \frac{4M}{Sl^2} \qquad (5.5)$$

Here the volume of the cuvette $V = Sl$. If we substitute the value of $d\rho/dx$ in equation (5.3) and allow for the diffusion of atomic vapour from the centre to both sides, we get:

$$\frac{dM}{M} = -\frac{8D}{l^2} d\tau \tag{5.6}$$

It follows from this equation that the relative losses of atomic vapour are in inverse proportion to the square of the length of the cuvette, and do not depend on its cross-section.

If we integrate equation (5.6), we get

$$M_\tau = M_0 e^{-(8D/l^2)\tau} \tag{5.7}$$

Here M_0 is the mass of the vapour at the initial instant of time.

From equation (5.7), the mean time spent by the atoms in the cuvette is

$$\tau_2 = \frac{l^2}{8D} \tag{5.8}$$

As an example, we shall estimate the diffusion of mercury vapour in nitrogen ($D_0 = 0.11$ cm²/s) from a cuvette 50 mm long, with a pressure of 1 atm and a temperature of 1500°K. According to equations (5.4 and 5.8), with $n = 1.6$ (see p. 287), $\tau_2 = 1.8$ sec.

In addition to diffusion through apertures, we must also consider diffusion through the porous walls of the cuvette. For an approximate estimate we use the diffusion coefficient measured experimentally[9] for CO_2 through graphite in nitrogen, $D_0^{gr} = 2.2 \cdot 10^{-3}$ cm²/s. In this case the relationship of the diffusion coefficient to temperature is expressed by the equation[9]

$$D^{gr} = D_0^{gr}(T/T_0)^{1.34}$$

so that with $T = 1500°K$, $D^{gr} = 2.1 \cdot 10^{-2}$ cm²/s. If we assume, for the purpose of simplicity, that the density of the atomic vapour varies linearly from the value of ρ within the cuvette to zero at its outer boundary, we get

$$M_\tau = M_0 e^{-(D^{gr} S_w/t_w V_w)\tau} \tag{5.9}$$

Here S_w is the area of the walls of the cuvette, V_w is its volume, and t_w is the thickness of the walls. For a tubular cuvette with an internal diameter of 2.5 mm and an external diameter of 6.0 mm,

$$\tau_2' = \frac{t_w V_w}{D^{gr} S_w} = 0.52 \text{ sec}$$

The amount of atomic vapour lost through the walls is thus much greater than the losses through apertures in a cuvette.

To prevent atomic vapour from diffusing through the walls, the inside can be lined with metal foil or the cuvette can be made of pyrolytic graphite.

Tantalum foil about 0.1 mm thick is very effective for lining cuvettes, since tantalum has a higher melting point (3270°K) than the other available metal (titanium, molybdenum, etc.). Since, however, metal screens rapidly form carbides their lives are very short, and they begin to crack after a few series of measurements. Moreover metal foil is contaminated with certain impurities (for instance iron), and as a result it is impossible to determine these elements. It is therefore better to make the cuvettes of pyrolytic graphite.

This type of graphite is obtained by the pyrolysis of hydrocarbons (usually methane) at a temperature of about 2000°C.[10] Cuvettes can be either made entirely of pyrolytic graphite or coated with a layer of this substance. The advantages of pyrolytic graphite are that it is impermeable to gases, has a higher sublimation point (3700°C) than standard graphite, is pure, and has a high thermal conductivity, so that the cell is uniformly heated. Pyrolytic graphite resists oxidation by oxygen tens of times better than standard graphite and cuvettes made of it thus have longer lives.

Experiments have confirmed these advantages. For instance, the losses of atomic vapour from a pyrolytic graphite cuvette investigated by atomizing iron and zinc were 33 per cent less than from unlined graphite cuvettes of the same dimensions. The cuvettes were tubular, with external diameters of 6 mm, internal diameters of 2·5 mm, and a length of 30 mm. Since the atomic vapour was better contained, the iron and zinc were determined in the pyrolytic graphite cell with almost double the sensitivity of determinations in unlined graphite cuvettes. Cuvettes made of standard graphite and coated with a layer of pyrolytic graphite 0·1–0·5 mm thick also provided good results.

The following experiments, conducted by Khartsyzov and the author, further confirmed the arguments stated above regarding the mechanism by which atomic vapour leaves cuvettes. The distribution after atomization of radioactive tracer elements was determined in the individual components of a special cell, the design of which is shown in Fig. 5.9. The chamber described in reference 34 was used for the experiments.

The cell consists of the two quartz half-cylinders (3), mounted coaxially with the cuvette (1) and the stainless steel circular end-plates (4). The electrode (2) is introduced through an aperture in the lower cylinder which is held between the coolers (5) in the half-rings (6).

The edges of the cylinders are ground, and the upper half-cylinders fitted to the lower. The thin quartz windows (8) are fitted into the steel sleeves (7), which are screwed against the cuvette with intermediate graphite washers (9).

The experiments were made with γ-active Na^{22}, Mn^{54} and Zn^{65} preparations. Measurements of the radioactivity of samples introduced into the cuvette as chlorides, and of the radioactivity of the different components of the cell after the atomization, were made with a γ-spectrometer. In these experiments, the initial radioactivity of the samples was about 1 μc.

The results of the experiments are tabulated, for cuvettes with different diameters and with or without pyrolytic graphite coatings, in Table 5.1. The samples were atomized at cuvette temperatures of about 2100°C, with argon in the chamber at a pressure of 3 atm.

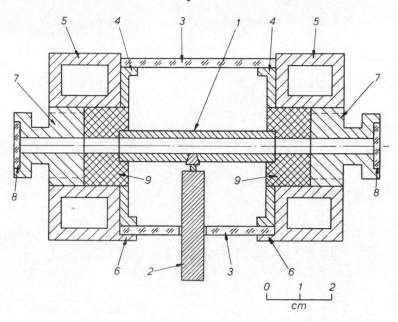

FIG. 5.9. Isolating cell. 1. Cuvette. 2. Electrode. 3. Quartz half-cylinder. 4. Stainless steel circular end-plates. 5. Cooler. 6. Supporting half-ring. 7. Steel sleeve. 8. Thin quartz window. 9. Graphite washer.

To find the effect on the amount of atomic vapour lost of the side aperture through which the electrode is introduced, an experiment was conducted with a cuvette with no aperture, the sample being introduced into the furnace on tungsten foil.

When the results of the experiments, given in the table, are considered we can draw the following conclusions:

1. A coating of pyrolytic graphite effectively localizes the atomic vapour radially within the cuvette. Actually, with the 'blind' cuvette (experiment 1), no atomic vapour was lost.

2. When a cuvette with no coating of pyrolytic graphite was used (experiment 5), the loss of vapour by diffusion through the walls was greater than the loss through the aperture; this accords with the theoretical assessment made earlier.

3. In every case in which cuvettes with a side aperture for the electrode were used, there was appreciable radial loss of atomic vapour, in spite of the fact that the cuvettes were coated with pyrolytic graphite (experiments 2–4).

This was evidently due to the diffusion of atomic vapour through the head of the electrode, which was made of standard impermeable carbon. The relative loss of atomic vapour through the side aperture for the electrode increases, as would be expected, as the diameter of the cuvette is reduced; the increase is from 20 per cent for a 4·5 mm diameter cuvette (experiments 2 and 3) to 40 per cent for a 2·5 mm diameter cuvette (experiment 4).

To eliminate, or at least reduce, the radial loss of atomic vapour through this

TABLE 5.1 Distribution of radioactivity in the components of the isolating cell

Experiment	Cuvette	Initial radio-activity (per cent)	Radio-activity at the electrode (per cent)	Radio-activity at the graphite washers (per cent)	Radio-activity at the metallic end sleeves (per cent)	Radio-activity at the cuvette ends (per cent)	Radio-activity at the quartz screen (per cent)	Radio-activity at the metal end-plates (per cent)	Total radio-activity (per cent)
1	4·5 mm diameter, coated with pyrolytic graphite, no aperture for introducing electrode	100		91·5	1·7	6·6			99·8
2	4·5 mm diameter, coated with pyrolytic graphite, aperture for introducing electrode*	100	5·0	50·5	4·0	21·3	18·5		99·3
3	4·5 mm diameter, coated with pyrolytic graphite, with aperture for introducing electrode*	100	5·0	51·2	7·5	13·0	21·6		98·3
4	2·5 mm diameter, coated with pyrolytic graphite, with aperture for introducing electrode*	100	1·5	44·2	3·5	2·0	36·0	8·6	95·8
5	4·5 mm diameter, not coated with pyrolytic graphite, with aperture for introducing electrode*	100	4·3	15·0	5·5	24·3	42·1	8·5	99·3

* Side aperture in cuvette ground to a 60° taper.

aperture, it is best for the electrodes to be made of graphite impermeable to gases and for the aperture to be small. Experiments showed that, other conditions being equal, electrodes tapered at 30° almost halved the loss from electrodes tapered at 60°.

4. In the experiments, it was found that the ends of the cuvette became radioactive owing to the partial condensation of atomic vapour on the colder ends of the cuvette, especially when the cuvette was made of uncoated graphite (experiment 5); for this cuvette was the least uniformly heated.

Let us now return to discussing possible methods of reducing diffusion through open apertures in cuvettes. It follows from equation (5.6) that the relative losses of atomic vapour are directly proportional to the diffusion coefficient. This coefficient in turn depends on the experimental conditions: the pressure and temperature of the medium. Lowering the temperature is a good method of reducing the diffusion coefficient only if volatile elements are being determined. For instance, when mercury is determined, the temperature of the cuvette can be lowered to 600–700°C. In other cases, the reduction in the temperature is limited by the necessity for atomizing samples rapidly, or at any rate by the necessity for maintaining a temperature at which the atomic vapour will not condense.

Increasing the pressure of the inert gas is the method most commonly used[4] for reducing diffusion. Its effectiveness can be seen at a glance from the oscillograms in Fig. 5.8, which indicate the change in absorbance in a cuvette, as time passes, with different argon pressures.

The atomic vapour diffusion coefficient does not only depend on the experimental conditions, but also on the type of gas in which diffusion takes place. According to kinetic theory, the diffusion coefficient is in inverse proportion to the square root of the mass of atoms and the square of the diameter of the gas molecules:

$$D \simeq \frac{1}{d^2 \sqrt{M}} \qquad (5.10)$$

Therefore the coefficients of diffusion for argon and helium, for instance, must differ by 3:1. Comparative measurements of diffusion made by Nikolaev[7] for zinc in argon and helium confirm this. The oscillograms are given in Fig. 5.10. Interpretation of the experimental data proves that the diffusion coefficients for zinc in these gases actually differ by 2·8:1, and this agrees well with the theoretical figure. In order, therefore, to reduce the diffusion of atomic vapour, inert gases with molecules of large diameters, such as argon or nitrogen, should be used.

It also follows from equation (5.6) that the loss of atomic vapour is inversely proportional to the square of the length of the cuvette, so that cuvettes of the maximum length should be used. However, cuvette length is increased only with considerable technical difficulty; for the power input increases in proportion to the length of the cuvette, while the luminous flux passing through the cuvette from the source decreases. For these reasons, it is not practicable to use cuvettes longer than 100 mm. A cuvette length of the order of 30–50 mm can be recommended from experience.

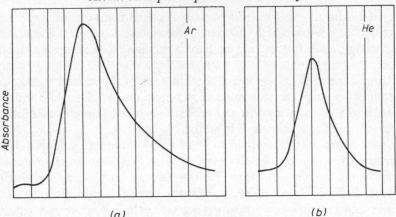

FIG. 5.10. Variation in the absorbance of the Zn 3075·9 Å line during the atomization of zinc in different gases.[7]

Finally, cuvettes of large diameter, fitted with end-caps in which there are small apertures for the beam of light, can be used to reduce the diffusion of atomic vapour through apertures (Fig. 5.11).

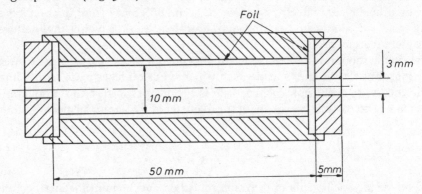

FIG. 5.11. Cuvette with caps.

The rate at which atomic vapour diffuses from a cuvette with caps can be calculated from

$$M_\tau = M_0 e^{-(2DS_c/t_c V)\tau} \qquad (5.11)$$

Here S_c is the transverse cross-sectional area of the apertures in the caps, t_c is the thickness of these caps, and V is the volume of the cuvette. When this equation was deduced, it was assumed that the atomic vapour is uniformly distributed throughout the cuvette, and that the density drops linearly in the apertures in the caps.

The length of time spent by mercury vapour in the capped cuvette shown in Fig. 5.11, under the same conditions as were used earlier for calculating the diffusion from a tubular cuvette, is $\tau_2 = 8\cdot1$ sec. If the pressure of the foreign gas is increased, for instance, to 3 atm, τ_2 increases to 24·3 sec.

The experiments described by L'vov[4] confirmed the advantages of capped cuvettes as regards the localization of atomic vapour. Typical recordings showing the diffusion of atoms of cadmium from a capped cuvette, for argon pressures of 1 and 9 atm, are reproduced in Fig. 5.12. The respective times for which the atoms of cadmium remained in the cuvette were 4·3 and 23 sec. The abrupt drops in the curves on the recordings for argon at 9 atm (at points C) correspond to the moments at which the electrodes were removed from the cuvette. Obviously, when this is done, intensive convection currents develop through the opened side apertures and the apertures in the caps, and as a result the remaining cadmium vapour rapidly leaves the cuvette.

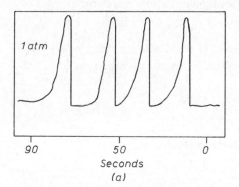
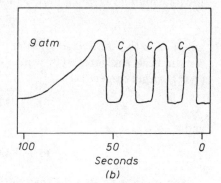

FIG. 5.12. Measurements of absorption recorded for cadmium at different pressures of argon.

The calculated and experimentally measured diffusion of metal vapour thus shows that a cloud of atomic vapour can be localized in a cuvette for a time τ_2 longer than the time τ_1 in which a sample is completely atomized.* At the moment when the element is fully atomized, the number N of atoms of the element in the cuvette must be equal to the number N_0 of atoms of this element in the sample. If the weight of the element concerned (in grams) is equal to M, and its atomic weight is A_w,

$$N_{\text{peak}} = N_0 = \frac{6 \cdot 02 \cdot 10^{23} M}{A_w} \qquad (5.12)$$

Here $6 \cdot 02 \cdot 10^{23}$ is Avogadro's number.

Similarly, the integrated number of atoms in the cuvette during the period τ_2 can be expressed according to (3.17) as follows:

$$\int_0^\infty N_\tau \, d\tau = N_0 \tau_2 = \frac{6 \cdot 02 \cdot 10^{23} M \tau_2}{A_w} \qquad (5.13)$$

Here we should draw attention to the following important fact. When different methods of measuring absorption were considered in Chapter 3, p. 122, we did not take into account the shape and dimensions of the analysis cell (we tacitly assumed these parameters to be the same in every case). Now that we have firm

* There is further experimental confirmation of this conclusion on p. 286.

information regarding cuvettes, we can introduce these parameters into the equations linking the measured signal with the amount of an element in a sample.

We know that absorbance is proportional, not to the total number of atoms in an analysis cell, but only to the effective length of the absorbing volume, $\int_0^l N(x)dx$, where $N(x)$ is the concentration of atoms at a point x in an absorbing volume with a total length l. For practical purposes, therefore, the ratio of the effective length of the absorbing volume to the amount of an element in a sample is important.

If we assume the atoms to be uniformly distributed over the cross-section of a cuvette (we shall consider later whether this assumption can be made safely), and that the cross-section is constant throughout the length of the cuvette, we can take the total number N_τ of atoms in a cuvette at a moment of time τ as being

$$N_\tau = S \int_0^l N_\tau(x)dx, \tag{5.14}$$

where S is the transverse cross-sectional area. If we substitute N_{peak} and $\int_0^\infty N_\tau \, d\tau$ for the peak and integrated values of the effective length in equations (5.12) and (5.13), we get respectively

$$\int_0^l N_\tau(x)dx = \frac{6 \cdot 02 \cdot 10^{23} M}{A_w S} \tag{5.15}$$

and

$$\int_0^\infty d\tau \int_0^l N_\tau(x)dx = \frac{6 \cdot 02 \cdot 10^{23} M \tau_2}{A_w S} \tag{5.16}$$

It is interesting to note that the peak value of the effective absorbing volume length does not depend either on the length of the cuvette or on the type of function $N(x)$, and is governed solely by the amount of the element in the sample and the transverse cross-sectional area of the cuvette.

Let us now return to discussing the assumption that atomic vapour is uniformly distributed over the cross-section. This is, in general, quite obvious, if we take into account the fact that the length of a cuvette is usually far greater than its diameter and that atomic vapour cannot penetrate its walls. Nevertheless this assumption was verified experimentally by measuring Cd 2288 Å line absorption in cuvettes with different diameters.[4] The results of these experiments are given in Table 5.2. The measurement conditions were: argon pressure 9 atm, temperature 1400°K, cuvette lined with foil and fitted with caps.

TABLE 5.2 Relationship of sensitivity to cuvette cross-sectional area

Diameter (mm)	S (cm²)	M (10^{-10}g)	A	$\dfrac{M}{AS} 10^{10}$
9.8	0.75	5	0.73	9.1
6.8	0.36	2.5	0.78	8.9
4.8	0.18	1.25	0.75	9.3

The results of the experiments show that the amount M of a substance necessary to produce the same absorption A in a cuvette with a different diameter is exactly proportional to the cross-sectional areas of the cuvettes. This relationship can be true only if the atoms are uniformly distributed over the transverse cross-section of the cuvette.

In connection with questions associated with the localization of the atomic vapour of a sample within a cuvette, it is interesting to consider Massmann's attempt at simplifying the procedure.[29] To simplify the apparatus, Massmann dispensed with the introduction of samples into the cuvette by means of an electrode. Instead, he applied the sample, in solution or as a solid, to the inner surface of a cold graphite tube through a small 1 mm diameter aperture in its wall. When the solution had been applied, the graphite tube was gently heated until the droplets had dried up. Impulsive atomization of the dried residue was effected by heating the cuvette rapidly, using a step-down transformer with a maximum output of 400 amp at 10 V. To increase the heating rate, the cuvette was thin-walled, its internal diameter being 8 mm and its wall thickness 1 mm; it was 55 mm long. The graphite tube was housed in a semi-closed chamber, through which argon was passed continuously in order to reduce the amount of graphite burned.

What, in my opinion, are the drawbacks of this variation of the method?

1. The sample atomizes from the cuvette wall much more slowly than from an additional electrode. In the first place, therefore, there is fractional atomization of elements with different volatilities, and, in the second place, the peaks corresponding to maximum absorption are lower; consequently, the absolute sensitivity with the peak method of measuring absorption is also lower. The correctness of using the integration method of measuring absorption is also dubious in this case, since the temperature of the cuvette is steadily rising while the measurements are being made, and τ_2, the time spent by atoms in the cuvette, is not constant.

2. The maximum temperature to which the sample is heated is governed by the temperature of the inner wall of the cuvette. In order, therefore, to atomize even elements with average volatility, such as iron or copper, a very powerful source of current must be used for heating the cuvette. It is also not satisfactory to raise the temperature of the cuvette owing to the increase in the continuous background from the cuvette walls. Both factors exert greater effects in proportion to the fourth power of the temperature.

3. When samples are introduced into the cuvette by this method, analysis is slower than with the normal variation of the method, since the sequence followed by all the stages of the analysis (weighing out the solution, drying it, the pulsed heating of the tube, cooling it to prepare for a fresh measurement) for each sample definitely takes longer than analysing by measuring series of several samples.

Definitely, therefore, this method of simplification is not advisable.*

DISSOCIATION OF COMPOUNDS

When we deduced the equation for the effective length of an absorbing volume, we assumed that all the atoms of an element being determined which are present in the

* It should be noted that in spite of these failings in the method Massmann[26] used for vaporization, he obtained high absolute sensitivities for volatile elements (see p. 226), and also obtained very valuable results when determining these elements in solid specimens.

form of atomic vapour within a cuvette are free, i.e. that they are not associated in compounds of any sort. Theoretical and experimental grounds are, however, required to back this assumption.

It was pointed out in Chapter 3 that the degree of dissociation of refractory MeO-type compounds in the gas phase depends on the partial pressure of the oxygen radical. Under conditions of thermodynamic equilibrium, the partial pressure of a component of this type must be governed by the reaction

$$C + O \rightarrow CO$$

Equilibrium constants for the reaction, and saturated vapour pressures for atomic carbon, are given in Table 5.3 for different temperatures.

TABLE 5.3 Equilibrium constants of the reaction of dissociation of CO, and saturated vapour pressures for C at different temperatures[11]

T (°K)	K_p (atm)	P_C (atm)
2000	$7.5 \cdot 10^{-22}$	$2.6 \cdot 10^{-11}$
2500	$3.7 \cdot 10^{-16}$	$1.4 \cdot 10^{-7}$
3000	$2.4 \cdot 10^{-12}$	$4.6 \cdot 10^{-5}$

The amount of molecules of CO in a cuvette is governed, on the one hand, by the amount of oxygen left within the chamber when it is blown out with inert gas and the amount of oxygen in this inert gas, and, on the other hand, by the atomization of the sample being analysed in the cuvette. In the ultimate case in which the volume of atomic vapour of the sample is about 0·1 times the volume of the cuvette, the pressure of the CO vapours may be as high as 0·1 atm. Even, however, in this ultimate case the partial pressure of the atomic oxygen

$$P_0 = \frac{K_p \cdot P_{CO}}{P_C}$$

does not exceed 10^{-8} atm (at 3000°K).

As is to be expected, the conditions in a cuvette and in reducing flames are very nearly the same and are conducive to the complete dissociation of compounds of the MeO or MeOH type.

In reference 4, these assumptions were verified by comparing the sensitivity with which certain pairs of elements, with different oxide and hydroxide dissociation energies, were determined (Table 5.4).

The other characteristics which govern the amount of absorption (except for the atomic weight) are nearly the same. The measurements were made in argon at 1 atm, in a 3 mm diameter cuvette, at 2200°C. The sensitivities, expressed in mols (mols/unit absorbance), are given in the penultimate column of the table. For purposes of comparison, the degrees to which these compounds are dissociated in an oxidizing air-acetylene flame, according to Hinnov and Kohn,[12] are given in the right-hand column.

TABLE 5.4 Sensitivities for elements with oxides or hydroxides with different dissociation energies

Resonance line (Å)	Compound	Dissociation energy (eV)	Sample (g) producing $A = 0.1$	Sensitivity (mols/unit absorbance)	Degree of dissociation in a flame (%)
K 4044	KOH	3.8	$6.0 \cdot 10^{-9}$	$1.5 \cdot 10^{-10}$	43
Cs 4555	CsOH	4.5	$5.0 \cdot 10^{-8}$	$3.8 \cdot 10^{-10}$	5
Sr 4607	SrO	4.9	$2.6 \cdot 10^{-10}$	$3.0 \cdot 10^{-12}$	11
Ba 5535	BaO	6.0	$1.7 \cdot 10^{-9}$	$12 \cdot 10^{-12}$	0.2
Mn 4031	MnO	4.2	$7.0 \cdot 10^{-10}$	$1.3 \cdot 10^{-11}$	100
Cr 4254	CrO	4.4	$7.0 \cdot 10^{-10}$	$1.3 \cdot 10^{-11}$	13
In 4102	InOH	3.7	$1.5 \cdot 10^{-9}$	$1.2 \cdot 10^{-11}$	
Al 3962	AlO	5.0	$3.5 \cdot 10^{-10}$	$1.3 \cdot 10^{-11}$	~1*

* In an oxidizing oxy-acetylene flame at 3200°K.[13]

The results given in this table show that the sensitivities expressed in mols for the pairs of elements concerned practically coincide, while the differences in sensitivity due to the difference in the degree of dissociation amount in some cases, in oxidizing flames, to several orders of magnitude. It may therefore be assumed that, in a graphite cuvette, the elements present in the form of vapour consist of free atoms.

CALIBRATION

In order to convert a measured value of absorption into the amount of an element in a sample, we must first know the relationship between these quantities. In principle, this relationship can be established either theoretically from constants for the element being determined, measured by independent methods, and the known experimental conditions or empirically by calibrating the instrument by means of standards containing known amounts of the element concerned.

The theory in its present state is adequate for the first method to be used for atomic absorption measurements. (We shall discuss this in greater detail in the section on the measurement of absolute oscillator strengths.) To do this, however, we must know the required physical constants accurately enough, eliminate any possible systematic errors in determining the absolute values of temperature, pressure and cuvette dimensions, and use a slightly more complicated procedure for measurement. For practical purposes, therefore, the empirical method for determining the relationship between an instrumental reading and the amount of an element in a sample is more convenient and reliable.

Usually, calibration is effected with standards prepared to cover the entire working ranges of values of the quantities measured. The following factors are extremely important:

1. The type of calibration function, or type of calibration graph.
2. The variations in calibration in time.
3. The effects of the composition of samples (third components) on the calibration graph.

We shall consider the last point in the next section, and shall confine ourselves here to discussing factors associated solely with the first two points.

To simplify and accelerate the calibration procedure, it is best that the absorption quantity measured should be proportional to the amount of an element in a sample, since the calibration procedure can then be reduced to the measurement of no more than a single point on a straight line passing through the coordinate origin. It was shown earlier, on p. 30, that this condition is provided by the method of measuring the absorbance of atomic vapours through which a monochromatic beam of light is passing.

In practice, the resonance lines emitted by sources of light are definitely not monochromatic owing to Doppler broadening, hyperfine structure, and—most important—line broadening from self-absorption in the light source.

The fact that emission lines are not monochromatic is bound to cause curvature of the calibration graph, and this will be particularly appreciable with high absorbances. Emission line broadening is not, however, the only cause of curvature of calibration graphs. As Menzies was the first to point out,[33] the heterogeneity of the atomic vapour (over the cross-section of the beam) and the background in the spectral region isolated by the monochromator are also bound to have a substantial effect on the curvature of the graphs.

The effects of all the causes given above can formally be reduced to the same pattern: the simultaneous measurement of the absorption from several beams of light which are absorbed to different extents; there is the sole distinction that, in the case of non-monochromatic emission, the difference between the beams is one of wavelength, while, in the case of a heterogeneous layer, it is associated with the different positions of the beams relative to the absorbing volume.

The immense amount of experimental data that has now been accumulated for atomic absorption spectrochemical analysis (mainly for the flame method of atomization) confirms this: the calibration graphs for most elements are curved with absorbances higher than 0·6–0·8. Owing to the curvature of the graphs, instruments must be calibrated for the entire range of measurements.

When samples are atomized in cuvettes the picture is different. In the first place, as we proved earlier (p. 212), the distribution of vapour over the cross-section of a cuvette is absolutely uniform. In the second place, if the pressure of the foreign gas in the chamber is above atmospheric, the absorption lines become broader owing to the Lorentz effect; consequently, the emission lines become effectively more monochromatic (i.e. in terms of the ratio of absorption line-width to emission line-width). For this reason, the curvature of the calibration graphs is bound to be less when samples are atomized in a cuvette in which the foreign gas is at a high pressure. As an illustration, Fig. 5.13 shows calibration graphs for cadmium obtained with argon at atmospheric and high pressure in the chamber. The calibration graph with argon at 9 atm deviates from a straight line by only 10 per cent at an absorbance of 1·5, while, with a pressure of 1 atm, the deviation is already 230 per cent at an absorbance of 1·0.

At the same time, when the peak method of recording absorption in a cuvette is used, slight distortion of the curves may result from the purely technical fact that a rapidly changing process is being recorded with automatic recorder in which the pen moves along the scale at a limited speed.

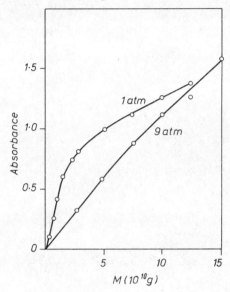

FIG. 5.13. Calibration curves for the Cd 2288 Å line, with argon at different pressures.

Naturally, the distortion in peak amplitude recordings must, in this case, increase as the peaks themselves increase; this may finally cause additional curvature of the calibration graphs at high absorbances. When choosing an automatic recording instrument therefore, it is best to select a fast one with the minimum pen travel time (0·5 or 0·25 sec) or to make measurements covering only a small region on the scale.

The integration method of measuring absorption does not suffer from this failing. It is possible to show that the integrated absorption does not depend on the time constant τ_c of the recording system; we can therefore select a sufficiently high value of τ_c to ensure that the mechanical limitations of the recording device do not affect the correctness of the recording.

Furthermore, the integrated absorption must in general be less susceptible to deviations from proportionality between the absorbance A and the amount M of an element present, since the relative importance of the pulse region with the maximum values of absorption is small by comparison with the total pulse area. Moreover, by regulating the speed at which a sample is atomized, it is possible to achieve a situation in which the maximum values of absorption in a pulse do not exceed the absorbance values at which the relationship between M and A is no longer proportional. The range in which there is proportionality between Q_A and M can therefore be greatly expanded to higher values of M. This is confirmed by the calibration graph for thallium given in Fig. 5.14, in which Q_A is plotted against M. Q_A remains proportional to M up to $M = 1·5 \cdot 10^{-9}$ g, and this is at least ten times the range of proportionality for $A_{\text{peak}} = f(M)$.

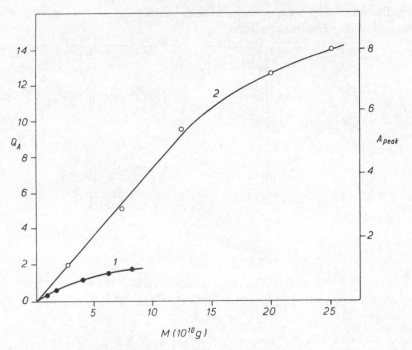

FIG. 5.14. Calibration graph for thallium. 1. $A_{peak} = f(m)$. 2. $Q_A = f(m)$.

Let us now consider the question of the stability of calibration for atomic absorption analysis using a flame and a cuvette. Here we must consider the effects of uncontrollable variations in the light source conditions, in the adjustment of the beam of light relative to the absorbing volume, and in the atomization parameters, which affect the measured concentration of atoms in the volume, and the effects of uncontrollable changes in the measurement conditions (temperature and pressure), which determine the shape and position of the absorption lines.

Although the current passing through a light source is usually controlled with great accuracy, changes in pressure in the lamp as time passes (not to mention increases in temperature while the lamp is heating up) may also cause changes in the width of the resonance lines being studied. These changes are particularly great if the lamp is used under critical conditions and line broadening is caused by self-absorption.

This has been pointed out many times in published works on flame atomic absorption.[43] The same effect is observed when measurements are made with a graphite cuvette, if the foreign gas pressure is close to atmospheric. For instance, if the current passing through the hollow cathode is increased from 5 mA to 10 mA, the sensitivity with which cadmium is determined in a cuvette at 1 atm is halved (Fig. 5.15). If the pressure of the argon is increased to 9 atm, however, the same change in the discharge conditions does not affect the sensitivity of the measurements. The reason for this is that, at 9 atm, the absorption line becomes so broad that the changes in the breadth of the emission line are not important. Increasing the pressure of the foreign gas in an atomic vapour thus reduces the effects on calibration of the excitation conditions in light sources. When the integration method of measuring absorption is used, according to what has been stated above calibration must depend even less on the light source conditions.

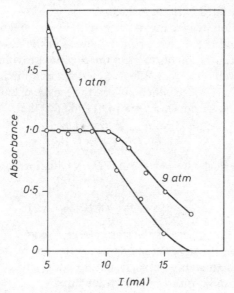

FIG. 5.15. Effect of the current passing through a hollow-cathode lamp on the sensitivity of cadmium determination at different argon pressures.

The heterogeneity of flames at different levels and different points across their width means that the analysis zone through which to pass the beam of light from the emitting source must be extremely carefully selected.[35] In view of this, any accidental changes in the positions of the separate components of the optical system, the light source, lens, diaphragms and flame, relative to one another may cause considerable changes in sensitivity, i.e. they may accordingly cause changes in the calibration. In the case of a cuvette this instability is entirely absent.

Let us next consider the effects of uncontrollable variations in the atomization parameters on the sensitivity of measurements. In this connection we shall return to the results obtained earlier (pp. 116–22) regarding atomization conditions which guarantee that there will be a simple connection between the measured concentration of atoms in a cuvette and the amount of an element in a sample.

When the equilibrium method of measuring absorption in a flame is used, these conditions are the constant rate of introduction of the mist into the flame and the constant time interval spent by atoms in the analysis zone in the flame. Constancy of the rate at which mist is introduced into the flame depends on how consistently the nebulizer works and whether the spray chamber functions steadily. Whether or not the atoms remain in the absorbing volume for a constant length of time depends on the composition and consumption of the components of the fuel mixture and the state of the burner nozzles. Some of the parameters which affect consumption and the length of time spent by atoms in the flame are either not controlled at all or controlled with insufficient accuracy for possible variations in the conditions to be ignored. Changes in the calibration of the spectrophotometer may, therefore, also be associated with these factors.

When peak absorption in a cuvette is measured, the parameters which govern the atomization time τ_1 and the time τ_2 spent by atoms in the cuvette cannot have any effect on the peak concentration of atoms in the cuvette, provided that the condition $\tau_1 \ll \tau_2$ is sufficiently well observed. When the integration method is used, the only important parameters are those which govern the length of time spent by atoms in the cuvette (τ_2). We know from equations (5.8) and (5.4) that

$$\tau_2 = \frac{l^2}{8D} = \frac{l^2}{8D_0}\left(\frac{T_0}{T}\right)^n \frac{P}{P_0} \tag{5.17}$$

If we substitute numerical values of T_0 and P_0 in equation (5.17), and assume that $n = 1 \cdot 6$ (see p. 287), we get:

$$\tau_2 = \frac{10^3 l^2 P}{D_0 T^{1 \cdot 6}} \tag{5.18}$$

Before we discuss the effects of the parameters P and T on the value of τ_2 in equation (5.18) and on possible variations in calibration caused by uncontrollable variations in these parameters, we shall consider the effects of temperature and pressure on the shape and position of absorption lines.

When measurements are made in a flame, we can ignore the effects of pressure and temperature on absorption lines, since uncontrollable changes in flame temperature are extremely small, and indeed the effects of these changes on line profiles, governed by the combined Doppler and Lorentz effects, are only small. Variations in barometric pressure, say between 710 and 790 torr, can only influence the Lorentz effect; at atmospheric pressure, however, the Lorentz contribution to the total line width is usually small (see Table 1.7).

When measurements are made in cuvettes containing foreign gas at a high pressure, matters are entirely different. In this case, the line profile is almost entirely governed by the Lorentz effect. Absorbance is associated with the effective length $(Nl)_{\text{eff}}$, according to (1.46) and (1.4), as follows:

$$A = \frac{0 \cdot 205 e^2 f}{mc \Delta \nu_s} (Nl)_{\text{eff}} \tag{5.19}$$

Taking into account equations (5.15) and (5.16) for the effective length of an absorbing volume in a cuvette, we get the following expression for the peak value of absorption, with an accuracy up to a constant multiple:

$$A_{\text{peak}} \simeq \frac{f}{A_w S \Delta \nu_s} M \tag{5.20}$$

while for integrated absorption we have

$$Q_A \simeq \frac{f \tau_2}{A_w S \Delta \nu} M \tag{5.21}$$

If we take into consideration the relationship of Δv_s to pressure and temperature given by equation (1.22), and the relationship of τ_2 to the same conditions according to equation (5.18), we find that

$$A_{\text{peak}} \simeq \frac{T^{0.7}}{SP} M \qquad (5.22)$$

and

$$Q_A \simeq \frac{l^2}{ST^{0.9}} M \qquad (5.23)$$

If we consider these expressions, we can see that in both methods of measuring absorption in a cuvette the results are inversely proportional to the cross-section and bear about the same relationship (though of opposite sign) to temperature. At the same time, there are substantial differences: when integrated absorption is measured, the effects of pressure are eliminated, but the length of the cuvette becomes of great importance.

Since the length of the cuvette can very easily be controlled with a high degree of accuracy, measures of Q_A are on the whole less prone to random experimental error than measures of A_{peak}. (Relative fluctuations in temperature and pressure were about the same throughout the experiments).

When measurements are made in a graphite cuvette, the temperature of the cuvette and the pressure of the foreign gas are constantly controlled with a very high degree of accuracy. Very effective allowances can also be made for variations in cuvette dimensions (when the cuvette is changed).

As opposed, therefore, to the flame, any changes in the atomization conditions, associated for instance with variations in absorbing volume dimensions, pressure or temperature, can be allowed for with rather high accuracy.

To summarize the results of the discussion, therefore, we can say that uncontrollable factors affect results less when cuvettes are used than when flames are used. Moreover, the factors which still affect results in the use of cuvettes can be allowed for quantitatively. For this reason, there are good prospects for developing an absolute method of analysis, based on cuvettes, with a calibration obtained once and for all time for each element (to be more precise, for each combination of analysis line and foreign gas). Taking into account the proportionality between the measured analysis signal and the content of an element, calibration over a sufficiently wide range of measurement can be reduced to determining the proportionality constants in equations (5.22) and (5.23).

INDEPENDENCE OF CALIBRATION FROM SAMPLE COMPOSITION

The effects of foreign constituents on results or on calibrations made in their absence inevitably limit most methods of spectrochemical analysis. Frequently in the history of spectrochemical analysis, methods suggested for eliminating composition effects have been found, when investigated carefully, to be defective in various respects or to be only approximate. The development of spectrochemical methods that are to be unaffected by sample composition must therefore start from soundly based theory for each particular method, reliable experimental data to confirm the theory, and, most important, extensive practical tests of the method.

As regards these requirements, the details at present available concerning the use of cuvettes are insufficient for drawing any categorical conclusions. The results given hereafter can therefore be considered only as preliminary results.

As we did earlier when discussing the effects of sample composition on the results of atomic absorption measurements in flames (p. 174), we shall confine ourselves to considering matters associated with the kinetics of atomizing samples. Obviously, the effects of foreign components on the process of dissociation of compounds in cuvettes are not important, since, whatever the amounts of foreign elements, the medium in the cuvette is still of a reducing nature (p. 214). The effects of sample composition on the partial ionization of certain elements can always, as in the use of a flame, be eliminated by adding an excess of an easily ionized component to the sample. The effectiveness of this method in the case of cuvettes has been confirmed in experiments carried out by Penkin[14] using a King furnace. In addition, cuvettes, unlike flames, provide unlimited possibilities as regards varying the temperature of the atomic vapour. When easily ionized and highly volatile elements are determined, therefore, and this applies primarily to alkali elements, we can use the cuvette temperature with which the corresponding elements are not ionized to a great extent.

Finally, we shall consider the effects or results of optical interference due to molecular absorption and scattering of light by the molecular vapour of the parent sample (p. 225).

Let us first dwell on the peak method of measuring absorption in a cuvette. The theoretical calculations and experimental measurements described on p. 211 have shown that it is correct to use the peak method of measuring absorption if the element is fully atomized in a time much shorter than the length of time spent by the atomic vapour in the cuvette. This impulsive atomization can be accomplished with most elements in pure form (in the absence of excess amounts of other components). We still, however, have absolutely no idea how the same elements are atomized in the presence of an excess of low-volatility components, which clearly cannot be atomized in the required time because their amounts in a sample may be as much as 10^{-4}–10^{-3} g (against the figure of 10^{-8} g which we used for the maximum content for the element being determined). Let us, in view of this, consider the results of the experiments in which Nikolaev[29] determined zinc and aluminium in the presence of excesses of elements with low volatility.

The results of determining $1 \cdot 10^{-10}$ g of aluminium in the presence of $1 \cdot 10^{-5}$ g of iron, nickel, cobalt, chromium, titanium, or copper are given in Table 5.5. It follows from this table that these elements have no effect on the results of determining aluminium. The same work also found that $5 \cdot 10^{-6}$ g of aluminium, iron, titanium, cobalt, copper, vanadium, magnesium, or silicon had no effect on the results of determining $6 \cdot 10^{-9}$ g of zinc when the samples were atomized with a cuvette temperature of 1470°C, at which temperature the electrode was insufficiently heated by d.c. arc to atomize its own material.

TABLE 5.5 Estimating the effects of foreign elements on the results of determining $1 \cdot 10^{-10}$ g of aluminium

Element (10 µg amounts introduced)	Al found (µg)
Fe	$1.03 \cdot 10^{-4}$
Ni	$0.98 \cdot 10^{-4}$
Co	$1.04 \cdot 10^{-4}$
Cr	$1.00 \cdot 10^{-4}$
Ti	$0.96 \cdot 10^{-4}$
Cu	$1.02 \cdot 10^{-4}$

Why are determinations made in a cuvette, unlike those made in a flame, free from the effects of low volatile foreign components?

1. As shown on pp. 178 and 202 it takes a thousand times longer to atomize a substance in a cuvette (10^{-1} sec) than in a flame (10^{-4} sec). Indeed, the time taken to heat and atomize a substance in a cuvette is so long that refractory compounds have ample time to decompose.[16] The layer of sample applied to the electrode is very thin (a sample weighing 10^{-5} g applied to an area of 0.01 cm² is only a few micrometres thick), and elements are removed extremely rapidly from this layer.

2. The atomization of samples in cuvettes takes place in a highly reducing medium (from the surface of incandescent graphite). The breakdown of the crystal lattice of the sample and the decomposition of compounds are therefore accelerated by the processes of reduction of compounds.

The conditions under which samples are atomized in a cuvette thus differ substantially from the conditions under which mists are atomized in flames. Obviously this is the reason why refractory components do not affect the results of measurements made in cuvettes. This advantage possessed by the peak method of measuring absorption in cuvettes enabled Nikolaev and Aleskovskii[5] and Nikolaev[6] to use the method successfully for determining aluminium and zinc in metallurgical specimens with complex compositions by plotting calibration graphs for pure solutions of the elements. The results they obtained are given in Tables 5.6 and 5.7.

The analysis results given in the tables are reproducible with an error of ~5 per cent.

In all the cases considered above the samples were added to the electrodes in solution form. The amount of residue after drying out was $<10^{-4}$ g; it was spread in a fairly uniform layer, well bonded to the graphite, over the tip of the electrode. The entire mass of the sample was therefore heated practically as rapidly as the underlying base.

The heating of samples weighing $>10^{-4}$ g, introduced as a powder or separate particles into a channel in the electrode, must take place in a slightly different manner. In this case, the sample is heated by heat transfer from the parts of itself that are in direct contact with the walls of the channel. The elements being determined may therefore be atomized more slowly than from a residue uniformly distributed over a surface (particularly if the samples are poor thermal conductors, such as oxides and salts). For this reason, systematic errors may be caused by using a calibration made for pure elements and results may be too low.

TABLE 5.6 Results of determining aluminium in samples of known composition[5]

Sample	Aluminium content (%)	Aluminium determined (%)
Steel	0·72	0·70
Steel	0·49	0·51
Nickel alloy	0·48	0·49
Silicocalcium	1·57	1·50
Copper	0·04	0·045
Brass	0·13	0·115
Ferrovanadium	3.00	2·93
Low-alloy steel	0·08	0·09
Low-alloy steel	0·14	0·13
Titanium	4·85	4·95
Titanium	1·07	1·11

TABLE 5.7 Results of determining zinc in samples of known composition[6]

Sample	Zinc content (%)	Zinc determined (%)
Brass	37·68	38·1
Bronze	5·28	5·18
Bronze	0·44	0·43
Complex composition aluminium alloys	5·55	5·51
	5·90	5·86
	1·01	1·04
	0·72	0·72
Silumin (11·79 per cent Si)	0·021	0·019
Cadmium	0·002	0·002
Antimony	0·0006	0·0005

In principle, the integration method of measuring absorption is free from this source of systematic error. With a view to verifying that results are independent of the time taken to atomize samples, Katskov, Lebedev, and I carried out the following experiments. The same amounts of an element were atomized in a cuvette with various amounts of additional heating for the electrode and with no such additional heating. The signals in unit absorbance were recorded automatically. In accordance with the procedure described on p. 201, the values of Q_A were found by weighing the recorder chart areas corresponding to the absorption pulses. The results of this weighing $M_{P(g)}$ were converted into values of Q_A.

Experimental conditions: Cuvette coated with pyrolytic graphite, diameter 2·5 mm, length 40 mm, $T = 1400°C$, $P(Ar) = 4$ atm. The absorption of $2·5 \cdot 10^{-10}$ g of thallium was measured at the Tl 2768 Å line.

TABLE 5.8 Results of measuring integrated absorption with samples atomized at different speeds

H (mm)	A_{max}	M_P (g)	Q_A (absorbance × sec)
191·5	0·72	0·089	1·7
177·6	0·66	0·090	1·7
135	0·51	0·098	1·8
85	0·32	0·105	2·0
65	0·24	0·088	1·7
44	0·16	0·091	1·7
35	0·13	0·103	1·9
25	0·09	0·090	1·7

The results of the experiments, given in Table 5.8, confirm that altering the rate at which samples are atomized does not affect Q_A. The rate of atomization is indicated indirectly by the maximum value of absorption H (mm), or by A_{max}, which varied by 8:1 during these experiments.

Integration is not yet used enough in cuvette methods. There is, however, no doubt that it would make a decisive step towards the entire elimination of foreign-component effects on results of atomic absorption determination of elements in samples of whatever composition or state of aggregation.

ABSOLUTE SENSITIVITY

By absolute sensitivity we shall mean the mass of an element corresponding to some chosen value of absorption. As with the equilibrium method of measuring absorption for a flame, we shall conditionally take as datum for the peak method of measurement the absorbance corresponding to 1 per cent absorption, i.e. $A°_{peak}=0·0043$. Similarly, we can take the value corresponding to 1 per cent absorption in 1 sec as the datum for the integration method of measurement, i.e. $Q°_A 0·0043$ absorbance × seconds. By definition, therefore, we can understand the peak method sensitivity as being

$$M°_{peak} = \frac{M}{A_{peak}} A°_{peak} \tag{5.24}$$

and for the integration method,

$$M°_{int} = \frac{M}{Q_A} Q°_A \tag{5.25}$$

It was established on p. 221 that equation (5.22) gives us:

$$\frac{M}{A_{peak}} \sim \frac{SP}{T^{0·7}} \tag{5.26}$$

and that equation (5.23) gives us

$$\frac{M}{Q_A} \sim \frac{ST^{0·9}}{l^2} \tag{5.27}$$

It follows from these relationships that the sensitivities depend on the experimental conditions (temperature and pressure) and the dimensions of the cuvette. Since these parameters may vary quite widely according to the problem concerned, we must relate the prescribed sensitivity figures to definite values of the parameters included in equations (5.26) or (5.27).

Further discussion will be confined to the peak method of measurement, since not much experimental data is yet available for the integration method. As standard values, we shall use the area S, corresponding to a cuvette diameter $r = 2\cdot 5$ mm and an argon pressure $P = 2$ atm. This selection is based on practical expediency. As regards the temperature T, we should prescribe the minimum cuvette temperature T_{min} for each element, with which absorption of the element can be measured.

Table 5.9 contains actual measured data for the amounts M of elements, and gives the experimental conditions corresponding to these amounts; it also contains the results of calculating $M°_{peak}$ from these data for the following standard conditions: $A°_{peak} = 0\cdot 0043$, $P = 2$ atm (argon), $r = 2\cdot 5$ mm, and T_{min}.

The mean absolute sensitivity with which the elements investigated were determined was 10^{-13}–10^{-11} g. The exceptions to this are boron, the sensitivity with which is, as in flames, surprisingly low (considering the quite high oscillator strength $f = 0\cdot 33$ of the resonance line and the low atomic weight), mercury and potassium (the most sensitive lines for which were not used in measuring), also iodine, phosphorus, scandium and titanium. The sensitivity figures were higher than 10^{-13} g in the case of beryllium, cadmium, magnesium, silicon, and zinc. It should be noted that some of the sensitivity values given were obtained during the early development of the method; these values could now be improved.

The sensitivities established by Massmann[26] for an 8 mm diameter graphite cuvette and 1 per cent absorption were very close to the figures given above. Nine of the elements (Zn, Ag, Cd, Mg, Pb, Sb, Bi, Hg and In) investigated gave mean

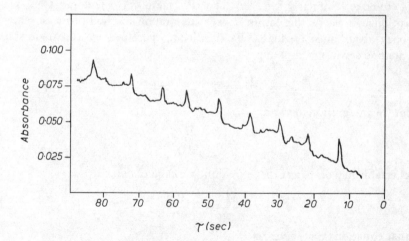

FIG. 5.16. Recordings of absorption produced by $6\cdot 25 \cdot 10^{-14}$ g of cadmium (nine successive determinations). The steady displacement of signal level is associated with instability of high frequency lamp radiation.

absolute sensitivities between 10^{-12}–10^{-10} g. In extrapolating for a cuvette with a diameter of 2·5 mm, these sensitivities must be increased by an order of magnitude, i.e. to 10^{-13}–10^{-11} g.

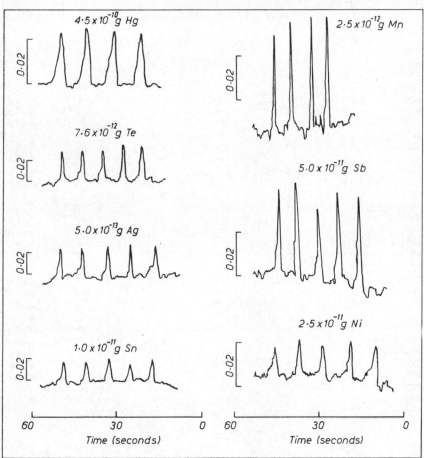

FIG. 5.17. Recordings of the atomic absorption of several elements. Recording conditions those of Table 5.9.

For most elements, the small amounts indicated in Table 5.9 can quite definitely be determined. This confirms, in particular, the recordings given in Figs. 5.16 and 5.17, which were obtained under the conditions given in Table 5.9.

It is interesting to compare the method described above with other methods of analysis. As an example, the absolute limits of detection achieved for cadmium by different modern methods of instrumental analysis are compared in Table 5.10. The results given in this table show that, as regards capacity for accurate determination, the cuvette method of atomic absorption analysis is far superior even to such methods as mass spectrometry and neutron activation.

Why does the method described above provide a much better sensitivity than other spectrochemical methods, in particular the flame method?

TABLE 5.9 Absolute sensitivities

Line (Å)	Cuvette diameter (mm)	Argon pressure (atm)	Temperature (°C)	A_{peak}	M (g)	$M°_{peak}$ (g)
Ag 3281	2.5	2	1800	0·020	$5·0 \cdot 10^{-13}$	$1 \cdot 10^{-13}$
Al 3093	4·5	1	2100	0·03	$2·5 \cdot 10^{-11}$	$1 \cdot 10^{-12}$
As 1973	2·5	2	1700	0·30	$5·6 \cdot 10^{-10}$	$8 \cdot 10^{-12}$
Au 2428	2·5	2	1700	0·28	$7·0 \cdot 10^{-11}$	$1 \cdot 10^{-12}$
B 2498	2·5	2	2400	0·105	$5·0 \cdot 10^{-9}$	$2 \cdot 10^{-10}$
Ba 5535	3·0	1	2200	0·05	$1·0 \cdot 10^{-10}$	$6 \cdot 10^{-12}$
Be 2349	4·5	6	2800	0·17	$2·6 \cdot 10^{-12}$	$3 \cdot 10^{-14}$
Bi 3068	2·5	2	1800	0·031	$2·5 \cdot 10^{-11}$	$4 \cdot 10^{-12}$
Ca 4227	2·5	2	2300	0·244	$2·5 \cdot 10^{-11}$	$4 \cdot 10^{-13}$
Cd 2288	1·2	1	1500	0·014	$6·0 \cdot 10^{-14}$	$8 \cdot 10^{-14}$
Co 2407	2·5	2	2200	0·020	$7·5 \cdot 10^{-12}$	$2 \cdot 10^{-12}$
Cr 3579	2·5	2	2200	0·125	$5·0 \cdot 10^{-11}$	$2 \cdot 10^{-12}$
Cs 8521	2·5	2	1900	0·072	$6·6 \cdot 10^{-12}$	$4 \cdot 10^{-13}$
Cu 3248	2·5	2	2100	0·044	$6·3 \cdot 10^{-12}$	$6 \cdot 10^{-13}$
Eu 4594	4·0	2·5	2450	0·13	$3·9 \cdot 10^{-10}$	$5 \cdot 10^{-12}$
Fe 2483	2·5	2	2100	0·11	$2·5 \cdot 10^{-11}$	$1 \cdot 10^{-11}$
Ga 2874	2·5	2	2100	0·12	$2·5 \cdot 10^{-11}$	$1 \cdot 10^{-12}$
Ge 2652	7·0	5	2400	0·19	$8·7 \cdot 10^{-10}$	$3 \cdot 10^{-12}$
Hg 2537	2·5	2	700	0·014	$5·0 \cdot 10^{-11}$	$2 \cdot 10^{-11}$
I 1830	4·0	1	1900	0·11	$2·0 \cdot 10^{-9}$	$3 \cdot 10^{-11}$
I 2062	2·5	2	2400	0.02	$5·0 \cdot 10^{-9}$	$1 \cdot 10^{-9}$
In 3039	2·5	1	1900	0·08	$8·0 \cdot 10^{-12}$	$4 \cdot 10^{-13}$
K 4044	2·5	2	1800	0·074	$6·3 \cdot 10^{-10}$	$4 \cdot 10^{-11}$
Li 6708	3·0	1	1900	0·05	$5·0 \cdot 10^{-11}$	$3 \cdot 10^{-12}$
Mg 2852	4·5	2	1800	0·10	$3·0 \cdot 10^{-12}$	$4 \cdot 10^{-14}$
Mn 2795	2·5	2	2000	0·048	$2·5 \cdot 10^{-12}$	$2 \cdot 10^{-13}$
Mo 3133	2·5	2	2500	0·08	$5·0 \cdot 10^{-11}$	$3 \cdot 10^{-12}$
Ni 2320	2·5	2	2200	0·012	$2·5 \cdot 10^{-11}$	$9 \cdot 10^{-12}$
P 1775	2·5	1	1900	0·10	$8·0 \cdot 10^{-11}$	$3 \cdot 10^{-12}$
P 2136	2·5	3	2400	0·01	$5·0 \cdot 10^{-10}$	$2 \cdot 10^{-10}$
Pb 2833	2·5	2	1900	0·063	$3·0 \cdot 10^{-11}$	$2 \cdot 10^{-12}$
Pd 2476	2·5	2	2100	0·05	$5·0 \cdot 10^{-11}$	$4 \cdot 10^{-12}$
Pt 2659	2·5	2	2300	0·086	$2·5 \cdot 10^{-10}$	$1 \cdot 10^{-11}$
Rb 7800	2·5	2	1900	0·030	$7·5 \cdot 10^{-12}$	$1 \cdot 10^{-12}$
Rh 3435	2·5	2	2300	0·034	$6·3 \cdot 10^{-11}$	$8 \cdot 10^{-12}$
S 1807	4·0	1	1900	0·12	$1·0 \cdot 10^{-8}$	$1 \cdot 10^{-10}$
Sb 2311	2·5	2	2000	0·040	$5·0 \cdot 10^{-11}$	$5 \cdot 10^{-12}$
Sc 3912	4·0	2·5	2350	0·08	$2·8 \cdot 10^{-9}$	$6 \cdot 10^{-11}$
Se 1961	2·5	2	1600	0·092	$2·0 \cdot 10^{-10}$	$9 \cdot 10^{-12}$
Si 2516	2·5	2	2250	0·23	$2·7 \cdot 10^{-12}$	$5 \cdot 10^{-14}$
Sn 2863	2·5	2	2000	0·022	$1·0 \cdot 10^{-11}$	$2 \cdot 10^{-12}$
Sr 4607	3·0	1	2200	0·05	$2·0 \cdot 10^{-11}$	$1 \cdot 10^{-12}$
Te 2143	2·5	2	2000	0·030	$7·6 \cdot 10^{-12}$	$1 \cdot 10^{-12}$
Ti 3653	2·5	2	2500	0·054	$5·0 \cdot 10^{-10}$	$4 \cdot 10^{-11}$
Tl 2768	2·5	2	1800	0·01	$2·5 \cdot 10^{-12}$	$1 \cdot 10^{-12}$
V 3184	4·5	2	2800	0·10	$2·5 \cdot 10^{-10}$	$3 \cdot 10^{-12}$
Yb 3988	4·0	2·5	2150	0·125	$5·0 \cdot 10^{-11}$	$7 \cdot 10^{-13}$
Zn 2138	4·5	4	1500	0·04	$1·0 \cdot 10^{-12}$	$3 \cdot 10^{-14}$

This question is very easily answered if we remember that the flame version of the atomic absorption method is based on recording equilibrium absorption, the

TABLE 5.10 Limits of detection of cadmium using different methods (g)

Atomic absorption		Emission			Neutron activation[18]		Mass spectrometry§
Cuvette	Flame*[15]	Arc[16]	Hollow cathode[17]	Copper spark†[17]	$2 \cdot 10^{14}$ neutrons cm²/s	$5 \cdot 10^{11}$ neutrons cm²/s	
$3 \cdot 10^{-14}$	$5 \cdot 10^{-10}$	$1 \cdot 10^{-9}$	$3 \cdot 10^{-8}$	$2 \cdot 10^{-7}$	$1 \cdot 10^{-10}$	$5 \cdot 10^{-8}$	$2 \cdot 10^{-10}$

* 0·05 ml of solution required
† With photographic recording of spectra for pure solutions of the element
§ This corresponds to a calculated absolute limit of detection of 10^{12} atoms[18]

cuvette version being based on recording peak absorption. It was established on p. 121 that the difference in absolute sensitivities for these methods of measuring absorption is governed by the ratio τ_2/τ_1, where τ_2 is the time spent by atoms in the flame zone through which the light passes, and τ_1 is the time taken to introduce the sample into the flame. Usually $\tau_1 \simeq 1$ sec, and $\tau_2 = 10^{-4}$ sec. The mean difference between the absolute sensitivities of the methods compared is therefore $10^4:1$, and this agrees well with the experimental data given in Table 5.10.

The high absolute capacity of the method for determining elements is extremely useful for micro-analysis, when the total amount of a sample may not be more than a few micrograms, and for analysing radioactive or toxic substances, when it is advisable in the interests of the experimenter that the sample should be as small as possible. The use of cuvettes for analysing toxic samples is also effective because the analysis space in which samples are atomized is effectively insulated from the outer atmosphere by the chamber itself. In order further to protect the interior of the chamber from the condensation of atomic vapour of toxic substances on the walls, the quartz insulating cell described on p. 206 (Fig. 5.9) can be used.

Khartsyzov and I used this method for analysing $\beta\gamma$-active preparations of Zn^{65}, Cd^{109}, $Te^{121,123}$, and Se^{75}. Although, in accordance with their category of 'no isotope carrier', the final products must not contain stable isotopes of these elements, the insufficient purity of the initial targets and of the reagents used for extracting the isotopes may cause them to be contaminated. The amount of a carrier present was estimated by comparing the total amount of an element measured by the atomic absorption method with the amount of isotope calculated from radioactivity measurements.

The results of analysing one of the batches of preparations are given in Table 5.11. When the figures in this table are considered, it is found that the preparation $Cd^{109}Cl_2$ actually corresponds to the designation 'no isotope carrier', while the preparation $Zn^{65}Cl_2$ contains a huge excess of stable zinc, exceeding the amount of the isotope Zn^{65} by three orders of magnitude.

Finally the high sensitivity of the method brings us close to solving one of the principal problems in analytical chemistry, that of analysing ultra-pure substances: semiconductor and atomic materials, the ores of rare metals, etc. At present, the analysis of pure substances is restricted by the insufficient absolute sensitivity of the methods; for the concentrate separated out after the chemical or physical thermal treatment of a sample contains amounts of the impurities being determined which are less than the absolute sensitivity. As an instance, if the absolute sensitivity with which any particular element can be determined is $1 \cdot 10^{-8}$ g, we have to use a sample weighing more than 100 g, to determine $1 \cdot 10^{-8}$ per cent of the element. In most cases these amounts cannot be used, and it is practically impossible to process them. Obviously a combination of the method described above with preliminary chemical or physical concentration will solve this problem, since samples 1000 times smaller will be used as the initial samples.

TABLE 5.11

Preparation	Specific activity (uc/ml)	Calculated amount of isotope present (g/ml)	Amount of the element measured by the atomic absorption method (g/ml)
$Cd^{109}Cl_2$	1·8	$7·1 \cdot 10^{-7}$	$7·9 \cdot 10^{-7}$
$Zn^{65}Cl_2$	1·85	$3·5 \cdot 10^{-7}$	$6·2 \cdot 10^{-4}$

OPTICAL INTERFERENCE IN ATOMIC ABSORPTION MEASUREMENTS

One of the principal generally recognized advantages of atomic absorption analysis, which applies equally to all methods of atomization, is that no optical effects are superimposed on the atomic absorption measured. This statement is only true, however, in cases in which the amount of foreign substances in the sample being analysed is small, and is not more than 4–5 orders of magnitude greater than the amount of the element being determined.

It was actually proved[3] as far back as 1959 that the atomization of comparatively large amounts of a substance in a cuvette greatly attenuated the beam of light from a light source. This attenuation was effective over quite a wide range of the spectrum, i.e. it was of a definitely non-selective nature. During the atomization of 50 μg of NaCl in a 3 mm diameter cell (with $T=2500°K$ and $P=1$ atm), the attenuation for the 4254 Å (chromium) line was about 0·2 absorbance unit. This effect was interpreted as follows. As the atomic vapour passes through the apertures in a heated cuvette, a cloud of NaCl particles is formed by condensation, and this cloud scatters the beam of light passing through.

Later on, it was established that the scattering of light is not the only process superimposed on atomic absorption when large amounts of samples are atomized in cuvettes. The beam of light is attenuated much more, particularly in the short-wave ultra-violet region of the spectrum, by molecular absorption of vapour of the parent substance in the sample.

As an example, the distribution of the optical interferences over a wide region of the spectrum, between 2000 Å and 10 000 Å, during the atomization of 100 μg of

Pb in the form of $Pb(NO_3)_2$ is shown in Fig. 5.18. The sources of light were a hydrogen lamp and an incandescent lamp.

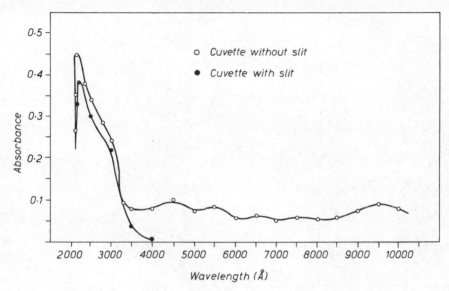

FIG. 5.18. Optical interferences during the atomization of 100 μg of Pb in a 2·5 mm diameter cuvette at 1900°C, with a pressure of 2 atm of argon.

This graph shows that there is a comparatively constant attenuation of the beam of light between 3500 and 10 000 Å; this, as we shall see later, is due to light scattering; between 2000 and 3000 Å the absorbance increases considerably, with a maximum at about 2150 Å; this is due to molecular absorption by Pb_2.

Let us discuss each of these effects separately.

We know (for example, from Ditchburn[37]) that, when light passes through a turbid medium the amount scattered depends on the ratio of the wavelength (λ) to the diameters (d) of the scattered particles. Fig. 5.19 shows the relationship in general form. Three regions can be isolated on this graph, in each of which scattering alters in a different manner as the wavelength changes. In the region $\lambda \ll d$, the scattered light energy is practically independent of the wavelength; with $\lambda \simeq d$ there is a low scattering maximum, and in the region $\lambda \gg d$ scattering rapidly decreases with the wavelength, in proportion to λ^{-4} (the Rayleigh scattering region).

It is absolutely clear that the scattering of a beam of light passing through a cuvette, observed when considerable amounts of sample are atomized (Fig. 5.18), should come in the region $d \gg \lambda$. The theory of scattering for this region has been given by Mie.[38] It follows from the curve plotted in Fig. 5.18 that the magnitude of the Pb particles in a cloud is not in any case less than 10 000 Å, i.e. 1 μm.

In order to reduce the effect of the scattering of light by an atomic vapour condensing at the point where it leaves a cuvette, I have suggested[3] that cuvettes with slits at their ends should be used (Fig. 5.20). The departure of atomic vapour

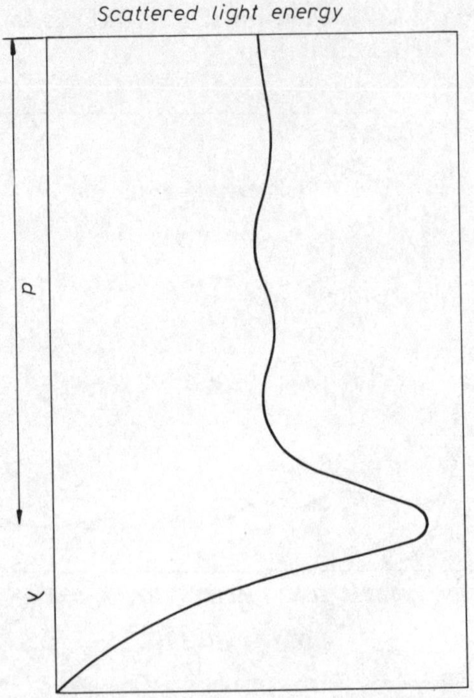

Fig. 5.19. Scattered light energy versus wavelength of light for scattered particles d in diameter.

through slits away from the beam of light greatly reduces the amount of the substance leaving the cuvette along the axis of the beam of light. It is easy to see that the part of atomic vapour condensing in the path followed by a beam of light with a cross-sectional area of πr^2, where r is the radius of the cuvette, must in the case of slits h deep and $2r$ high be:

$$\frac{\pi r^2}{\pi h 2r} = \frac{r}{2h} \qquad (5.28)$$

For a cuvette with a radius $r = 1\cdot 5$ mm and a slit depth $h = 5$ mm, therefore, for instance, the scattering must be reduced to $0\cdot 15$, i.e. to 15 per cent, of that in a

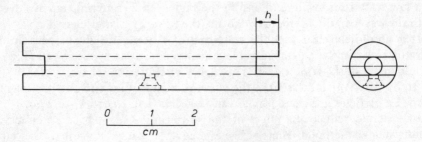

Fig. 5.20 Cuvette with slits at its ends

cuvette without slits. Actually experiments[3] proved that, in a cuvette with slits of these dimensions, the atomization of 100 μg of NaCl caused practically no scattering of the 4254 Å line.

Let us now consider the attenuation caused by molecular absorption by vapour of the parent substance. Many substances, when they become vapour, retain or form molecules which are stable at high vapour densities. These molecules include, in particular, the halides of many metals and diatomic and polyatomic molecules consisting of atoms of one type (for instance Bi_2, Se_2, Te_2, etc.). Molecular absorption is revealed not only in cuvettes, but also in flames, and particularly in the spent gases from flames.[39]

Fig. 5.21 shows the absorption spectrum for the KI molecule recorded by us at intervals of 50 Å with a hydrogen lamp. In order to eliminate the effect of scattering on the results, a cuvette with 3 mm deep slits was used. For comparison, the KI absorption spectrum, observed by Koirtyohann and Pickett[39] when measuring the absorption in the spent gases from an oxy-hydrogen flame, is plotted on the same graph. When the curves are compared we find that, in spite of the entirely different measurement conditions, in both cases the effect is of exactly the same nature. The fact that this effect occurs is further confirmed by the fact that the maxima and minima of the curves in Fig. 5.21 correspond to the positions of the extreme values for the KI absorption spectrum investigated earlier[40] in heated quartz cells.

The stability of the molecules, and therefore also the molecular absorption, decreases as the temperature of the surrounding medium increases. Fig. 5.22 shows

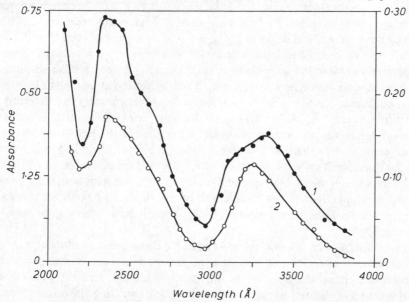

FIG. 5.21. Absorption spectrum for KI vapour. 1. Graphite cuvette, 2·5 mm in diameter, at 1700°C, 2 atm of argon (1μg KI). 2. Flame adaptor for spent gases.[39]

the effect of cuvette temperature on the magnitude of absorption of the 2150 Å spectrum region during the atomization of 4 μg of LiI in a cuvette with slits. Raising the temperature from 1700°C to 2000°C about halves the molecular absorption.

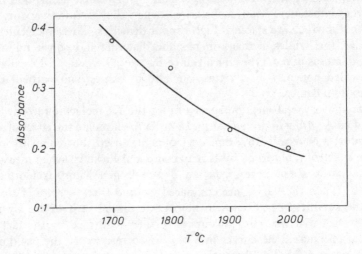

FIG. 5.22. Effects of cuvette temperature on molecular absorption of LiI vapour.

The interference of scattering and molecular absorption on atomic absorption sets a limit to the increase in the relative sensitivity with which elements are determined in a sample. For this very reason Massmann, therefore, when he atomized large amounts of samples (up to 2 mg in a graphite cuvette), used a multi-channel spectrometer, one of whose channels was adjusted to a non-resonance line comparatively close to the analysis line; this channel was used for the simultaneous recording of non-selective interference. This system was successfully used for making automatic allowance for interference developing during the determination of volatile elements (Ag, Cd, etc.) in metallic zinc, lead and copper. Interference, corresponding to the absorption of up to 80 per cent of the light, was allowed for with an error of less than 0·3 per cent. This system of recording was tested on metal samples, with no slits in the cuvette. Attenuation of the beam of light was principally associated not with molecular absorption, but with scattering, which is practically independent of the wavelength of the light. The fact that the analysis and comparison lines were spectrally non-homologous was not of any great importance.

In general, however, the effectiveness of using this system of allowing for optical interference is quite dubious, in view of the possible abrupt variation in the molecular absorption along the spectrum. In actual fact our attempts at allowing for optical interference superimposed on the Al 3092 Å and Cu 3248 Å lines compared with the Ne 3520 Å line by means of a two-channel system were not successful. This is confirmed by results obtained later by Massmann,[41] who found a considerable difference (of 2–3:1) in the degree to which optical interference associated with the atomization of large amounts of copper was superimposed on the As 1890 Å line and the 2020 Å comparison line.

The use of a two-channel system for allowing for molecular absorption is not, therefore, in general advisable (for any sample). In principle, of course, superimposed interference can be taken into account by means of a non-homologous comparison line, adjusting the difference between the lines by altering the particulars of the recording channels. This method is, however, cumbersome and not sufficiently accurate.

Koirtyohann and Pickett[52] proposed a more efficient and elegant technique of allowing for non-selective interference. Their method consists of using the beam of light from a continuous spectrum source, isolated by the same channel as that used for isolating the resonance line, as a comparison. Since, with the slit widths customary for spectrophotometric measurements, the magnitude of the atomic absorption for continuous emissions is a mean two orders of magnitude smaller than for line emission (see p. 30), the effect of atomic absorption on the reference beam of light can be ignored. The advantage of this method is that the wavelengths of the main and reference light beams almost coincide, so that a possible effect resulting from the two signals being spectrally non-homologous is eliminated. The other advantage of this last method is that it is simple: there is no need to use a two-channel monochromator or two separate recording systems, and the problem of selecting comparison lines in the light source spectrum does not exist.

This principle was used to allow for the molecular absorption of H_2SO_4 and NaCl superimposed on the atomic absorption of zinc (Zn 2139 Å) and cadmium (Cd 2288 Å) when these were being determined in the spent gases from an oxy-hydrogen flame.[32] Since the amount of superimposition (in absorbance) did not exceed 0·28 and the measurement conditions were extremely stable, Koirtyohann and Pickett[32] used the successive measurement of absorption with resonance and continuous emission sources.

In the case of cuvettes, the successive measurement of absorption is not suitable owing to the non-reproducible nature and considerable magnitude of the non-selective interference. In view of this, a spectrophotometer in which automatic allowance was made for non-selective interferences while simultaneously recording the absorption from both sources was used; this method was described earlier.

Experiments were made, using this spectrophotometer, in recording small atomic absorptions against a background of greater non-selective interferences. The results of the measurements for bismuth (Bi 3068 Å) and thallium (Tl 3776 Å) are given in Tables 5.12 and 5.13. The cuvette temperature was 1600°C for determining thallium and 1860°C for bismuth, the internal diameter of the cuvette was 2·5 mm, and the pressure of the argon in the chamber was kept at 2 atm. The error representing the reproducibility of the results of making determinations was not, with the smallest amounts of elements, more than 20 per cent.

The second column of each table gives the amounts of optical interference caused by the atomization of the pure parent substance during measurement with a single source of light. All the other results were obtained by automatic measurement of the difference between the absorption of light from two sources: line and continuous.

Within the limits of measurement error, the atomic absorption in the presence of excesses of foreign substances proves to be the same as for the element being determined in pure form. If we compare the maximum values of optical interference for bismuth in the presence of LiI, and for thallium in the presence of KH_2PO_4, with the values of atomic absorption for the minimum amounts of the elements being determined, we find that atomic absorption can, at least, be measured against a background of interference 40 times greater.

TABLE 5.12 Atomic absorption of bismuth (in absorbance units) in different parent substances

Bismuth (g)		Parent substance (g)		
		KH_2PO_4 $6 \cdot 10^{-5}$	LiI $4 \cdot 10^{-5}$	LiCl $1 \cdot 10^{-4}$
0	0	0·2*	1·13*	0·43*
$2·5 \cdot 10^{-11}$	0·027	0·028	0·030	0·031
$5 \cdot 10^{-11}$	0·053	0·055	0·056	0·060

* Molecular absorption (a cuvette with a slit was used).

TABLE 5.13 Atomic absorption of thallium (in absorbance units) in different parent substances

Thallium (g)		Parent substance (g)		
		KH_2PO_4 $1 \cdot 10^{-4}$	LiI $2 \cdot 10^{-5}$	LiCl $2 \cdot 10^{-5}$
0	0	0·37*	0	0
$5 \cdot 10^{-12}$	0·0095	0·0095	NM	NM
$1·5 \cdot 10^{-11}$	0·026	0·024	NM	NM
$5 \cdot 10^{-11}$	0·076	NM	0·063	0·070

NM = not measured.

* Non-selective attenuation resulting from scattering and molecular absorption (cuvette with no slits).

It must be emphasized that this system of recording can only be effective when allowing for interference with a continuous spectrum. The formation of continuous molecular absorption spectra is associated with transition between two states, of which at least one is unquantized, i.e. it has a continuous series of energy values.[42] These transitions are primarily transitions to a continuous upper state, resulting in the dissociation of molecules (the most typical case is that of diatomic molecules of metal halides), and secondly transitions from a continuous lower state corresponding to collisions between free atoms in metal vapours. The latter type of continuum is only observed with sufficiently high vapour densities. In addition to this, the continuous nature of molecular absorption, typical of measurements in a cuvette (with high foreign gas pressure and high parent substance vapour densities), can also partly be associated with impact broadening of the components of the band structure, this broadening resulting in the disappearance of the discrete spectrum.

We have not in our practical work, up to now, encountered any cases in which this method of allowing for optical interference has proved ineffective. From now on, therefore, when we consider the relative sensitivity of measurements, we shall not take optical interference associated with the presence of excess amounts of foreign substances in samples into consideration.

RELATIVE SENSITIVITY

When we consider relative sensitivity, it is convenient to consider two cases, different in principle:

1. All the sample being analysed is converted into atomic vapour form, so that the ratio of atoms of the element being determined to atoms of foreign elements in the vapour phase and the initial sample are identical.

2. The sample is not completely atomized, only fractionally, so that the ratio of atoms of the element to atoms of the parent substance in the vapour phase may be different from their ratio in the sample.

Let us first consider the more general and widespread first case. Assuming that the composition of the sample and atomic vapour are equivalent, the relative sensitivity is governed by the absolute sensitivity on the one hand, and the limiting magnitude of the sample atomized on the other.

The limiting magnitude of the sample atomized depends in turn on the method by which the analysis signal is measured. When peak absorption is measured, the volume of sample atomic vapour cannot exceed the volume of the cuvette, since otherwise the partial expulsion of atomic vapour of the parent substance from the cuvette will cause a loss of the element being determined.

Increasing the internal volume of the cuvette makes it possible to atomize correspondingly larger samples. The volume can only, however, be increased and the absolute sensitivity remain unchanged if the length of the cuvette, and not its internal diameter, is increased.

Absolute sensitivities measured by experiments were related to a 2·5 mm diameter cuvette; we shall therefore assess relative sensitivities for a cuvette with the same diameter and 40 mm long, a foreign gas (argon) pressure of 2 atm, and a mean cuvette temperature of 2000°C. The weight of a sample with a mean atomic weight A, which when fully atomized at a temperature T and a foreign gas pressure P has a volume equal to the cuvette volume $V_c = Sl$, can be estimated using the equation:

$$M = \frac{SlAPT_0}{V_0 P_0 T} \quad (5.29)$$

Here V_0 is the volume of a gram-molecule with $T_0 = 273\cdot15°K$ and $P_0 = 1$ atm.

If we substitute the values given above for the parameters in (5.29) and assume $A = 50$, we get $M = 1 \cdot 10^{-4}$ g. Since the mean absolute sensitivity is 10^{-13}–10^{-11} g, under the same measurement conditions the relative sensitivity must be 10^{-7}–10^{-5} per cent.

The results of determining bismuth and thallium in certain salts, given on p. 236, can be used for illustrating these possibilities. The lowest measured concentrations were $2\cdot15 \cdot 10^{-5}$ per cent for bismuth (absolute sensitivity $4 \cdot 10^{-12}$ g), and $5 \cdot 10^{-6}$ per cent for thallium (absolute sensitivity for the Tl 3776 Å line—$2 \cdot 10^{-12}$ g).

The results of determining aluminium in pure metals have been described by Nikolaev and Aleskovskii.[5] Very small concentrations of aluminium in samples were measured with a high margin of sensitivity: with an aluminium concentration of $7 \cdot 10^{-4}$ per cent the absorbance was about 0·25 (absolute sensitivity for determining aluminium $1 \cdot 10^{-12}$ g).

The integration method of measuring absorption has not yet been used for determining microscopic amounts of elements. Nevertheless certain advance conclusions have already been drawn regarding the possibilities of this method. When absorption is measured by the integration method, the samples analysed can be far larger than when the peak method is used. Actually the amounts of sample consumed depend on the rate n at which the sample is atomized and the atomization time τ_1. The condition which limits increase in n is that the atoms of the sample accumulating in the cuvette must not exceed a critical value N_{crit}, with which the concentration of sample atoms is commensurate with that of atoms of the foreign gas. When equilibrium is reached between the amount of atomized sample (n_{crit}) and of sample leaving the cuvette (N_{crit}/τ_2), the following condition must be fulfilled:

$$n_{\text{crit}} = \frac{N_{\text{crit}}}{\tau_2} \tag{5.30}$$

Hence the maximum amount of sample which can be atomized in a time τ_1 (under steady conditions) is governed by the equation

$$n_{\text{crit}} \tau_1 = N_{\text{crit}} \frac{\tau_1}{\tau_2} \tag{5.31}$$

Taking into account the fact that the mean value of $\tau_2 \simeq 1$ sec, and the recording time τ_3 (where $\tau_3 \geqslant \tau_1 + 4\tau_2$) must not be more than a reasonable length, i.e. $10-10^2$ sec, we can count on increasing the sample dimensions until they are 10–100 times N_{crit} used with the peak method of measuring absorption.

The maximum anticipated gain in relative sensitivity may therefore, provided that the absolute sensitivities of the methods are the same, also be 10–100 times.

It is interesting to compare the relative sensitivity in a cuvette, when peak absorption is measured, with the relative sensitivity when equilibrium absorption is measured in a flame. Provided that the compositions of the sample and the atomic vapour are equivalent, in both cases the maximum relative sensitivity is governed by the ratio of the minimum concentration of atoms of the element recorded to the maximum permissible concentration of the parent substance in the atomic vapour. Since the minimum concentration of atoms of an element found is about the same in the case of both variations, the difference between the relative sensitivities can only be governed by the maximum concentration of particles of the parent substance.

Let us consider what is the maximum concentration of atoms of the parent substance in a flame. It is quite easy to see that the ratio of V_{par}, the volume of vapours of the parent substance, to V_g, the volume of gas, can be represented as follows:

$$\frac{V_{par}}{V_g} = \frac{V_0 C w_{sol} \epsilon}{V_g M} \qquad (5.32)$$

Here, as before, $V_0 = 22.4 \cdot 10^3$ cm³/g · mol (the volume of 1 gram-molecule of vapour), C is the concentration of the solution atomized (g/ml), w_{sol} the rate of atomization of the solution (ml/min), ϵ the efficiency of atomization, V_g the total volume of gas (cm/min), M the molecular weight (g/mol).

Assuming that C = 0.01 g/ml, $w_{sol} = 2$ ml/min, $V_g = 10^4$ cm³/min, and A = 50 g/mol, we get the following for a flame with the gases premixed ($\epsilon = 0.1$):

$$\frac{V_{par}}{V_g} \simeq 10^{-4} = 0.01 \text{ per cent}$$

and for the flame from direct injection burners ($\epsilon = 1.0$):

$$V_{par}/V_g \simeq 10^{-3} = 0.1 \text{ per cent}$$

Since all flames burn at atmospheric pressure, the partial pressures of the vapour of the parent substance in a flame cannot exceed 10^{-4} atm and 10^{-3} atm for premix and direct injection burners respectively. In a cuvette the concentration of particles of the parent substance can be raised to a few tens of per cent with relation to the foreign gas with which the chamber is filled. If, therefore, the pressure of the foreign gas is raised to several atmospheres, the partial pressure of the vapour of the parent substance in the cuvette can be increased to 1 atm.

Hence the ratio of the maximum permissible concentration of atoms of the parent substance to the concentration of atoms of the element, and consequently also the relative sensitivity for a cuvette, must, in principle, be 3-4 orders of magnitude greater than that for a flame. The mean relative sensitivity for a flame is 10^{-3} per cent, while for a cuvette (see above) it is 10^{-7}–10^{-5} per cent. This difference agrees with our estimate.

The criterion of the ultimate concentration of particles of a parent substance in an atomic vapour can also be used for assessing the relative sensitivities of other methods of spectrochemical analysis, for instance the emission method with a carbon arc. Let us consider the photographic variation of the method, which is the variation most used in practice for the spectrochemical analysis of pure substances.

Problems associated with estimating sensitivity for emission analysis in an arc theoretically were first investigated in detail, for both the photoelectric and photographic methods of recording spectra, by Mandel'shtam and Nedler.[19,20] According to Mandel'shtam and Nedler estimating the limiting concentration of atoms which can be measured in arc discharge plasma amounts to the following task:

1. Finding the connection between the concentration of atoms in plasma and the ratio of the analytical signal to the background (I_L/I_B).

2. Determining the lowest value of I_L/I_B which can be found by calculating statistical fluctuations in the photocurrent or the grain size of the photographic emulsion.

The first part of the task can be solved comparatively easily, but the second part requires data, which are not always known, concerning the characteristics of the light source, the spectroscopic device, and the detector. The final equations for calculating I_L/I_B obtained by Mandel'shtam and Nedler are rather cumbersome,[19,20] and not very convenient for comparing with the results of experiments.

It seems more rational to carry out this estimation omitting the second stage of the calculation. To do this we estimate the concentration of atoms in plasma for a certain value of I_L/I_B selected in advance. This value can be the criterion, often used for estimating the sensitivity of photographic methods, of excess of line blackening on the background of a continuous spectrum per 0·1 unit of absorbance. We thus get

$$\gamma \log \frac{I_L + I_B}{I_B} = 0 \cdot 1 \qquad (5.33)$$

With a contrast factor $\gamma = 1$, equation (5.33) corresponds to*

$$\frac{I_L}{I_B} = 0 \cdot 26 \qquad (5.34)$$

Emission intensity for resonance (which are the most sensitive) lines for atoms can be expressed by:

$$I_L = \frac{hc}{1 \cdot 51} \frac{f}{\lambda^3} N_0 e^{h\nu/kT} \text{ (ergs. sec.cm}^3\text{)} \qquad (5.35)$$

where f is the oscillator strength, N_0 is the concentration of atoms in the unexcited state, which can be taken for elements with high ionization potentials and a single ground level, as being equal to the total concentration of atoms of the element concerned.

The following factors cause background on spectrograms: continuous background from incandescent particles and electrodes, molecular bands, electronic continuum due to free-free and free-linked electron transitions taking place during collisions with positively charged ions. Of these only the last cannot in principle be eliminated. Following Mandel'shtam and Nedler,[19,20] therefore, we shall assume that the background is associated solely with the electronic continuum. The coefficient of emission of the electronic continuum can be calculated theoretically for a carbon arc and the 3500–2000 Å range, using the equation

$$\epsilon_\nu = 6 \cdot 8 \cdot 10^{-38} \frac{N_e^2}{\sqrt{T}} e^{h(\nu_g - \nu)/kT} \text{ (ergs/cm}^3\text{)} \qquad (5.36)$$

which agrees well with the results of experiments.[21] Here N_e is the concentration of electrons, ν_g the boundary frequency beyond which the intensity of emission drops by an exponential law. For a carbon arc, $\nu_g \simeq 0 \cdot 9 \cdot 10^{15}$ sec^{-1}.[21]

---

* When the frequency selection method, described in references 27 and 28, is used for the photoelectric recording of spectra, the limits of detection achieved are governed by statistical fluctuations of the photo-current. With a luminous flux of $\sim 10^{-8}$ lm and an exposure of 60 sec, the limit of detection of I_L is $\sim 0 \cdot 003$ times I_B. Therefore, the frequency selection method has great advantages over photographic methods of recording.

The intensity of the background I_B in the spectral region occupied by the line being measured can be expressed as:

$$I_B = \epsilon_\nu d\nu \qquad (5.37)$$

The magnitude of $d\nu$ is determined by the equation

$$\Delta\nu = \frac{c\Delta\lambda}{\lambda^2} = \frac{cDs}{\lambda^2} \text{ (sec}^{-1}\text{)} \qquad (5.38)$$

where D is the spectrograph dispersion (cm/cm) and s the slit width (cm).

If we substitute (5.35) and (5.37) in (5.34), and also use (5.36) and (5.38), by rewriting (5.37) relative to N_0 we get:

$$N_0 = 4\cdot0 \, . \, 10^{-12} \frac{N_e^2 Ds\lambda}{\sqrt{(T)}f} e^{-h\nu_g/kT} \qquad (5.39)$$

This equation enables us to estimate the concentration of atoms in plasma necessary for determining elements when spectra are excited in carbon arcs. We calculate N_0 for a particular element, with meaned parameters (λ and f), using a medium dispersion spectrograph for recording spectra. Using the values of the parameters given in Table 5.14, we obtain a value of $N_0 = 5\cdot5 \, . \, 10^7$ atoms/cm$^3 \simeq 10^8$ atoms/cm^3.

According to experimental measurements,[22,23] the concentration of particles of the parent substance in an arc plasma is less than 1 per cent of the total number of particles in the plasma, i.e. does not exceed 10^{16} particles/cm^3. With the mean parameters indicated above and uniform atomization of the sample, therefore, the relative sensitivity in a carbon arc cannot exceed 10^{-6} per cent.

It should be emphasized once more that, in the calculation, no allowance was made for ionization or the partial excitation of atoms, nor for the existence of other sources of background besides the electronic continuum. The calculated value can therefore only be used as the limit for the ideal and most favourable case of analysis. Allowing for this, the agreement between the value obtained and the limits known from numerous experimental results can be considered good.

The examples given above prove convincingly that the criterion of the limiting concentration of particles of the parent substance, introduced in order to assess the relative sensitivity of the method in which a cuvette is used, is equally effective for considering other methods of spectrochemical analysis based on the uniform atomization of samples (provided that the composition of the sample is equivalent to that of the vapour phase).

This is not so for fractional vaporization of samples. Here a more favourable ratio of atoms of the element being determined to atoms of the parent substance than in the initial sample can be obtained by vaporizing elements from materials, which are less volatile. We know that this method is extensively used in emission spectrochemical analysis of pure sustances: the method of fractional distillation and the vaporization method are founded on it.[16]

Fractional vaporization of elements is particularly effective in atomic absorption measurements made with a cuvette, since as opposed to the method of fractional distillation in an arc the vaporization of elements can be regulated independently of the process of measuring concentrations, while, on the other hand, as opposed to the vaporization method, both analysis stages occur in the same absorbing volume and are performed simultaneously.

TABLE 5.14 Parameters for calculating the limiting concentration of an element which can be determined in a carbon arc

$N_e = 10^{15}$ electrons/cm^3 $\qquad f = 0\cdot 1$
$T = 6000°K$ $\qquad s = 20\,\mu m = 2 \cdot 10^{-3}$ cm
$\lambda = 3000$ Å $= 3 \cdot 10^{-5}$ cm
$D = 13\cdot 5$ Å/mm $= 1\cdot 35 \cdot 10^{-6}$ cm/cm (for the ISP–28 spectrograph at 3000 Å)

The most important condition for using the peak method of measuring absorption in fractional vaporization is that it should be possible to isolate elements from the parent substance in a very short space of time. It was proved on p. 222 that the presence of up to 10^{-5} g of foreign elements with low volatilities does not affect the process of atomization of aluminium or zinc.

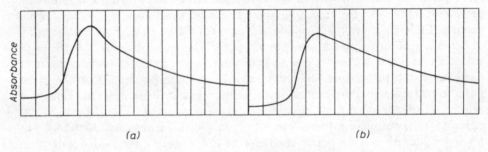

FIG. 5.23. Variation in the absorbance of the 3092·7 Å line for aluminium when it is atomized (a) from pure aluminium and (b) from tungsten powder.[7] Argon pressure 4 atm, $T = 2600°K$, arc current 50 amp. Interval of time scale 0·2 sec.

The rate at which atoms of aluminium are atomized from samples of tungsten weighing about 10^{-2} g was measured oscillographically by Nikolaev and Aleskovskii.[7] One of the oscillograms is shown in Fig. 5.23. For purposes of comparison there is an oscillogram of the atomization of pure aluminium under the same experimental conditions. We can see by comparing the oscillograms that in both cases the atomization times were the same.

The results of determining the concentration of aluminium in tungsten using samples with different weights (Table 5.15) also indicate that the rate at which elements are atomized does not depend on the sample size. In the fractional vaporization of elements from materials with low volatilities, therefore, it is possible to use calibration graphs plotted for the pure element. Results of the determination of aluminium and zinc in graphite and certain metals with low volatilities are given in Table 5·16, and the results of determining the iron, lead and manganese in graphite powder are given in Table 5.17. In all cases the weighed samples were inserted into a channel in the electrode.

TABLE 5.15 Results of determining aluminium in tungsten samples with different weights[5]

Weight (mg)	Absorbance	Al content (10^{-5} per cent)
0·20	0·08	5·0
0·25	0·13	6·4
0·30	0·16	6·6
0·45	0·30	8·3
0·50	0·29	7·1
0·65	0·30	5·8
0·80	0·40	6·3
0·90	0·54	7·5
1·10	0·63	7·2
1·65	0·75	5·6
1·70	0·90	6·6

Mean $6·5 . 10^{-5}$ per cent

TABLE 5.16 Determination of aluminium and zinc in metals with low volatilities and graphite[29]

Specimens	Sample weights (mg)	Absorbance	Amounts determined (per cent)
Molybdenum	0·35	0·56	$2 . 10^{-4}$ Al
Graphite	0·8	0·51	$8 . 10^{-5}$ Al
Tungsten	0·8	0·40	$6 . 10^{-5}$ Al
Tantalum	2·5	0·60	$3 . 10^{-5}$ Al
Niobium	4·5	0·60	$1·7 . 10^{-5}$ Al
Tungsten	10·4	0·03*	$5·8 . 10^{-5}$ Zn
Niobium	0·2	0·60†	$1·5 . 10^{-5}$ Zn

* Measurements made with the Zn 3076 Å line
† Measurements made with the Zn 2139 Å line

TABLE 5.17 Determination of elements in graphite

Samples	Iron		Lead		Manganese	
	C (%)	w (%)	C (%)	w (%)	C (%)	w (%)
Normal graphite	$1·8 . 10^{-4}$	36	$3·7 . 10^{-6}$	14	$3·3 . 10^{-6}$	27
Purified graphite	$7·0 . 10^{-5}$	27	$4·3 . 10^{-6}$	7	$1·3 . 10^{-6}$	23
High purity graphite	$2·1 . 10^{-5}$	33	$3·0 . 10^{-6}$	17	$2·2 . 10^{-7}$	14

Finally, Table 5.18 gives the result of determining cadmium, lead, bismuth, antimony and tin in standards of zirconium dioxide. This table indicates the error representing the reproducibility of the results of determinations. On the whole the results agree satisfactorily. The slight discrepancy in the data, which in three cases is (2·3–2·4):1, and in eight cases is (1·3–1·6):1, is possibly due to the error in producing standards and the insufficient accuracy with which the calibration solutions (pure solutions of the salts of elements being determined) were prepared.

Attention should be drawn to the margin of sensitivity existing in all the cases given in Tables 5.16, 5.17 and 5.18; this margin can be realized by using large amounts of samples, reducing cuvette diameters, measuring smaller absorbances, and in some cases using more sensitive lines. For instance cadmium was determined in zirconium dioxide, from the Cd 3261 Å line, in a 3·5 mm diameter cuvette. The use of a cuvette with a diameter of 2·5 mm, and of the 700 times more sensitive Cd 2288 Å line, makes it possible, other conditions being equal, to determine down to 10^{-8} per cent cadmium effectively.

When we estimate the relative sensitivity of the method in which elements are fractionally vaporized from materials with low volatilities as a whole, we should count on samples weighing about 10^{-3} g for the peak method of measurement and up to 10^{-1} g for the integration method. As both variations have absolute sensitivities of 10^{-13}–10^{-11} g, therefore the peak method must provide a relative sensitivity of 10^{-8}–10^{-6} per cent, the integration method a relative sensitivity of 10^{-10}–10^{-8} per cent.

TABLE 5.18 Determination of elements in zirconium dioxide

Element	Sample number	Certificate (per cent)	Determined (per cent)	Coefficient of variation, w (per cent)
Cadmium	2	$3 \cdot 1 \cdot 10^{-4}$	$1 \cdot 3 \cdot 10^{-4}$	6·2
	3	$1 \cdot 1 \cdot 10^{-5}$	$4 \cdot 9 \cdot 10^{-5}$	6·5
	4	$2 \cdot 4 \cdot 10^{-5}$	$1 \cdot 5 \cdot 10^{-5}$	4·9
Lead	2	$5 \cdot 3 \cdot 10^{-4}$	$4 \cdot 2 \cdot 10^{-4}$	18
	3	$1 \cdot 9 \cdot 10^{-4}$	$1 \cdot 9 \cdot 10^{-4}$	17
	4	$6 \cdot 5 \cdot 10^{-5}$	$7 \cdot 4 \cdot 10^{-5}$	6·4
Bismuth	2	$1 \cdot 2 \cdot 10^{-4}$	$0 \cdot 9 \cdot 10^{-4}$	22
	3	$4 \cdot 0 \cdot 10^{-5}$	$2 \cdot 9 \cdot 10^{-5}$	10
	4	$1 \cdot 6 \cdot 10^{-5}$	$1 \cdot 1 \cdot 10^{-5}$	9
Antimony	2	$1 \cdot 1 \cdot 10^{-3}$	$0 \cdot 7 \cdot 10^{-3}$	19
	3	$3 \cdot 8 \cdot 10^{-4}$	$5 \cdot 7 \cdot 10^{-4}$	25
	4	$1 \cdot 7 \cdot 10^{-4}$	$1 \cdot 6 \cdot 10^{-4}$	10
Tin	2	$4 \cdot 3 \cdot 10^{-4}$	$1 \cdot 8 \cdot 10^{-4}$	9
	3	$1 \cdot 6 \cdot 10^{-4}$	$1 \cdot 0 \cdot 10^{-4}$	18
	4	$4 \cdot 7 \cdot 10^{-5}$	$3 \cdot 1 \cdot 10^{-5}$	12

The opposite method can be used for concentrating elements when a cuvette is employed: the fractional distillation of the parent substance, which is more volatile than the elements being determined. This procedure can actually be employed with the same apparatus as that for analysis immediately before the determination. A particular case of this method is that of determining the impurity elements in liquid substances, water, acids, organic compounds, etc., when the simple evaporation of the solution or the ashing of the organic substance perform the function of preliminary enrichment. To illustrate the possibilities of this method, the results of determining aluminium in water and acids are given in Table 5.19.[5]

Table 5.19 Results of determination of aluminium in water and acids by atomic absorbtion[5]

Specimen	Volume of solution analysed (ml)	Absorbance	Aluminium content (per unit)
Mains water	0·005	0·30	$8 \cdot 10^{-6}$
Distilled water	0·010	0·12	$1·6 \cdot 10^{-6}$
Doubly distilled water	0·020	0·12	$8 \cdot 10^{-7}$
Hydrochloric acid	0·005	0·45	$1 \cdot 10^{-5}$
Nitric acid	0·005	0·65	$1·1 \cdot 10^{-5}$

If only 10 μl of solution (with a dry residue content of $<10^{-5}$ per cent) is introduced into the electrode, a relative sensitivity for the solution of about 10^{-9}–10^{-7} per cent can be achieved.

In general, the use of a graphite cuvette combined with the preliminary enrichment of samples, which can be carried out by any chemical and thermal means, is extremely effective, since owing to the high absolute sensitivity of the cuvettes this combination makes it possible to take full advantage of the preliminary enrichment of samples, even when the amount of the initial substance is small. For instance, although the method of vaporizing elements from substances with low volatilities, as a means of preliminary enrichment of samples, makes it possible to achieve a coefficient of enrichment in excess of 100, the gain in the analysis of the concentrate in an arc over the direct determination of elements in an arc is not usually more than 10:1.[16] The reason for this lies in the insufficient sensitivity of the emission method of spectrochemical analysis; for most elements this is 10^{-8}–10^{-7} g.[16] These figures correspond to $2 \cdot 10^{-5}$–$2 \cdot 10^{-4}$ per cent of an element in the initial sample (optimum sample weight 50 mg). Naturally if the concentration of an element is lower than this, the method of analysing concentrate in an arc is not sufficiently sensitive. (The procedure by which concentrate is analysed in a hollow cathode[17] affords slightly better possibilities, since with this method the absolute sensitivity is about an order of magnitude higher than the sensitivity with an arc.)

ACCURACY AND REPRODUCIBILITY OF ANALYTICAL RESULTS

The basic concepts representing analysis error are those of the precision and accuracy of analysis. If we follow the classification of errors given by Nalimov,[24] analysis accuracy means the deviation of the mean result of determinations from the true amount of an element in a sample, while the reproducibility of analysis is the scatter of the results relative to the mean value.

Let us first consider precision of the method. The reproducibility error builds up during the two principal stages of the analysis: the production of an atomic vapour and the measurement of absorption with a spectrophotometer. The sources of error in making measurements with spectrophotometers were discussed earlier (p. 95), where it was proved that these errors can, in principle, be reduced to the detector shot noises, which are of statistical origin.

Random errors associated with the production of an atomic vapour in a graphite cuvette are governed by the following factors: specimen heterogeneity, introduction of weighed amounts of samples into electrodes, and fluctuations in cuvette temperature and foreign gas pressure. Let us assess the effects of these factors on the magnitude of the analysis signal.

We know from Nalimov[24] that the error σ_z in measuring any quantity, $z = f(x_1 x_2 \ldots x_n)$, which is a function of the series of independent parameters $x_1, x_2 \ldots x_n$, can be expressed as follows in terms of the error σ_{x_i} of x_i parameters:

$$\sigma_z^2 = \left(\frac{\partial f}{\partial x_1}\right)^2 \sigma_{x_1}^2 + \left(\frac{\partial f}{\partial x_2}\right)^2 \sigma_{x_2}^2 + \ldots + \left(\frac{\partial f}{\partial x_n}\right)^2 \sigma_{x_n}^2 \tag{5.40}$$

Using equation (5.40), which is called the law of accumulation of errors, we can express the coefficient of variation w in measuring the analysis signals A_{peak} and Q_A in terms of coefficients of variation of the parameters included in equations (5.22) and (5.23),

$$w_A = \sqrt{(0.49\, w_T^2 + w_P^2 + 4 w_r^2 + w_M^2)} \tag{5.41}$$

$$w_Q = \sqrt{(0.81\, w_T^2 + 4 w_r^2 + 4 w_l^2 + w_M^2)} \tag{5.42}$$

Let us estimate the values of the coefficients of variation for each of the parameters. The fluctuations in temperature do not exceed a mean 20–30°K during measurements; the relative fluctuations in the temperature between 2000 and 2500°K do not therefore exceed 1 per cent. It is quite easy to keep the pressure of the foreign gas in the chamber constant to within 0·1 atm. Between 2 and 5 atm, therefore, the relative fluctuations in pressure do not exceed 5–2 per cent. It must, moreover, be pointed out that changes in the pressure near 1 atm have, in general, less effect on absorption than would appear from equation (5.22), since the line contour at atmospheric pressure is largely governed by the Doppler effect, which does not depend on the pressure.

When cuvette dimensions are measured the coefficients of variation can be reduced to 0·1 per cent when determining length and to 1 per cent when determining diameter.

Next let us estimate the error associated with the procedure of dosing samples.

It was stated earlier that errors of dosing of a few microlitres of solution can easily be reduced to 0·025 µl, and when 0·05–5 µl of solution is introduced this corresponds to 5–0·5 per cent. The situation is much worse when solids are introduced by weighing the electrodes before and after introduction of samples. In this case the absolute error is 0·1 mg, and this corresponds to relative errors of between 1 and 100 per cent when introducing samples weighing 10–0·1 mg. It is in general impossible to introduce weighed batches of less than 0·1 mg for this reason.

In view of this it is interesting to use the technique, normally employed for emission spectrochemical analysis, of introducing an internal standard into the substance being analysed, and simultaneously recording the signals from the element being determined and the comparison element, for calculating the amount of a sample introduced into a cuvette. This method was first used for cuvettes by Massmann.[26] In his experiments he used a multi-channel spectrophotometer. The scatter due to inaccurate introduction of samples was slightly reduced for certain combinations of elements, but with other combinations of elements the use of an internal standard increased the measurement error. For instance, in the case of silver-zinc, the mean square deviation in the values of A_{Ag}/A_{Zn} was about 20 per cent as against 12–15 per cent for each of the quantities A_{Ag}/A_{Zn} separately. Obviously the poor correlation between the absorbances of different elements found by Massmann[26] was associated with the ineffective method used for atomizing samples; the samples were introduced into a cold cuvette and atomized by heating the cuvette itself, not by means of a separately heated additional electrode (see p. 213).

The introduction of an internal standard was also used by Katskov and L'vov for determining impurity elements in certain powdered samples with low volatilities, in particular zirconium dioxide (Table 5.18) and powdered graphite (Table 5.19). When zirconium dioxide was being analysed, zinc was used as the internal standard (Zn 3076 Å), and when graphite was analysed, silver was used (Ag 3281 Å). These elements were introduced into the samples being investigated in the form of solutions. The results given in Table 5.20 show that the use of internal standards makes it possible to allow reliably for the weight of the sample. Actually the standard deviation ($\bar{C} = 1\cdot33 \cdot 10^{-4}$ per cent) in the results of determining cadmium was only 6·2 per cent. The apparatus used in these experiments was somewhat cumbersome, since it included two separate monochromators; this can be eliminated by using a polychromator with two exit slits and a two-pen recorder.

As well as the procedure for introducing samples into the electrodes there is another factor which causes a scatter in the amount present in the electrodes of the element being determined; this is heterogeneity of the substances analysed. Let us as an example consider the case in which the element being determined is present in the form of particles, the individual grains about 3 μm, in a powdered sample. The volume of a grain will then be $1\cdot4 \cdot 10^{-11}$ cm³, and its weight, with a density of 3 g/cm³, will be about $4 \cdot 10^{-11}$ g. It is quite easy to calculate that a sample weighing 10^{-3} g will, when the concentration of the element being determined is $1 \cdot 10^{-4}$ per cent, contain about a mean 25 grains of the element. The coefficient of variation in the number of grains n in individual samples will correspond to the equation:

$$\frac{\Delta n}{n} = \frac{1}{\sqrt{n}} \qquad (5.43)$$

With $n = 25$, the coefficient of variation due to the statistical distribution of the grains in samples will be 20 per cent, and with a sample weight of 0·1 mg and $n = 3$ it will be about 60 per cent.

TABLE 5.20 Determination of cadmium in zirconium dioxide
(Sample 2)

A_{Cd}	M_{Cd} ($\times 10^{-9}$ g)	A_{Zn}	M_{Zn} ($\times 10^{-8}$ g)	Sample weight (mg)	C (10^{-4} per cent)
0·054	0·50	0·099	1·40	0·35	1·43
0·063	0·58	0·125	1·77	0·43	1·49
0·091	0·48	0·179	2·54	0·63	1·33
0·109	1·00	0·206	2·92	0·73	1·37
0·126	1·16	0·255	3·61	0·91	1·28
0·183	1·68	0·353	5·00	1·25	1·34
0·196	1·80	0·395	5·60	1·40	1·29
0·280	2·58	0·580	8·21	2·06	1·25
0·296	2·73	0·619	8·77	2·19	1·24

Experiments I conducted in which $1 \cdot 10^{-4}$ per cent of bismuth was determined in granite standards prepared by the successive dilution of a standard containing a large amount of the pure base substance confirm this. Fig. 5.24 shows the results of measuring the peak absorption of the Bi 3068 Å line. The scatter of results, in terms of the mean arithmetical deviation from the curve, comprise respectively 43 per cent, 34 per cent and 10 per cent in the regions for weighed batches of 0·05–0·15 mg, 0·15–0·25 mg and 0·25–0·35 mg.

It is interesting to note that it is even possible to establish a quantitative correlation between the coefficient of variation w, which represents the reproducibility of the results within a series of measurements of a particular element in samples, and the mean size P of the samples used for analysis. Fig. 5.25 shows the correlation between the error and the mean sample size, found by Katskov and L'vov when analysing powdered zirconium dioxide (Table 5.18). As would be expected, the error distribution is subject to the relation

$$w \sim \frac{1}{\sqrt{P}} \quad (5.44)$$

By comparing relationships (5.43) and (5.44), we can determine the number (n) of grains of an impurity element or internal standard per unit of sample weight. We get a mean figure of about 11 grains per mg of zirconium dioxide.

In order to reduce the errors associated with sample heterogeneity, it is advisable that the samples analysed should be ground as finely as possible. If this is done the number n of grains of an impurity element in the sample weight must increase in inverse proportion to the cube of the dimensions of the particles. As an instance, if a substance is ground down to particles with dimensions not exceeding 1 μm, a sample weighing 1 mg, with a concentration amounting to 10^{-4} per cent of an impurity element, must contain not less than 700 grains of the element being determined (instead of 25 in the case of 3 μm particles).

The most radical means of preventing heterogeneity in solid substances is, however, to convert comparatively large amounts of the substances (about a few tens of milligrammes) into solutions, and then select the aliquot part. It is important

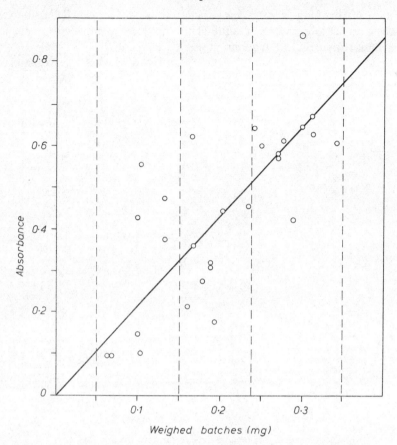

Fig. 5.24. Scatter of results of determination of bismuth in granite as a function of sample weight.

that this procedure at the same time also reduces the error in weighing solid samples.

In the atomic absorption analysis of solutions in graphite cuvettes, therefore, the random error is made up basically of the spectrophotometric measurement error, fluctuations in the pressure of the foreign gas, and the error in dosing volumes of solutions less than 1 μl. When solid samples are analysed the weighing error increases, and there is an additional source of errors associated with sample heterogeneity.

The distribution of the total random error in atomic absorption measurements in a cuvette was investigated by Nikolaev,[6] who used the determination of zinc in an aluminium alloy sample converted into a solution as his example. The concentration of zinc in the alloy was 1 per cent. The measurements were made with an argon pressure of 4 atm, using the Zn 3076 Å line. Altogether 616 determinations were made. It was proved, by means of the χ-criterion,[24] that the distribution observed was close to normal. Graphical comparison of the results of the experiments and a theoretical curve corresponding to the normal distribution provides a visual confirmation of this (Fig. 5.26). The parameters of the recorded distribution (in absorbances) are as follows: mean value of the random quantity 0·911, dispersion $94 \cdot 10^{-6}$. Thus the coefficient of variation was 5·1 per cent.

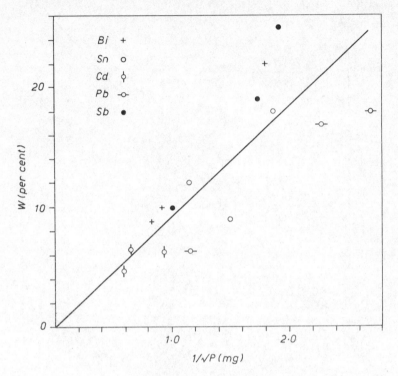

FIG. 5.25. Coefficient of variation versus sample weight when determining impurities in zirconium dioxide (see Table 5.18).

Experience gained in the author's laboratory has shown that, even under the best experimental conditions, the coefficient of variation when determining peak absorption cannot be reduced below 3 per cent.

Let us now dwell on the possible sources of systematic error in atomic absorption measurements in graphite cuvettes, which may result in the mean determined figure varying from the true amount of an element in a sample. The first factors to relate to this are those associated with the excessively high absolute sensitivity of the method: firstly the possibility of the materials used in the analysis (electrodes and cuvettes), also of the samples, being contaminated with the elements being determined, and secondly the instability of very dilute solutions.

The first problem is particularly important when determining elements which are much encountered in nature: iron, zinc, aluminium, calcium, sodium, potassium, etc., which are usually present not only in the substances and reagents used for analysis, but also in the air. The greatest risk is presented by the possibility that dust may fall on the electrodes containing the sample, since no allowance can be made for this as opposed to contamination of the reagents or water. According to our data, contamination from the air while samples are being introduced into electrodes (in about 5 minutes) becomes appreciable when determining $<10^{-11}$ g of iron, zinc or aluminium.

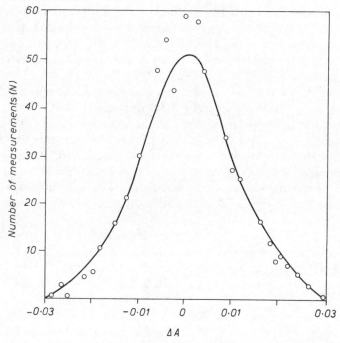

FIG. 5.26. Distribution of the results of determining zinc in solution in terms of the deviation from the mean absorbance.[6] (N is the number of measurements.)

The instability of very dilute solutions is associated principally with the hydrolysis of certain salts and the adsorption of elements by the walls of vessels. It was proved, for instance, that the concentration of $1 . 10^{-3}$ per cent of a solution of Ni, Mn, Mo, V, Au, Pt, Ru and Ti in 6 per cent mineral acid decreases by ten times if it is kept in a glass vessel for seventy-five days.[18] When the adsorption in a cuvette is measured, for certain elements the solution concentration must be $1 . 10^{-7}$ per cent. The instability of the concentration of an element may therefore be even greater.

Adsorption of an element by the vessel walls may develop not only while a solution is being kept, but even while it is being introduced with a micro-pipette; in this case the risk of adsorption is particularly great owing to the high specific surface area of glass with relation to the volume of solution. The author has observed this effect when determining the lead in a solution containing 0·1 μg of Pb per ml. A certain volume of the solution was taken in a glass pipette, and fed into five different electrodes in identical successive amounts. The results of the determinations (in absorbances) decreased, in the order in which they were introduced into the electrodes, as follows: 0·23–0·20 0·18–0·15–0·13. The reason why the last batches of solution contained less lead was that they were in the pipette for longer, and made contact with a larger pipette area than the first batches.

Elements are less readily adsorbed at vessel walls when quartz is used, and even less when polyethylene is used. For instance, when a solution containing a concentration of 10^{-4} per cent of manganese was kept in a polyethylene vessel for a year, the concentration of manganese did not change.[18] The other advantage of polyethylene over glass is its purity. Polyethylene contains the following amounts of common elements: $2 \cdot 10^{-4}$ per cent Si, $6 \cdot 10^{-5}$ per cent Fe, $(2-3) \cdot 10^{-5}$ per cent Al, Ca, Cr and Pb. When dilute solutions are investigated, therefore, polyethylene vessels and micro-pipettes with polyethylene tips should be used.

The other possible reasons for discrepancies between the results of determination and the true amounts of elements in samples include the differences in the rates at which impurity elements are vaporized from substances with low volatilities, or in pure form, since these differences may cause the results of measuring peak absorption to be too low (see p. 224) when elements in solid materials are determined. The integration method of measuring adsorption does not suffer from this limitation.

To conclude, let us estimate the time taken for analysis with a graphite cuvette. The total analysis time is governed by the durations of the following procedures:

1. Preparation of samples for analysis: application of five different solutions to the electrodes, drying droplets, and installing the electrodes in the chamber—7 minutes.

2. Preparation of the chamber for making measurements: blowing out the chamber, filling it with gas up to the increased pressure, heating the cuvette—1 minute.

3. Measurement procedure: the successive introduction of the five electrodes into the cuvette—1 minute.

4. Switching on heating, releasing the excess pressure, and processing the results—1 minute.

A series of five determinations thus takes about 10 minutes, and up to thirty determinations of any particular element can be made in an hour.

When solid samples are introduced into electrodes by weighing, the procedure for preparing samples for analysis is lengthened to 15–20 minutes, so that the total duration of a series of determinations is lengthened to 20–25 minutes. When a two-channel spectrophotometer is used for simultaneously measuring the content of an internal standard, the time taken to introduce solid samples into electrodes is reduced to 2–3 minutes.

The estimates given above only indicate the instrumental possibilities of the method. In practice, as when the flame method of atomizing samples is used, the procedure for preparing the solution makes the greatest contribution to the total analysis time. The true analysis time, which represents the output and speed of the method which can be realized in practice, is therefore similar to the time for atomic absorption flame analysis (at any rate for analysing solid samples).

Chapter 6

SPECIAL FIELDS OF APPLICATION FOR THE ATOMIC ABSORPTION METHOD OF MEASUREMENT

DETERMINATION OF HALOGENS AND CERTAIN NON-METALS IN THE VACUUM ULTRA-VIOLET REGION

The resonance lines for certain non-metallic elements, and also halogens and gases, are in the vacuum ultra-violet region of the spectrum, which covers wavelengths shorter than 2000 Å. This boundary is quite clearly defined by two factors: the molecular absorption of oxygen, which begins to develop appreciably at less than 1950 Å (Figs. 6.1 and 6.2), and the absorption of quartz, the material most used in spectroscopy.

In spite of these difficulties, the vacuum ultra-violet region has now been sufficiently well mastered. Vacuum grating spectroscopic devices are used for observing the spectra. Lithium fluoride (LiF) or fluorspar (CaF_2) which transmit up to 1050

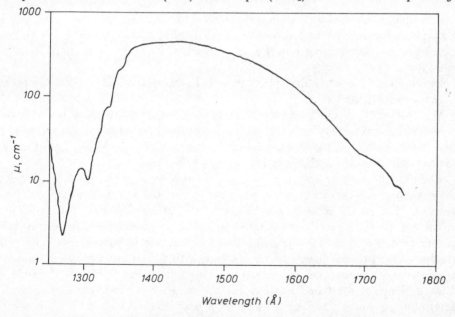

FIG. 6.1. Variation of the coefficient of oxygen absorption in the vacuum region (1250–1750 Å).

and 1250 Å are used for the windows and lenses. Above 1500 Å, sapphire (α-Al_2O_3) or very thin layers of optical quartz can be used for making the windows. Photomultipliers with quartz or glass windows, coated with a thin layer of luminescent sodium salicylate, which has an approximately constant quantum yield over a very wide region, between 400 and 3400 Å, can be used for the light detectors.

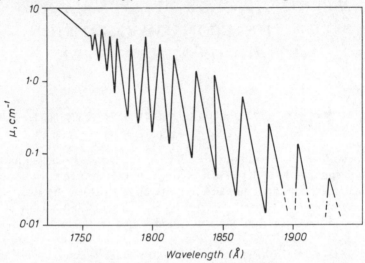

FIG. 6.2. Variation of the coefficient of oxygen absorption in the vacuum region (1750–1950 Å).

The vacuum ultra-violet region is much used for emission spectroscopic analysis. For this purpose different types of multi-channel photoelectric device have been developed for determining non-metallic elements, halogens, and gases in metals and alloys.

Development of atomic absorption analysis in the vacuum ultra-violet region has been much more modest.

We shall deal with the analysis of gases separately (p. 256), and confine ourselves here to setting forth the principal results obtained in the atomic absorption determination of halogens and certain non-metallic elements. The principal resonance lines for these elements are given in Table 6.1.

The basic obstacle to entering the region of wavelengths shorter than 2000 Å is the oxygen absorption in the method of atomizing samples most used in analysis practice, the flame. According to Allan,[33] even at 1960 Å the absorption in a 10 cm long air-hydrogen flame is 50 per cent, and in an air-acetylene flame it is about 70 per cent. At 1890 Å, air-propane flame absorption is 89 per cent. The more vigorous absorption of flames by comparison with the air atmosphere is evidently associated with absorption of oxygen molecules from excited levels. In spite of these difficulties, the flame method has been used comparatively effectively for determining selenium and tellurium;[33,34] it was later also used for determining arsenic.[35] The determinations were made from the analysis lines of these elements lying above 1900 Å.

TABLE 6.1 Resonance lines of the halogens and some non-metallic substances[32]

Element	Line (Å)	Element	Line (Å)
F I	955, 952	P I	1788, 1783, 1775
Cl I	1380, 1347	S I	1826, 1821, 1807
Br I	1576, 1489	As I	1973, 1938, 1890
I I	1830	Se I	1961
C I	1657, 1561	Hg I	1849

The first step into the actual vacuum region of the spectrum was taken by Sullivan and Walsh,[36] who tried to determine phosphorus using the 1775 Å line. They used a hollow cathode as the means of atomizing. The details of these experiments were not published; it was merely made clear that phosphorus contents of 0·01–0·05 per cent P in copper were measured quantitatively.*

The next step in mastering the region shorter than 2000 Å was taken by Massmann,[37] who used a graphite cuvette for making measurements. The cuvette was housed in a chamber,[38] through which a continuous current of argon was passed. In other ways the apparatus was a normal 'non-vacuum' apparatus. This, however, proved sufficient for determining arsenic satisfactorily using the 1890·4 Å and 1972·6 Å lines. When a cuvette with an internal diameter of 8 mm was used, the absolute sensitivity of determination (for 1 per cent absorption) was $1 \cdot 10^{-9}$ g. The peak method was used for recording absorption with the sample to be analysed first introduced into the cuvette which was then heated to 2400°C by the pulsed method.

Preliminary experiments in which a cuvette was used for determining sulphur, phosphorus and iodine were recently carried out by Khartsyzov and L'vov. Since the resonance lines of these elements come in the 1700–1900 Å region, a vacuum grating monochromator was used for recording emissions. The source of light and the two-lens illuminating system were housed in two cylindrical jackets, firmly held against the chamber containing the cuvette (see Plate 3), and fixed to the spectrophotometer rails. The windows and lenses were made of lithium fluoride. The chamber and cuvette functioned in the normal manner. A gentle current of argon was blown continuously through the casing; this argon passed through the unsealed joints between the casings and the chamber. The sources of light were high-frequency spherical lamps with the pure elements introduced into them, made in the laboratory. The thickness of the walls of the quartz envelopes of these lamps was about 0·2 mm, and in this region of the spectrum quartz absorption can in practice be ignored.

The experiments proved that the elements investigated can effectively be determined using all the resonance lines known for them. A temperature of about 1900°C is sufficiently high for measuring absorption. There is absolutely no background from the cuvette, and a d.c. amplifier can be used for signal recording, without modulating the latter.

* During the six years which have elapsed since Sullivan[36] published his findings, no fresh data have appeared regarding the use of hollow cathodes for making measurements in a vacuum. This has necessitated a certain caution in estimating the prospects of the method for practical analysis work.

The highest absolute sensitivities were recorded for the P 1775 Å, I 1830 Å and S 1807 Å lines. The measurements were made by the peak method, in a 4 mm diameter cuvette, and the absolute amounts of the elements were 10^{-9}–10^{-8} g. The elements were introduced into the electrodes in the form of dilute solutions of salts and acids. Sensitivities (for 1 per cent absorption) were $3 \cdot 10^{-12}$ g P, $3 \cdot 10^{-11}$ g I, and $1 \cdot 10^{-10}$ g S.

This procedure can obviously be used for determining chlorine or bromine without any difficulty whatever. Measurement of the absorption of fluorine is made more difficult by the fact that no substances which are transparent in the region shorter than 1050 Å are available for light sources. As regards carbon, though it is quite possible to measure the absorption of this element, the method of atomizing samples in a cuvette made of graphite cannot be used. In this respect tubes with hollow cathodes have better prospects.

The possibility of using the Hg 1849 Å line, which is two orders of magnitude more sensitive than the Hg 2537 Å intercombination line, is definitely interesting for determining small amounts of mercury.

ANALYSIS OF GASES

All the resonance lines of gases lie in the vacuum ultra-violet region of the spectrum (Table 6.2), while with the exception of helium and neon they are above 1050 Å, the transmission boundary for lithium fluoride. There is no doubt, therefore, that the measurement of the absorption of gases, except for helium and neon, is technically possible. By comparison with the analysis of solid and liquid substances, the analysis of gases is simpler, since it is not necessary first to convert specimens into gaseous form. When inert gases are determined, and under normal conditions these are in the form of free atoms, it is not necessary to dissociate the molecules. In view of this, it is tempting to use the atomic absorption method of measurement for analysing gases. The spectra emitted by inert gases in the vacuum region are extremely simple: basically only the resonance lines are emitted. In some cases, therefore, when light detectors with a restricted range of spectral sensitivity are used (for instance ionization chambers filled with the vapours of organic compounds),[32] no spectroscopic apparatus at all is necessary. As an example, the Xe 1295 Å resonance line for xenon can be recorded by means of an ionization chamber, with a lithium fluoride window, filled with nitric oxide. Cells with lithium fluoride windows can be used as absorbing volumes. Obviously the atomic absorption measurement of gases must be more accurate than for any other elements, since measurement of the analysis signal may last for as long as is necessary.

When molecular gases (oxygen, hydrogen and nitrogen) are determined, the molecules of gas can be dissociated into atoms either by heating the gas or by means of an electrical discharge. These facilities have not yet been used in practice for gas analysis, but we know of several research projects, somewhat similar as to experimental technique, in which certain gases have been determined from their molecular absorption. As an instance, the concentration of water vapour in air was measured from the absorption of the L_α 1216 Å resonance line for hydrogen.[40] This line was also used for determining up to 10^{-4} per cent of water vapour in nitrogen.[39]

A number of research projects[41,42] have been devoted to determining oxygen in gas mixtures from absorption in the Schumann region. Khartsyzov and I developed a simple non-dispersion procedure for determining up to 10^{-3} per cent O_2 in different gases by means of a compact gas analyser made of quartz, consisting of a hydrogen tube with a hollow cathode, a 100 mm long cell filled with the mixture being analysed, and an emission detector; this was an ionization chamber containing dimethylaniline vapour. The apparatus functioned as follows. The beam of light passes from the hydrogen tube through the cell and reaches the detector, where it ionizes the dimethylaniline vapour. The current in the ionization chamber is proportional to the beam of light reaching the detector. Since the ionization potential of dimethylaniline is 7·14 eV, only light with wavelengths shorter than 1740 Å has the ionizing effect. On the other hand, the thin quartz windows separating the source, the cell and the detector pass the spectrum region with wavelengths of over 1500 Å. The light detector thus only reacts to radiation in the 1500–1740 Å region. Since the coefficient of absorption of oxygen in this range alters by more than 10:1, the A:C calibration graph is a curve. The sensitivity (for 1 per cent absorption) corresponds to a partial pressure of oxygen in the cell of $9·3 . 10^{-3}$ torr. When the cell is filled with the mixture being analysed to a pressure of 600 torr, the relative sensitivity is $1·5 . 10^{-3}$ per cent. The coefficient of variation, indicating the precision of measurement, is 2–3 per cent (with $A > 0·1$). The atomic absorption of gases is not confined to the vacuum region of the spectrum alone. When inert gases are excited there is considerable absorption of certain lines in the visible part of the spectrum. It was proved that absorption is maximum for lines corresponding to transitions to the metastable levels $^3P_0$ and $^3P_2$, and also the excited level $^3P_1$.[43-45] The gas is excited in electrical discharges. Table 6·3 gives the inert gas lines corresponding to maximum absorption (over 40 per cent) when a gas is excited in a continuous hollow cathode.[46] In these experiments the gas pressures were 1 torr for helium, 0·8 for neon, and 0·5 for argon, krypton and xenon. The current passing through the tube was 75 mA. Spectrally pure gases were used for the research.

The absorption is very reproducible from one experiment to the next. For instance, when the absorption of the Kr 7601·54 Å line was measured, the coefficient of variation was 1·7 per cent.

In spite of the ease of observing the absorption of these lines, their use is quite restricted in gas analysis. The fact is that the population of the metastable levels depends greatly on the purity of the gas. If an impurity element with an ionization energy less than the metastable-level excitation energy is present, the concentration of excited atoms decreases owing to second-order collisions with atoms of the impurity element, and also to reduction in the electron temperature of the plasma. Reduction in the concentration of metastable atoms in turn results in reduced absorption.

This relationship was used by Bochkova[3] for determining the amounts of impurities in inert gases. The absorption $(1 - T)$ was calculated from measured values of the intensity of emission I_e from an emission discharge lamp, the intensity of emission I_a from an absorption discharge tube, and the total intensity I_Σ of the tubes during simultaneous use, by the equation

$$(1-T) = \frac{I_e + I_a - I_\Sigma}{I_e} \text{ 100 per cent} \qquad (6.1)$$

Groups of lines corresponding to transitions to metastable levels concentrated in different regions of the spectrum were used for the measurements. The regions were: for neon 6143–6506 Å, for argon 6965–8015 Å, and for helium 10830 Å.

TABLE 6.2 Resonance lines for gases[32]

Element	Line (Å)	Transition	f_{abs}
He I	584·33	$1s^2\ {}^1S_0 - 2\ {}^1P_1$	0·28
Ne I	735·89	$2p^6\ {}^1S_0 - 3s\ {}^1P_1$	0·16
Ne I	743·72	$2p^6\ {}^1S_0 - 3s^3\ P_1$	—
Ar I	1048·22	$3p^6\ {}^1S_0 - 4s\ {}^1P_1$	0·28
Ar I	1066·66	$3p^6\ {}^1S_0 - 4s\ {}^3P_1$	0·036
Kr I	1164·87	$4p^6\ {}^1S_0 - 5s\ {}^1P_1$	0·135
Kr I	1235·84	$4p^6\ {}^1S_0 - 5s\ {}^3P_1$	0·158
Xe I	1295·59	$5p^6\ {}^1S_0 - 6s\ {}^1P_1$	0·23
Xe I	1469·61	$5p^6\ {}^1S_0 - 6s\ {}^3P_1$	0·28
Rn I	1451·56	$6p^6\ {}^1S_0 - 7s\ {}^1P_1$	—
Rn I	1786·07	$6p^6\ {}^1S_0 - 7s\ {}^3P_1$	—
H I	1215·67	$1\ S - 2\ P$	0·416
N I	1199·54	$2p^4\ S_{3/2} - 3s^4\ P_{5/2}$	0·18
N I	1200·2	$2p^4\ S_{3/2} - 3s\ {}^4P_{1/2}$	0·11
N I	1200·71	$2p^4\ S_{3/2} - 3s^4\ P_{1/2}$	0·062
O I	1306·03	$2p^4\ {}^3P_0 - 3s\ {}^3S$	0·033
O I	1304·87	$2p^4\ {}^3P_1 - 3s\ {}^3S$	0·033
O I	1302·17	$2p^4\ {}^3P_2 - 3s\ {}^3S$	0·033

The apparatus used for analysis purposes (Fig. 6.3) consists of an emission discharge lamp with a narrow capillary tube, an absorption discharge tube filled with

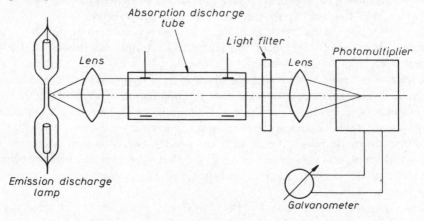

FIG. 6.3. Apparatus for the atomic absorption analysis of gases using non-resonance lines.[3]

Special Fields of Application for the Atomic Absorption Method of Measurement

the mixture to be analysed, an interference light filter, a photomultiplier, and a d.c. amplifier. As an example, Fig. 6.4 contains calibration curves for determining the nitrogen in helium, neon and argon.[3]

TABLE 6.3 The inert gas lines which are most sensitive to absorption in the 3900–8500 Å region[46]

Line (Å)	Intensity (relative units)	Absorption (per cent)	Level energy (cm^{-1})
He 5875·62	1000	45·5	169 081·111–186 095·90
He 3888·65	1000	51·5	159 850·318–185 558·92
Ne 6402·25	2000	73·5	134 043·790–149 659·000
Ne 6382·99	1000	45·4	134 461·237–150 123·551
Ne 6143·06	1000	52·3	134 043·790–150 317·821
Ar 8115·31	5000	73·7	93 143·800–105 462·804
Ar 8014·79	800	59·1	93 143·800–105 617·315
Ar 7635·11	500	72·6	93 143·800–106 237·597
Ar 7514·65	200	53·1	93 750·639–107 054·319
Ar 7383·98	400	48·3	93 750·639–107 289·747
Kr 8190·05	3000	55·5	80 917·561– 93 124·140
Kr 8112·90	5000	47·0	79 972·535– 92 295·199
Kr 8104·36	5000	43·7	79 972·535– 92 308·177
Kr 8059·50	1000	60·0	85 192·414– 97 596·718
Kr 7601·54	5000	52·6	79 972·535– 93 124·140
Kr 7587·41	1000	48·9	80 917·561– 94 093·662
Xe 8280·12	5000	47·1	68 045·663– 80 119·474
Xe 8231·63	5000	51·1	67 068·049– 79 212·970

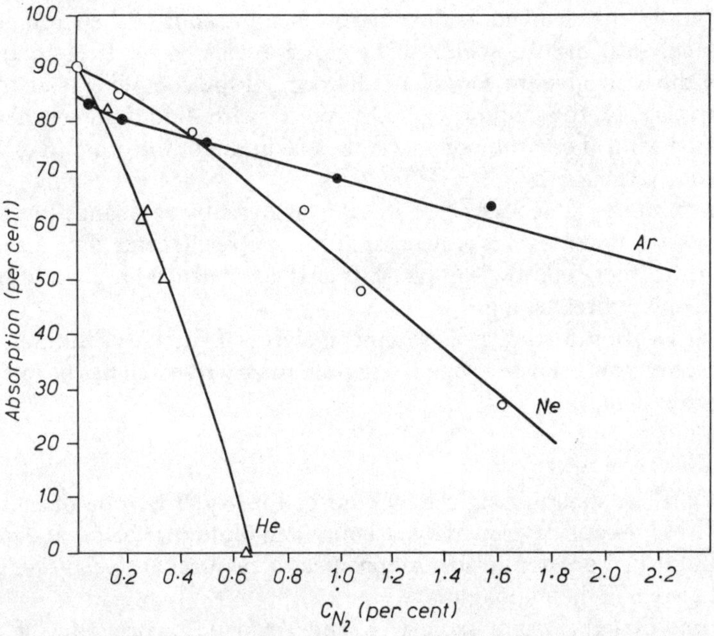

FIG. 6.4. Calibration curves for determining nitrogen in helium, neon and argon.[3]

The absorption of spectral lines corresponding to transitions to metastable levels can also be used for determining the concentration of the most inert gas, provided that the other components in the mixture have ionization energies higher than that of the metastable level. For instance, absorption in argon can be used for determining the amount of argon mixed with neon or helium.

Absorption of the He 10830 Å line group has also been used in the isotopic analysis of helium (p. 270).

ANALYSIS OF ISOTOPES

The advantage of the atomic absorption method for analysing isotopic compositions is that no spectral apparatus with great resolving power, such as is necessary in emission analysis for recording the separate components of the isotopic structure of the analysis lines, is required for determination. To be able to determine the concentration of any isotope in a mixture of isotopes by the atomic absorption method it is sufficient to measure the absorption of the resonance line, emitted by a single-isotope source of light, by a gaseous mixture of isotopes. Analysis is of course only possible if the isotope shift of the components of the line exceeds the line width in the source of light and the atomic vapour.

As opposed to the normal analysis of elements, in isotopic atomic absorption analysis the width of the absorption lines must be reduced. The best methods of measurement are therefore those in which atomic vapours with low foreign gas pressures and low temperatures are produced.

In order to estimate the possibility of using the atomic absorption method for isotopic analysis, the elements with sufficient isotopic shift of their resonance lines to enable analytical measurements to be made are given in Table 6.4. The last columns of the table give the Doppler half-width of the components at 190°K (the gas temperature in the hollow cathode cooled with liquid nitrogen), and the superimposed wing of one component on the maximum of the other (β_{ik}) calculated using equation (1.11).

The figures in the table show that isotopic analysis by resonance lines and lines with a metastable lower level is only possible for twelve elements. For the remainder of the elements, there is little isotopic shift of the resonance lines, and this is overlapped by Doppler broadening.

Atomic absorption methods of isotopic analyses of mercury, lithium, uranium and helium have now been developed, and preliminary research has been conducted for hydrogen and boron.

Isotopic analysis of mercury

A procedure for determining the amount of the Hg^{202} isotope of mercury in a mixture of isotopes was developed by Osbourn and Gunning[4] before the first papers had been published on atomic absorption spectrochemical analysis. Fig. 6.5 shows the optical system of their apparatus.

The source of light was a radio-frequency lamp (1), consisting of an 8 mm diameter tube containing practically pure Hg^{202} isotope (98.3 atomic per cent).

TABLE 6.4 Isotopic shift of resonance lines

Element	Isotopes	Wavelength (Å)	Lower level energy (cm^{-1})	Oscillator strength	$\Delta\nu_{isot}$² (cm^{-1})	$\Delta\nu_D$ (cm^{-1})	β_{ik} (per cent)
He	3–4	10830	159850·32	0·54	1·118	0·046	0
Li	6–7	6708	0	0·71	0·350	0·056	0
B	10–11	2497·7		0·33	0·168	0·125	0·79
Ne	20–22	6929·5	135890·67	NK	0·070	0·030	5 . 10^{-5}
Ar	36–40	7503·9	95399·87	NK	0·031	0·021	0·19
Re	185–187	3460·5		0·20	0·066	0·021	0
Os	188–190	4260·9		0·10	0·062	0·017	0
Pt	194–196	2659·5	0	0·12	0·052	0·027	3·6 . 10^{-3}
Hg	200–202	2536·5	0	0·03	0·179	0·027	0
Tl	203–205	3775·7	0	0·13	0·060	0·018	0
Pb	206–208	2833·1	0	0·21	0·083	0·024	0
U	235–238	5915·4	0	NK	0·300	0·011	0

NK denotes not known.

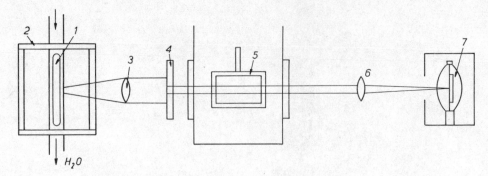

Fig. 6.5. Diagram of the apparatus for the isotopic analysis of mercury.[4]
1. Radio-frequency lamp. 2. Resonator. 3 and 6. Lenses. 4. Diaphragm.
5. Cell. 7. Detector.

Power was supplied to the lamp, housed in the resonator (2), from a micro-wave generator (2450 Mc/s). The lamp was cooled with running water, the temperature of which was maintained thermostatically at 25°C. The beam of light was passed by a lens (3) and the diaphragm (4) through the cell (5), and then after passing through a second lens (6) reached the detector (7), which was sensitive between 2000 and 3000 Å. Since there are no other mercury lines in this region, spectral devices are not necessary for isolating the Hg 2537 Å resonance line. The quartz cells, 0·1–5·0 cm long, were kept at between 12 and 40°C, with an accuracy of 0·1°C, thermostatically. After each analysis the cell was heated to dark red and at the same time exhausted by means of an oil diffusion pump; the next droplet specimen of the mercury to be analysed was then introduced into it.

The Hg I 2537 Å resonance line consists of five components (Table 6.5).

TABLE 6.5 The hyperfine structure of the Hg 2537 Å line

Displacement from the central component in cm^{-1}	Isotopes	Designations
+0·39	Hg^{199}, Hg^{201}	I
+0·15	Hg^{198}, Hg^{201}	II
0·0	Hg^{200}	III
−0·18	Hg^{202}	IV
−0·34	Hg^{199}, Hg^{201}, Hg^{204}	V

The concentration of the Hg^{202} isotope was determined by Ladenburg's method from the linear absorption (p. 21). The total pressure of the mercury vapour in the cell was calculated using the equation:

$$\log P_{(atm.)} = \frac{3 \cdot 185 \cdot 10^3}{T} + 5 \cdot 109$$

It was established, by means of a sample of mercury with the natural composition, that $\alpha = 1 \cdot 7$ corresponds to the linear relationship of k_0 to N (α is the ratio of emission line half-width to absorption line half-width). Since the width of the absorption line for the Hg^{202} isotope is governed solely by the Doppler effect, it is easy to compute that, for the absorption line, $\Delta\nu'_D = 3.4 \cdot 10^{-2}$ cm^{-1} and, for the emission line, $\Delta\nu''_D = \alpha\Delta\nu'_D = 5.7 \cdot 10^{-2}$ cm^{-1}. With this emission line width absorption by other isotopic components is practically eliminated.

Comparison of the results of analysing several samples containing between 30 and 95 per cent of Hg^{202} with the results of determination by mass spectroscopy shows that in five cases there is a deviation of 2–3 per cent, and in one case a deviation of 14 per cent. Only a few microgrammes of mercury are necessary for the analysis.

This method can also be used for determining the concentration of the other mercury isotopes.

The isotopic analysis of lithium

The flame variation of the atomic absorption method was first used by Zaidel' and Korennoi for the isotopic analysis of lithium.[5]

The resonance line of lithium, Li I 6707·8 Å ($2^2S_{1/2} - 2^2P_{1/2,3/2}$), has a doublet separation practically equal to the isotope shift of the components (Fig. 6.6). The result of this is that the line actually consists of three components, the central one being absorbed by both isotopes. In order, therefore, to determine the isotopic composition, the total amount of lithium atoms in the atomic vapour must be controlled.

A lamp with a water-cooled hollow cathode was used as the source of light; enriched Li^6 (99·8 atomic per cent) or Li^7 (99·1 atomic per cent) isotopes were introduced into the cathode. The beam of light, modulated by a rotating disk, was passed through a 20 mm long air-acetylene flame 10 mm wide. The Li 6708 Å analysis line was isolated by means of a grating monochromator and recorded by means of a photomultiplier for the infra-red region, a selective amplifier, and a millivoltmeter.

The total lithium content in the flame was controlled by means of an additional device for the simultaneous measurement of emission of the Li 6708 Å line; this device consisted of a monochromator, a photomultiplier, and a millivoltmeter.

The isotopic composition was determined from calibration graphs plotted by means of standards with a known isotopic composition; on these graphs $\log(I_7/I_{07})$ or $\log(I_6/I_{06})$ was plotted against the relative concentration of Li^6 atoms $[C_6 = n_6/(n_6 + n_7)]$.

If the line contours are not superimposed, assuming the ratio between the probabilities of transitions for the component doublets to be 2:1, the parameters of the calibration graphs are bound to be associated by the following equations:[5]

$$\log \frac{I_7}{I_{07}} = \log \frac{e^{-\beta(1+C_6)}[2e^{-\beta(1-3C_6)} + 1]}{3} \qquad (6.2)$$

$$\log \frac{I_6}{I_{06}} = \log \frac{2e^{-\beta} + 1}{3} - 0.434\beta C_6 \qquad (6.3)$$

where β is a coefficient depending on the length of the flame and the total concentration of lithium atoms $(n_6 + n_7)$ in the flame.

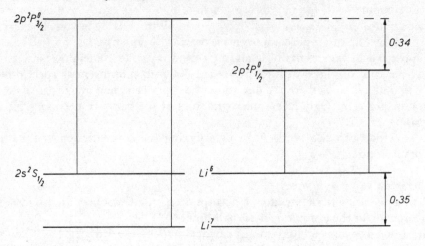

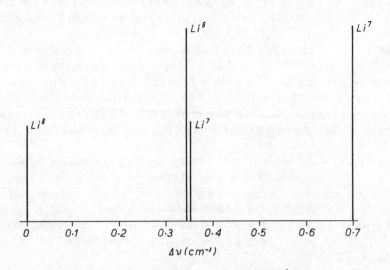

Fig. 6.6. Fine structure of the Li 6707·8 Å line.

It is interesting to note that, as opposed to the relationship of log (I_7/I_{07}) to C_6, the relationship of log (I_6/I_{06}) to C_6 is expressed by a linear function. Calibration curves plotted from the results of experiments follow the calculated curves sufficiently well. With low concentrations of the isotope Li⁶, a source with the isotope Li⁶ should be used for the analysis, while if the concentration of Li⁶ is large, the source should contain the isotope Li⁷.

It has been established that the presence of acids, alkalis, $AlCl_3$, or up to 0·25 g-mol/l of Na, K, Ca, Mg or Ba does not affect the results of analysis. The mean square error (absolute) between 2 and 45 per cent of Li^6 is 0·4–0·5 per cent.

Zaidel' and Korennoi[6] described the procedure which they developed for determining the isotopic composition of lithium from the emission of the Li 6708 Å line component due solely to the isotope Li^6. This component was isolated from the total line emission in the source of light with an isotopic filtration system; Walsh was the first to point out the possibility of using this.[7] The beam of light from the flame was passed through a tube containing Li^7 vapour, which absorbs the other two components of the line. The column of lithium vapour was formed in an iron tube about 30 cm long, filled with argon to 20 torr and heated to 500°C. The isotopic composition was determined from a calibration graph in which intensity was plotted against C_6 concentration. Between 2 and 91 per cent Li^6, the graph was a straight line. The standard error was not more than 2 atomic per cent.

The flame variation of the isotopic analysis of lithium was also developed by Manning and Slavin.[47] Two different methods were used for determining isotopic compositions. The first method consisted of measuring the amount of light absorbed from a source into which first one, then another isotope were introduced. Oxy-hydrogen flames were used as the source of light. An air-acetylene flame was used for the atomic vapour. Isotopic compositions were calculated using a calibration graph in which A_{Li^6}/A_{Li^7} was plotted against C_{Li^6}. Variation between $1·5 \cdot 10^{-4}$ and $3 \cdot 10^{-4}$ per cent Li in the total lithium content of solutions did not affect the position of the calibration curve.

The second method of determination consisted of measuring the amount of light absorbed from a source containing pure Li^6 isotope. The powerful component coinciding with the weak component of the line for the isotope Li^7 (see Fig. 6.6) was absorbed by adding an excess ($5 \cdot 10^{-3}$ per cent) of Li^7 to the flame. The absorption by the excess Li^7 was subtracted from the absorption by the sample being analysed with the added excess of Li^7, and the difference (in absorbance units) was compared with the concentration of the isotope Li^6. Although, as before, the source of light was a flame, and the emission lines were quite wide, the calibration curve was almost a straight line for between $1 \cdot 10^{-1}$ and $1 \cdot 10^{-4}$ per cent Li^6 in solution.

Goleb and Yokoyama[8] used the method of producing an atomic vapour in a hollow cathode for the isotopic analysis of lithium. The apparatus consisted of the source of light (a hollow-cathode lamp), a hollow-cathode absorption lamp, a grating monochromator produced by JACO, and an IP21 photomultiplier. The beam of light was modulated by means of a disk rotating at 60 c/s. The amplified signal was recorded using a chart recorder.

The hollow cathodes in the source lamp and the absorption lamps were cooled with running water. Emission in the lamp was most intense and stable when the lithium was introduced in the form of LiF, with a current of 10–20 mA and a helium pressure of 2 torr. The samples to be analysed in the form of LiOH were placed in a 6 mm diameter absorption lamp made of copper, 25 mm long. To obtain good reproducibility the LiOH solutions, 50 per cent diluted with acetone, were dried while the cathode was rotated at 60 rev/min. Altering the amount of specimen introduced from 50 to 1000 μg of Li does not affect absorption. The current in the absorption lamp was kept at 40 mA, the helium pressure at 2 torr. The working conditions for the absorption lamp were the same when the samples and standards were analysed; this was controlled by the total intensity of the emission of the Li 6708 Å line in the same apparatus immediately after measuring absorption. Obviously the absorption lamp emission was also modulated during this process.

The calibration graph was plotted for $\log(I_6/I_{06})/\log(I_7/I_{07})$ against C_6 (in atomic per cent). According to equations (6.2) and (6.3) the graph should have contained curves, but it can be considered practically to consist of a straight line between 2·5 and 10 per cent Li^6, the narrow range of concentrations used by Goleb and Yokoyama.[8] In their opinion, this method of plotting the graph is more accurate than plotting $\log(I_6/I_{06})$ against C_6, since more data are used in the former than in the latter case.

The reproducibility of the analysis results, including changing samples, is indicated by the error of 3·3 per cent.

Hollow-cathode lamps have the advantage over flames, for isotopic analysis, that the weight of the samples is measured in micrograms, also that no toxic substances are sprayed and the absorption line profiles are narrower.

The isotopic analysis of uranium

Goleb[9,29] described how he used both variations of the atomic absorption method for determining amounts of the isotopes U^{235} and U^{238}: measurement of the absorption of light from single-isotope sources, with the light passing through an atomic vapour of the sample being analysed, and measurement of the emission from the sample being analysed, with the light passing through a single-isotope filter.

Demountable water-cooled hollow-cathode lamps were used as the light sources and for producing atomic vapour. In order to atomize the uranium more intensively the lamps were filled with krypton (1 torr) or xenon (0·5 torr). By comparison with xenon, krypton has a simpler spectrum, and is therefore preferable for producing resonance emission. The 40 mm long cathode, hollow for all its length and with a diameter of 6 mm, in the absorption lamp was made of metallic uranium. In order to produce a homogeneous atomic vapour a lamp was designed with two ring-type anodes on either side of the cathode. The current was kept at 100 mA. The samples were inserted into the source lamp in the form of metal chips, oxides, or other compounds of uranium. When 50 mg of metallic uranium was introduced, with a current of 20 mA emission was steady after 30 minutes. The absorption and emission were recorded with the same apparatus as was used for analysing the isotopic composition of lithium.

Preliminary experiments proved that absorption was maximum for the 5915 Å, 5027 Å, and 4154 Å uranium lines (Table 6.6).

The 5027 Å line proved the best for the absorption method, and the 4154 Å line for the isotopic filtration method. Both lines have less developed hyperfine structures (for the isotope U^{235}) than the 5915 Å line.

The procedure for determining concentration by the absorption method of analysis consists of measuring the absorbance of the beams of light from two

TABLE 6.6 Absorption of uranium lines with relation to the isotopic composition of the uranium in the light source

Uranium lines (Å)	Isotopic composition (per cent)				Absorption (per cent)
	Source of light		Absorption tube		
	U^{238}	U^{235}	U^{238}	U^{235}	
5915·40	99·3	0·7	99·3	0·7	60
	7·0	93·0			5
5027·38	99·3	0·7	99·3	0·7	42
	7·0	93·0			3
4153·97	99·8	0·2	99·8	0·2	73
	5·4	94·6			2·5

sources containing practically pure U^{238} and U^{235} isotopes, when light is passed through an atomic vapour from the sample being analysed. When the isotopic filtration method is used, the percentage of light from the source containing the sample being analysed which passes through an atomic vapour containing atoms of pure U^{238} isotope is determined.

The isotopic filtration method has the following advantages over the absorption measurement method:

1. The calibration graph is linear.
2. Not only metallic samples of uranium, but also compounds of uranium, can be analysed directly; this also applies to alloys of uranium and other metals, for instance the rods consisting of 20 per cent U + 80 per cent Th which are used in nuclear reactors.
3. There is greater freedom as regards the selection and variation of experimental conditions.
4. The measurement procedure is simpler.

According to measurements of the concentration in the same specimen (51·5 per cent U^{238}), the discrepancy between the results and the mean concentration is 0·1 per cent U^{238}. The mean discrepancy between the results of mass spectrometric and atomic absorption analysis of specimens of uranium containing 0·58–78·8 per cent U^{238} is about 2 per cent.

Isotopic analysis of hydrogen

The resonance lines for hydrogen are in the vacuum region of the spectrum, and absorption can therefore only be measured in excited gas if the lower level corresponding to the analysis line is sufficiently populated. Ostrovskaya observed absorption of the Balmer series of lines for hydrogen, starting with the ($n = 2$) level with an energy of 10·15 eV.[3] A method of isotopic filtration was used: the emission from a discharge emission lamp filled with the mixture of isotopes being analysed (protium and deuterium), modulated with a frequency of 1600 c/s, was passed through another discharge lamp filled with protium.

Power was supplied to the water-cooled emission lamp from a VG-2 high-frequency generator. The gas was excited in an 8 mm diameter Wood absorption lamp 50 cm long by means of a high voltage transformer ($V_{max} = 10\,000$ V). The pressure of the protium in the absorption lamp was 0·9 torr, and the current density was ~0·4 amp/cm².

In order to determine the isotopic composition, transmittance factors, T_1 and T_2 respectively, were found when emission from pure protium or from the mixture analysed was passed through the absorption lamp, i.e.

$$T_1 = \frac{I_H}{I_H^0} \qquad (6.4)$$

and

$$T_2 = \frac{T_1 I_H^0 + I_D^0}{I_H^0 + I_D^0} \qquad (6.5)$$

Since

$$\frac{I_D^0}{I_H^0} = \frac{C_D}{C_H} \qquad (6.6)$$

we get

$$\frac{C_D}{C_H} = \frac{T_2 - T_1}{1 - T_2} \qquad (6.7)$$

or

$$T_2 = C_D(1 - T_1) + T_1 \qquad (6.8)$$

With the measurement conditions indicated, the value of T_1 for the H_α (6562·73 Å) line was about 0·1. The slope of the calibration graph (T plotted against C_D) proved to be not 0·9, as would be expected from equation (6.8), but ~0·5. The reason for the smaller slope was probably that the light characteristic of the photomultiplier was not linear when it received the total luminous flux from the light source and the absorption lamp. The considerable scatter of results (15–30 per cent) is evidently also due to the photomultiplier being exposed to the emission from the absorption lamp.

Isotopic analysis of helium

Isotopic filtration was used by L'vov and Mosichev[10] for the isotopic analysis of mixtures of $He^3 + He^4$. Since the resonance lines for helium lie in the vacuum region of the spectrum, gas excitation was used; this made it possible to work using lines in the visible and infra-red regions of the spectrum.

Fig. 6.7 is a diagram of the apparatus. The mixture to be analysed was placed in a glass discharge lamp (1) with a ballast capacitance, and emission was created in the capillary tube by means of the high-frequency generator (3), which had a working frequency of 145 Mc/s; the helium pressure in the lamp was $\simeq 3$ torr. The emission isolated by the diaphragm (2) from a region (5 mm in length) of the capillary tube was passed through the 30 cm long, 20 mm diameter absorption lamp (4) by means of a two-lens system, and then reached the input slit of the

Specail Fields of Application for the Atomic Absorption Method of Measurement 269

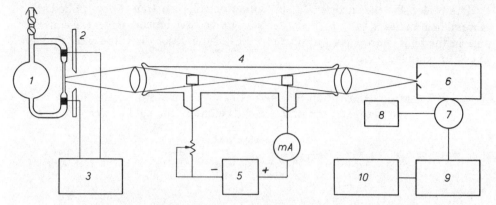

FIG. 6.7. Diagram of the apparatus for the isotopic analysis of helium.
1. Glass discharge lamp. 2. Diaphragm. 3. High-frequency generator.
4. Absorption lamp. 5. UIP-1 all-purpose power source. 6. ZMR-3
monochromator. 7. FEU-22 photomultiplier. 8. VS-22 rectifier. 9. Ul-2
electrometric d.c. amplifier. 10. EPP-09 potentiometer.

ZMR-3 monochromator (6). The absorption lamp was filled with pure helium He^4 (natural helium) to a pressure of 2 torr, and d.c. was supplied to it from an UIP-1 all-purpose power source (5). The electrodes in the lamp were made of tantalum foil, and were open-ended cylinders. A ballast resistance of up to 50 kΩ was included in series with the lamp, and the discharge was excited by means of a high-frequency pulse. The length of the absorbing volume (the positive column of the discharge) was about 160 mm. Emission was recorded with an FEU-22 photomultiplier (7) and an UI-2 electrometric d.c. amplifier (9), and was recorded by means of an EPP-09 potentiometer (10). Power was supplied to the photomultiplier from a VS-22 rectifier (8).

The intensity of emissions from the source lamp (I_e) and the absorption lamp (I_a) were measured, also the total intensity with both lamps emitting simultaneously (I_Σ). The absorption was calculated using the equation

$$A = 1 - \frac{I_\Sigma - I_a}{I_e} = 1 - \frac{I}{I_0} \qquad (6.9)$$

where I_0 and I are the intensities of the emission reaching the absorption lamp and passing through it.

The absorption was investigated for a number of lines. The results are tabulated in Table 6.7.

The group of three lines in the 10 830 Å ($2^3P_{0,1,2} - 2^3S$) region, ending at the metastable level 2^3S, proved to be the most convenient for analysis purposes. The structure of the line is shown in Fig. 6.8. The isotope shift of the lines (1·18 cm^{-1}) practically coincides with the multiplet splitting (1·07 cm^{-1}); the central component is therefore, as in the case of lithium, absorbed by both isotopes. In the case of pure He^3, the absorption is 0·12, and approximately corresponds to the specific gravity of this component (0·11).

The reason for the incomplete absorption of emission from a lamp filled with natural helium lies in the fact that self-absorption causes broadening of the emission line in the light source. As a result the wings of the absorption line pass through the absorption lamp unattenuated.

TABLE 6.7 Absorption of different helium lines with relation to the isotopic composition of the helium in the light source

Wavelength (Å)	Transition	Isotopic compositions (per cent)		Absorption lamp (He^4)	Absorption (per cent)
		Emission lamp He^3	He^4		
10830	$2^3S–2^3P$	100	—	100	12·0
		69·0	31·0	100	38·6
		—	100	100	92·5
3888·6	$2^3S–3^3P$	100	—	100	1·3
		69·0	31·0	100	24·6
		—	100	100	72·0
5015·7	$2^1S–3^1P$	100	—	100	2·8
		69·0	31·0	100	20·7
		—	100	100	58·0

The absorption remained practically constant when the current through the absorption lamp was altered over a wide range (between 2 and 20 mA). This made it possible to obtain highly reproducible results. With a current of about 5 mA used for further experiments, the absorption lamp emission did not exceed by more than 2–4 per cent that of the emission line.

Let us estimate the effects of self-absorption on the shape of the calibration curve in which A is plotted against C_{He}. Spektorov[11] used Milne's theory to obtain the following relationship of spectral line intensity I to the concentration N of atoms in a discharge with self-absorption taking place:

$$I = \frac{aN}{1+bN} \quad (6.10)$$

Here a and b are constants and do not depend on N. The value of b governs the degree of self-absorption in the discharge. With $b=0$, there is no self-absorption. Hence the intensity of the component He^4 in the emission lamp can be represented as follows:

$$I_{He^4} = \frac{aN_{He}\dfrac{N_{He^4}}{N_{He}}}{1+bN_{He}\dfrac{N_{He^4}}{N_{He}}} \quad (6.11)$$

Here $N_{He} = N_{He^4} + N_{He^3}$ corresponds to the total number of atoms of He. If we

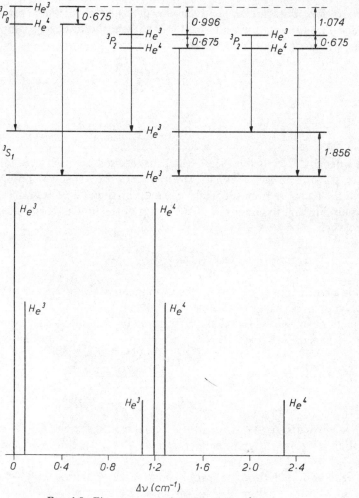

FIG. 6.8. Fine structure of the He 10 830 Å line.

replace N_{He^4}/N_{He} by C_{He^4}, the partial concentration of He⁴, and assume that if the pressure in the discharge is constant, N_{He} = constant, we get:

$$I_{He^4} = \frac{aN_{He}C_{He^4}}{1 + bN_{He}C_{He^4}} \tag{6.12}$$

Similarly we get the following equation for the intensity of the component He³ in the emission tube:

$$I_{He^3} = \frac{aN_{He}(1 - C_{He^4})}{1 + bN_{He}(1 - C_{He^4})} \tag{6.13}$$

It follows from equation (6.10) that aN_{He} corresponds to the total intensity of both the helium components when there is no self-absorption ($b=0$). We can therefore designate it I_{He}^0. The product bN_{He}, which represents the amount of self-absorption, does not depend on the partial concentrations of He³ and He⁴ (provided that the total pressure of the helium in the discharge lamp is constant). We shall designate this α. We thus get:

$$\left.\begin{aligned} I_{He^4} &= I^0_{He} \frac{C_{He^4}}{1 + \alpha C_{He^4}} \\ I_{He^3} &= I^0_{He} \frac{1 - C_{He^4}}{1 - \alpha(1 - C_{He^4})} \end{aligned}\right\} \quad (6.14)$$

The intensity I of the emission which has passed through the lamp consists of that part γ of the emission which corresponds to the non-overlapping components He³, and the emission from the wings of the components appertaining to the isotope He⁴. Assuming that the intensity of the wings of the line increases in proportion to $\sqrt{C_{He}}$, we get:

$$I = I^0_{He} \beta \sqrt{C_{He^4}} + I^0_{He} \frac{\gamma(1 - C_{He^4})}{1 + \alpha(1 - C_{He^4})} \quad (6.15)$$

Here β is a coefficient representing the proportion of the intensity of the component He⁴ applying to the wings of the line.

If equations (6.14) and (6.15) are substituted in equation (6.9), we finally get:

$$A = 1 - \frac{\beta \sqrt{C_{He^4}} + \dfrac{\gamma(1 - C_{He^4})}{1 + \alpha(1 - C_{He^4})}}{\dfrac{C_{He^4}}{1 + \alpha C_{He^4}} + \dfrac{1 - C_{He^4}}{1 + \alpha(1 - C_{He^4})}} \quad (6.16)$$

The coefficients α, β and γ can be found from experimental measurements of absorption for three different concentrations of He⁴ (including 0 and 100 per cent He⁴).

In Fig. 6.9 the experimental results (points) are compared with a curve calculated using equation (6.16) for the following values: $\alpha = 0.5$, $\beta = 0.05$, and $\gamma = 0.08$. Natural helium and the pure isotope He³ were used as the initial samples for making up the mixtures. With all the mixtures the pressure of the gas in the emission lamp was 3 torr. When the curve is examined, the following points are noted:

1. The serpentine form of the calibration curve is due to the self-absorption of emission in the light source. The shape of the curve is therefore largely governed by the discharge conditions in the emission lamp: the pressure of the gas and its purity, the dimensions of the capillary tube, and the discharge power. Obviously self-absorption in the light source can be eliminated, or at any rate reduced, by selecting different spectrum excitation conditions, but in view of the special features of the path followed by the curve (see point 2), self-absorption is conducive to making measurements more accurate. When small concentrations of He³ are determined, therefore, it is advisable to make use of the curvature of the curve, taking every possible care to ensure that spectrum excitation is the same in every case.

2. The slope of the calibration curve is close to 1:1 in the low ($<$10 per cent) and

high (>80 per cent) He^4 concentration regions. The relative errors in measuring concentration therefore correspond to the errors in measuring absorption.

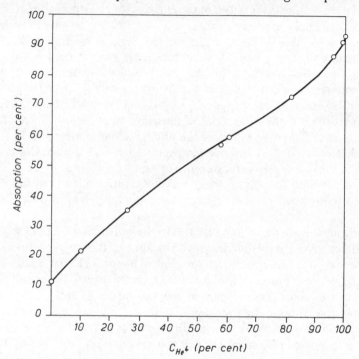

FIG. 6.9. Comparison of the results of measurements (points) with the calibrated curve for the isotopic analysis of helium.

3. In the region of medium concentrations of He^4 (40–70 per cent), the slope of the graph is very different from 1:1, and is about 0·67:1. The errors in measuring mean concentrations of He^4 are accordingly 50 per cent greater than the errors in measuring absorption.

The error indicating the scatter of measured absorptions from the mean absorption is about 0·2 per cent in absorbance units.

The isotopic analysis of boron

Goleb[48] attempted to analyse the isotopic composition of boron by the atomic absorption method. Hollow-cathode lamps cooled by means of running water were used, as they were when uranium was analysed.[29] The light sources contained the pure isotopes B^{10} and B^{11}. The emission lamps were filled with krypton at up to 1 torr, the absorption lamps with xenon at up to 0·5 torr. The length of the cathode in the absorption lamp was 40 mm, its diameter 12 mm. The current through the source lamp was 25 mA, the current through the absorption lamp 100 mA. In order to increase the light flux, both the B 2496 Å and B 2497 Å lines were used at once for measurement; these have the same isotopic shifts (the spectral slit width was 5 Å).

Under these measurement conditions, the absorption reached 70 per cent, but no differences were recorded between the amounts of light passing from the different single-isotope sources through the absorption lamp using the two different boron isotopes. This means that the width of the emission or absorption lines (and possibly of both) exceeded the isotopic shift. In my opinion the negative results of these experiments were due to the comparatively high temperature of the hollow cathodes, and to the fact that the best atomizing gas was not used in the absorption lamp. Actually, when the cathodes were water-cooled, the temperature of the gas within them was, owing to the poor thermal conductivity of boron and the high current density, no less than 400–500°C. In this case the coefficient β_{ik}, representing the superimposition of the wing of one of the components on the centre of the other, is 8–13 per cent. The other reason is the line broadening in xenon. It was proved on p. 49 that, when hollow-cathode lamps are filled with xenon, the line width is several times the line width of, for instance, neon.

It can be confidently asserted that, if cathodes are cooled with liquid nitrogen and the lamps filled with neon or argon, the isotopic analysis of boron will be possible.

It follows from the above review that the isotopic filtration method has considerable advantages over the method in which absorption is directly measured. Actually, in the latter case it is necessary strictly to control not only the nature of the emission in the light source, but also the conditions under which the atomic vapour is produced in the absorbing volume. When mercury was analysed, this involved maintaining the volume thermostatically at the correct temperature with an accuracy of about 0·1°C, while when lithium was analysed the emission of lithium vapour in a flame or in a hollow cathode was measured. The necessity for making additional measurements complicates the analysis procedure and apparatus, and also introduces additional errors into the results of determinations.

When the isotopic filtration method is used, the variation in the conditions under which the atomic vapour is produced (in an iron tube in the case of lithium, in a hollow cathode for uranium, and in a Geissler tube for helium) is not important, or is at least less important. This is quite natural, since the variations in the conditions under which the atomic vapour is produced amount, in this case, to a change in the resolving power of the 'isotopic filter'. These changes become more perceptible when mixtures of isotopes including a small amount of one of the isotopes are analysed.

Earlier we considered cases in which atomic absorption was used for determining the isotopic compositions of elements from resonance lines with appreciable isotopic shifts. The possibilities of atomic absorption analysis are not, however, confined to this. In many cases isotopic analysis can be performed for the components of the hyperfine structure of a line (see p. 15). Table 6.8 gives figures for the hyperfine resonance line shift of the odd numbered isotopes of certain elements, the concentrations of which mixed with even numbered isotopes can be determined by atomic absorption.

The possibility of isolating the separate components of a line, appertaining to one particular isotope, by means of isotopic filtration was used for the optical pumping of the $F=2\ ^2S_{1/2}$ sub-level—the Rb^{87} state.[49] For this purpose emission of the Rb^{87}

7800 Å resonance line was first passed through an atomic vapour containing Rb^{85}. The line component stimulating transitions from the $F=2$ sub-level was practically fully absorbed. High-frequency spherical lamps were used as the emission source, and the temperature of the atomic vapour was 61–69°C. Naturally this method can also be used for the isotopic analysis of a mixture of Rb^{85} and Rb^{87}.

The other possibility of performing isotopic analysis consists of using the resonance lines of ions, for which as a rule the isotopic shift is greater than for the resonance lines of atoms. The principal experimental difficulty here is that of producing beams of ions with a particle density sufficient for measurement purposes.

Finally, when isotopic compositions are analysed by the isotopic filtration method, the gap between the two Zeeman σ-components of a highly saturated absorption line can be used. It has been proved that a transmission band of 0·011 cm^{-1} for these filters can be obtained for the Hg^{198} 2537 Å line.[50] In another published work,[51] a transmission band of about 0·007 cm^{-1} was obtained for a filter for the same line. Emission corresponding solely to the isotope in which we are interested can be separated from the total emission from a two-isotope mixture by means of a Zeeman filter. The components of the second isotope are absorbed by the wings of the absorption line.

The atomic absorption method therefore has extremely good prospects for isotopic analysis, and even with the technical equipment at present available this method can be used for a comparatively large number of elements.

MEASURING THE CONCENTRATION OF NEUTRAL ATOMS IN THE GAS PHASE

The atomic absorption method was used by Cole, Hall, Kemp and Sellen for determining the concentration of neutral atoms of mercury and cesium in plasma beams produced in an ionic rocket engine.[12,13]

Ladenburg's linear absorption method (p. 21) was used for making the measurements. Fig. 6.10 is a diagram of the experimental apparatus. In both cases the source of light was a high-frequency spherical lamp. The beam of light, modulated by means of a rotating disk, passed successively through the vacuum chamber containing the beam of atoms, a quartz cell (used for calibrating the readings) and a filter isolating the Hg 2537 Å or Cs 8521 Å resonance line, and was then recorded by means of a photomultiplier and an oscillograph.

The experiments with mercury were performed in an experimental ion accelerator. The beam of light was passed, by means of a system of mirrors, through the engine exhaust plasma. To prevent mercury vapour from settling on the surfaces of the mirrors, the latter were protected by seals which only opened when signals were being measured.

The beam of caesium atoms simulating the jet from the ion engine was produced by means of a heated copper vessel covered with an 0·08 mm thick stainless steel grid with 706 apertures, their diameter being 0·25 mm. Traps cooled with liquid air were used for removing the cesium vapour.

The 3 cm long thermostatically controlled quartz cell, with plane windows and a tube containing metallic mercury or caesium, made it possible to produce a column of vapour in which the concentration of atoms was known; this concentration was calculated using the equations for saturated vapour pressure.

The absorptions measured with the cell in the experiments enabled it to be established that the value of α, representing the relationship of emission line half-width to absorption line half-width, is equal to 2 for both metals. The theoretical calibration curves calculated for $\alpha = 2$ and allowing for the hyperfine structure of the lines were confirmed by the experiments. The fuel required by the engine can be calculated on the basis of spectroscopic measurements of the concentration of mercury in the beam. The results of the measurements agree with those of determinations by weight.

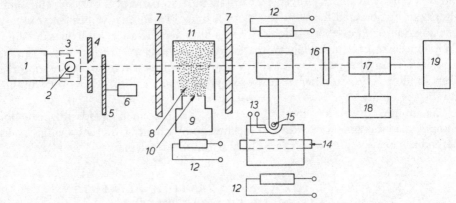

FIG. 6.10. Diagram of the apparatus for measuring concentrations of neutral atoms in beams of ions.[12] 1. High-frequency generator. 2. Spherical lamp. 3. Protective casing. 4. Diaphram. 5. Disk. 6. Synchronous motor. 7. Vacuum chamber. 8. Beam of atoms. 9. Effusion cell. 10. Cell apertures. 11. Cooled trap. 12. Heating element. 13. Thermocouple. 14. Heat elimination channel cooled with liquid nitrogen. 15. Tube containing droplets of metal. 16. Filter. 17. Photomultiplier. 18. Power supply unit. 19. Oscillograph.

The caesium atom concentrations measured in the beam were compared with the results of calculating beam density on the basis of the effusion model for the flow of vapour from the apertures in the chamber. The calculated results proved to be 40 per cent higher than the results of the atomic absorption measurements; the authors considered that this was due to the approximate nature of the effusion model of gas flow.

Absorption of the measured signal amounting to 10 per cent corresponds to a product Nl equal to $2 \cdot 10^{10}$ atoms/cm² for cesium and $1 \cdot 5 \cdot 10^{12}$ atoms/cm² for mercury. Sensitivity can be increased by passing the beam of light a number of times through the beam of atoms.

The possible sources of error in the measurements are: the formation of deposits of metal on the chamber windows through which the light passes, and the interference of emission from the atoms in the beam.

The atomic absorption method is also effective for measuring the concentration

TABLE 6.8 Hyperfine shift of the resonance lines of the odd numbered isotopes

Isotope being determined	Spin moment	Isotopes with zero spin	Wavelength (Å)	Lower level energy (cm⁻¹)	Oscillator strength	$\Delta\nu_{hfs}^2$ (cm⁻¹)	$\Delta\nu_D$ (at 190°K) (cm⁻¹)	β_{ik} (per cent)
Zn^{67}	5/2	64, 66, 68	3075·9	0	1·7 . 10⁻⁴	0·064	0·039	6·8 . 10⁻²
Kr^{83}	9/2	78, 80, 82, 84, 86	8059·5	85192·4	NK	0·106	0·013	0
Rb^{85}	5/2	NK	7800·2	0	0·80	0·083	0·014	0
Rb^{87}	3/2	NK						
Sr^{87}	9/2	84, 86, 88	6892·6	0	0·0014	0·079	0·015	0
Sn^{115}	1/2	116, 118	2863·3	0	0·23	0·178	0·032	0
Sn^{117}	1/2	120, 122				0·181	0·032	0
Sn^{118}	1/2	124				0·181	0·032	0
Xe^{129}	1/2	124, 126	8206·3	76197·3	NK	0·148	0·011	0
Xe^{131}	3/2	128, 130, 132, 134, 136				0·109	0·010	0
Os^{189}	1/2	184, 186, 187, 188, 190, 192	4260·9	0	0·014	0·207	0·017	0
Pt^{195}	1/2	190, 192, 194, 196, 198	2830·3	0	0·04	0·398	0·025	0
Bi^{209}	9/2	208, 210	3067·7	0	0·077	0·588	0·022	0
Po^{209}	—	208	4611·4	0	NK	0·157	0·015	0

Note: NK denotes not known.

of neutral atoms when studying the thermodynamic characteristics of substances, in particular the saturated vapour pressure of elements. It must be noted that, in spite of the large number of methods for determining vapour pressures and the heats of evaporation of chemical elements, the available data are not sufficiently accurate, and in many cases no data are available at all. According to Nesmeyanov,[15] who estimated on the basis of the generalization and critical consideration of all data published up to 1961, reliable values of vapour pressure had only been obtained for twenty of the 70–75 most frequently encountered elements. The data for eighteen elements were known to be unreliable, and absolutely no details of the vapour pressures of twenty elements, including elements in the platinum group and rare earth elements, had been obtained.

One of the first projects on the use of atomic absorption spectroscopy for determining the density of atomic vapour was the research carried out by Füchtbauer and Bartels in 1921.[30] In their paper the authors described a procedure for determining the concentration of caesium vapours from the integral absorption of resonance lines; the concentration was measured photographically by means of a continuous spectrum source. To make recording more convenient, the absorption lines were broadened by means of a foreign gas (nitrogen at 4 atm).

In spite of the relative simplicity of the atomic absorption method of measuring concentrations, up to now it has not been used on any extensive scale, owing to the absence of data on absolute oscillator strengths which are necessary for calculating absolute concentrations of atoms in a gaseous phase.

The position has now greatly improved. Absolute values of oscillator strengths have been measured, for many elements, accurately enough using methods which do not depend on the saturated vapour pressures of the elements. In the case of most other elements they can be measured comparatively easily (see below). More intensive use can therefore now be made of atomic absorption spectroscopy for investigating the physical and chemical characteristics of substances.

In this connection attention should be drawn to Vidale's research cycle;[31] he determined the saturated vapour pressure of sodium and that of sodium vapour above glass, and also measured the concentration of silicon vapour above silicon carbide. A modified King furnace, with the diffusion of vapour restricted by an inert gas, was used in these experiments.

MEASUREMENT OF ABSOLUTE OSCILLATOR STRENGTHS

Most of the experimental methods of determining absolute oscillator strengths are based on measuring the product of the oscillator strength f and the effective length Nl of the absorbing volume. To find f, we must know Nl; this is the main limitation of the use of absorption methods for absolute measurements.

Sealed quartz cells are not suitable for obtaining the atomic vapour of elements with low volatility. The column of atomic vapour in a King furnace cannot be accurately calculated owing to the uneven temperature distribution in the furnace and the movement of vapour towards the open apertures. Research by Ostrovskii, Parchevskii and Penkin has proved that the effective dimensions of the column of vapour in a furnace differ from the calculated dimensions (for a homogeneous column) by (6–10):1.[14] In addition to this, if the equations for the saturated vapour pressures of elements are used for calculating the concentration of atoms in a gas phase, the results of measuring oscillator strengths inevitably include the determination errors involved in using the vapour pressure equations. With many elements, measurement of their saturated vapour pressure is not sufficiently reliable, or just is not done.[15]

The effective length of a flame can only be approximately estimated quantitatively, owing to the gradient temperature and to the variation in the reducing conditions of flames from the centre to the edges.

In view of this, the method of atomizing samples in a graphite cuvette is of considerable interest; with this method an atomic vapour with a known effective length is produced. I used this method to obtain experimental measurements of absolute oscillator strengths for resonance lines.[16,17]

The combined method of measuring the total and line absorption, described on p. 20, was used for determining Nlf. We recall that, according to (1.43)

$$Nlf = \frac{mc}{\pi e^2} \frac{\Delta \nu}{k\phi(A_s)} \left(1 - \frac{I}{I_0}\right) \quad (6.17)$$

where $\phi(A_s)$ depends on the shape of the absorption line.

The differences between the magnitudes of $\phi(A_s)$ for extreme cases in which the line profile is due solely to the Doppler, or solely to the Lorentz, effects, increase as the absorbance A_s increases (see Fig. 1.5). For instance, with $A_s = 1\cdot 0$ this difference is 6 per cent, while with $A_s = 4\cdot 0$ it is 50 per cent.

Allowing for the contribution of both broadening factors to the true line profile, we attempted to confine ourselves to the smallest absorbances A_s with which $(I_0 - I)/I_0$ could be sufficiently accurately measured; we also selected experimental conditions under which the shape of the line profile was closest to being a Lorentz shape.

These requirements are not, in general, compulsory for using the method, since the magnitude of $\phi(A_s)$ can be calculated in advance for any mixed line profile. On the other hand, the contribution of both broadening factors to the total line width can be assessed on the basis of the calculated Doppler half-width and the total line half-width measured by the same combined method (p. 282).

The final equation, obtained from (6.17) and (5.15), for calculating oscillator strengths, takes the form:

$$f = 0\cdot 188 \cdot 10^{-11} \frac{AS\,\Delta \nu}{M\phi(A_s)k} (1 - I/I_0) \quad (6.18)$$

The experimental apparatus is absolutely the same as the apparatus used for analysis and described in detail on p. 206.

The procedure for determining f consists of the successive measurement of peak values of absorbance A_s from a line spectrum source, and of total absorption $1 - (I/I_0)$ from a continuous spectrum source, under identical experimental conditions, except that the line absorption is measured with 3–5 times smaller quantities of the element introduced into the cuvette than the amounts M introduced for measuring total absorption. The absorbance values found for line absorption are extrapolated to the quantities M. As an instance, $A_s = 1\cdot30$ for $5\cdot 10^{-9}$ g of lead (see Table 6.9) was found by extrapolating the measured absorbance of $0\cdot26$ for $1\cdot 10^{-9}$ g of lead. This method makes it possible to eliminate the effect of the spectrum source not being monochromatic, which is appreciable when high absorbances are measured (see Table 1.9).

The values of $1-(I/I_0)$, the total absorption, were calculated from measured values of $\log(I_0/I)$, since it is easier and more accurate to record small values of absorption on a logarithmic than on a linear scale.

Table 6.9 contains a full summary of the conditions and the results of the measurements made by L'vov in collaboration with Lebedev. The values of the function $\phi(A_s)$ for all the elements except for beryllium were read from the curve for purely Lorentz profiles, since the Doppler half-width of the lines measured was considerably smaller than the Lorentz half-width (p. 283). Even with an argon pressure of 10 atm, in the case of the Be 2349 Å line both effects make about the same contribution to line width. For the calculations, therefore, we used the mean value of $\phi(A_s)$ from both curves.

The mean coefficient of variation representing the reproducibility of the measurements was 10 per cent. Taking the mean of several tens of measurements, the random error in the final values of f did not exceed 2–3 per cent.

The last column of Table 6.9 gives oscillator strengths found by other research workers; these are the figures which, in our opinion, are most reliable. Most of the results were obtained by the Rozhdestvenskii 'hooks' method, or by measuring the total absorption in a beam of atoms. The exception to this is the value of f for the Be 2349 Å line, which was measured by the method of emission in an arc.[52] Besides this single experimental value, we only know of a number of calculated values. According to one recent work, the calculated value of f(Be 2349 Å) is $1\cdot59$.[53]

When the measured and known values of f are compared, we find that on the whole the method of determining absolute oscillator strengths described above provides results which agree with the most reliable values. At the same time, one must note that there is a tendency for the results to be consistently too low; this is particularly appreciable in the case of elements which were measured at higher temperatures (Ga, Ag, Sn, Mn, Fe, Cu and Be). The most likely reason for this is that a proportion of the element introduced is lost by the diffusion of vapour through the electrode head (p. 207).

The confirmation, obtained in this section, of the correctness of the method of measuring absolute oscillator strengths not only provides hope that it can be effectively used for other elements, which have not yet been investigated, but also provides proof of the argument that when the theory of the processes and phenomena on which the method is based have been sufficiently developed, it will be possible to arrive at an *a priori* coupling equation in each particular case of analysis.[27] This point of view is diametrically opposed to the assertion that all variations of spectrochemical analysis involve the use of standards.[28] Up to now many analysts have maintained that this last point of view is correct, and it is one of the principal

TABLE 6.9 Results of measuring the absolute oscillator strengths of resonance lines by the combined method in a graphite cuvette

Line (Å)	Transition	T (°K)	P (atm)	S (cm^2)	M (g)	$\Delta\nu$ (cm^{-1})	$1-(I/I_0)$	k	A_s	$\phi(A_s)$	f_{meas}	f_{publ}	
Zn 3076	$4\,^1S_0 - 4\,^3P_1$	1870	6	0·330	5·10^{-7}	14·8	0·048	0·98	1·67	0·446	1·3·10^{-4}	1·7·10^{-4}	(1)
Pb 2833	$6\,^3P_0 - 7\,^3P_1$	1840	5	0·365	5·10^{-9}	21·8	0·143	0·95	1·30	0·498	0·19	0·21	(18)
Cd 2288	$5\,^1S_0 - 5\,^1P_1$	1870	3	0·365	1·10^{-10}	28·1	0·032	0·99	1·03	0·550	1·3	1·2	(1)
Bi 3068	$6\,^4S_{3/2} - 7\,^4P_{1/2}$	1910	3	0·132	2·10^{-9}	19·5	0·088	0·97	0·84	0·593	0·077		
Sb 2311	$5\,^4S_{3/2} - 6\,^4P_{1/2}$	2130	3	0·365	5·10^{-9}	28·3	0·045	0·99	1·25	0·507	0·042		
Te 2259	$5\,^3P_2 - 6\,^5S_2$	1900	3	0·365	1·10^{-7}	19·0	0·051	0·98	1·40	0·482	1·8·10^{-3}		
Ga 2874	$4\,^2P_{1/2} - 4\,^2D_{3/2}$	2370	6	0·160	1·1·10^{-9}	28·6	0·078	0·97	1·01	0·551	0·19	0·23	(54)
In 3039	$5\,^2P_0 - 6\,^2D_{3/2}$	1870	6	0·160	2·1·10^{-9}	25·4	0·155	0·96	2·86	0·347	0·27	0·28	(54)
Tl 2768	$6\,^2P_{1/2} - 6\,^2D_{3/2}$	2000	6	0·146	2·1·10^{-9}	30·8	0·133	0·96	2·08	0·402	0·29	0·27	(55)
Ag 3281	$5\,^2S_{1/2} - 5\,^2P_{3/2}$	2350	10	0·162	4·10^{-10}	21·6	0·0868	0·98	1·57	0·457	0·34	0·45	(54)
Sn 2863	$5\,^3P_0 - 6\,^3P_1^0$	2700	6	0·162	5·3·10^{-9}	28·7	0·0749	0·97	1·96	0·414	0·11	0·19	(54)
Mn 2795	$a\,^6S_{5/2} - y\,^6P_{7/2}^0$	2400	6	0·160	1·6·10^{-10}	30·3	0·053	0·98	2·97	0·341	0·50	0·58	(56)
Fe 2967	$a\,^5D_4 - ^5F_5$	2900	4	0·122	2·5·10^{-9}	27·2	0·049	0·98	1·06	0·541	0·030	0·035	(25,26)*
Cu 3248	$4\,^2S_{1/2} - Y\,^2P_{3/2}^0$	2600	4	0·220	5·10^{-10}	22·6	0·068	0·98	1·17	0·522	0·16	0·31	(25)
Cu 3274	$4\,^2S_{1/2} - 4\,^2P_{1/2}^0$	2600	4	0·135	1·10^{-9}	22·3	0·096	0·97	1·76	0·435	0·082	0·16	(25)
Be 2349	$2\,^1S_0 - 2\,^1P_1^0$	2900	8	0·160	5·3·10^{-11}	43·1	0·0749	0·97	3·36	0·275†	0·62	0·24	(52)

* The value of f_{publ} for the Fe 2967 Å line was calculated using the equation $f(\text{Fe 2967 Å}) = 1 \cdot 1 \, 0 \, f(\text{Fe 3720 Å})$ taken from reference 26.
† The arithmetical mean of the values of $\phi(A_s)$ corresponding to the Doppler and dispersion contours.

THE LORENTZ WIDTH OF RESONANCE LINES

The Lorentz widths of the spectral lines in atomic spectra are determined either by direct methods involving the direct measurement of line contours, or by indirect methods based on measuring some quantity which is theoretically related to the width. In many cases indirect methods are less complicated as regards the apparatus (in particular there is no necessity for instruments with high resolving powers) and less laborious; in addition they enable the validity of the theoretical assumptions on which the methods are based to be assessed.

Of the indirect methods of determining the collision width of lines, the method based on plotting 'growth curves' has been most developed.[57-59] When a growth curve (the line intensity plotted against the concentration of atoms in a source of light) is plotted in absolute units, the ordinate of the point at which asymptotes produced to regions of the curve for small and large absorbance intersect, is linked simply with the quantity

$$a = \frac{\Delta v_L}{\Delta v_D} \sqrt{\ln 2} \qquad (6.19)$$

With this method, the source of radiation is a flame. The set of lines investigated is therefore confined to the visible and neighbouring ultra-violet (up to 3000 Å) region of the spectrum. The amount of the foreign gas being investigated in the products of combustion (for instance an inert gas), added to dilute the combustible gases, does not exceed 70–75 per cent, and the results of measurement can therefore only apply approximately to the pure gas. If the measured intensities are calibrated in absolute units, this involves errors which Behmenburg and Kohn[59] assessed at about 10 per cent. The variations in temperature in the flame, disturbances due to emissions from neighbouring lines, and the recognized arbitrary nature of producing asymptotes, introduce additional errors into the measurement results.

In view of these limitations in applying the 'growth curves' method, the possibility of determining line width by the combined method of measuring the total and line absorptions, which was used earlier (p. 279) for measuring Nlf, is interesting.

Although, with this method, there is no obligation to produce atomic vapour by any particular means, we shall now discuss the use of the method as applied solely to graphite cuvettes. By comparison with flames, the method by which an atomic vapour is produced in a cuvette opens up good prospects as regards variations in the pressure of the foreign gas, the temperature, and the great homogeneity of the atomic vapour.

Since, when measurements are actually made, not only collision broadening, but also the Doppler effect contribute to line width, we shall assess the effects of each of these factors on the results of measurement. In the case of a line broadened by the Doppler effect alone, according to equations (1.12) and (1.4) the absorbance at the centre of the line can be expressed as follows:

$$A_D = \log e \; \frac{2e^2 \sqrt{(\pi \ln 2)}}{mc} \cdot \frac{Nlf}{\Delta \nu_D} \tag{6.20}$$

In the case of a line broadened by the foreign gas alone, the absorbance in the region displaced by $\Delta \nu_s$ from the centre will, according to (1.45) and (1.4), be

$$A_L = \log e \; \frac{2e^2}{mc} \cdot \frac{\Delta \nu_L^2}{\Delta \nu_L^2 + 4\Delta \nu_s^2} \cdot \frac{Nlf}{\Delta \nu_L} \tag{6.21}$$

Taking (1.18) into consideration,

$$A_L = \log e \; \frac{2e^2}{mc} \; 0.658 \; \frac{Nlf}{\Delta \nu_L} \tag{6.22}$$

According to (1.43), the total absorption is

$$1 - \frac{I}{I_0} = \frac{\pi e^2}{mc} \; k\phi(A) \frac{Nlf}{\Delta \nu} \tag{6.23}$$

where as before the form of $\phi(A)$ will depend on the nature of line broadening (Fig. 1.6).

If we eliminate Nlf from equations (6.21)–(6.23) and (6.22)–(6.23), for a line with a purely Doppler profile we get:

$$\Delta \nu_D = \frac{2 \log e \sqrt{\ln 2}}{\sqrt{\pi}} \; \frac{\Delta \nu}{k\phi(A_D)} \; \frac{1 - I/I_0}{A_D} \tag{6.24}$$

and for a line with a purely Lorentz profile we get:

$$\Delta \nu_L = \frac{2 \log e \cdot 0.658}{\pi} \; \frac{\Delta \nu}{k\phi(A_L)} \; \frac{1 - I/I_0}{A_L} \tag{6.25}$$

When we compare equations (6.24) and (6.25), we find that, if $\Delta \nu_D$ and $\Delta \nu_L$ are equal, the ratio of total absorption to line absorption is 3–4 times less for the Doppler profile than for the Lorentz profile (we assume that, with $A > 4.0$, the ratio $\phi(A_L)/\phi(A_D) > 1.5$, and that the numerical coefficients in the equations are respectively equal to 0.408 and 0.182). The quantity $(1 - I/I_0)/A$, measured in the experiment for actual lines under conditions in which $\Delta \nu_L \geqslant \Delta \nu_D$, is governed practically solely by the quantity $\Delta \nu_L$.

The above arguments were confirmed by experimental research carried out by L'vov and Lebedev.[60,61] The accurate measurements were made in the same way as when absolute oscillator strengths were determined (p. 279). The conditions and results of the experiments are given in Table 6.10. The penultimate column contains values of $\Delta \nu_L$ reduced to a foreign gas pressure of 1 atm without allowing for the relationship of line half-width to temperature and hyperfine structure. Calculated values of $\Delta \nu_D$ are also given in Table 6.10 for purposes of comparison.

When we consider the results of the measurements, the following conclusions can be drawn:

1. Doppler broadening only has a slight effect on the results of determinations made under the assumption that the line contour is due solely to the collision effect. Actually the results of measuring $\Delta\nu_L$ for the Ag 3281 Å line with pressures of 6 and 10 atm only differ by 7 per cent. In some cases (with the exception of beryllium), the Doppler effect has even less influence.

2. In the case of the Be 2349 Å line, the Doppler half-width is much greater than the Lorentz half-width (at 3 atm). The Doppler half-width found using equation (6.24) is only 50 per cent greater than the calculated value. Since the Lorentz effect has little influence on the measured width, and the emission line in an uncooled hollow cathode is not monochromatic, the results can be considered to agree sufficiently well to confirm that the method is correct.

3. Line half-width is slightly less in nitrogen than in argon.

The results of measuring Lorentz half-widths by different methods are compared in Table 6.11. Comparison of the figures shows that the results of measuring by the combined method agree well with those of direct research into absorption line profiles,[62,63] conducted with high argon pressures, and also with the results of measuring $\Delta\nu_L$ by the indirect method,[64] by simultaneously studying total absorption and the anomalous dispersion close to absorption lines (the Rozhdestvenskii 'hooks' method).

The results of measuring $\Delta\nu_L$ by the 'growth curves' method, by Hinnov and Kohn,[58] deviate considerably from the results obtained by other methods, particularly for silver and thallium. This is obviously due to errors in the measurements made by Hinnov and Kohn. The determination of $\Delta\nu_L$ by the combined method is more correct than when f is determined (p. 279), since the results obtained do not depend on consistent errors in measuring the cross-sectional areas of cuvettes or the amounts of elements introduced into the cuvettes. When the mean value for several tens of independent measurements is taken, the standard deviation for the final result is not more than 1 per cent.

MEASURING COEFFICIENTS OF DIFFUSION OF ATOMS

Measurement of the coefficients of diffusion of atoms in gaseous media at high temperatures is, from the experimental point of view, a comparatively complex matter. This explains why, up to now, no accurate data have been available regarding the coefficients of diffusion of atoms, except for mercury, the diffusion of which has been studied at low temperatures. It is true that a number of works have recently been published in which the coefficients of diffusion of certain metals have been measured in arc discharge plasma.[19,20] The arc model used for the calculation has, however, been based on a number of assumptions, and these require better backing. One particular theory giving rise to doubts is the theory that radial electric fields and convection do not affect the process by which vapour 'leaks' by diffusion from the arc discharge zone.

Experimental research into the relationship of coefficients of diffusion to tem-

TABLE 6.10 Conditions for measuring line half-widths, and the results

Lines (Å)	Transitions	T (°K)	$\Delta\nu_D$ (cm^{-1})	P (atm)	$\Delta\nu_L$ (cm^{-1}) Argon	$\Delta\nu_L$ (cm^{-1}) Nitrogen	$\Delta\nu_L/P$ Argon	$\Delta\nu_L/P$ Nitrogen
Tl 2768	$6\ ^2P^o_{1/2}$–$6\ ^2D_{3/2}$	2000	0·081	6	0·935	0·865	0·156	0·144
Tl 3776	$6\ ^2P^o_{1/2}$–$7\ ^2S_{1/2}$	2000	0·059	6	0·935	0·850	0·156	0·142
In 3039	$5\ ^2P^o_{1/2}$–$5\ ^2D_{3/2}$	1900	0·096	6	0·796	0·790	0·133	0·132
Ga 2874	$4\ ^2P^o_{1/2}$–$4\ ^2D_{3/2}$	2400	0·146	6	0·764		0·128	
Al 3093	$3\ ^2P^o_{3/2}$–$3\ ^2D_{5/2}$	2800	0·236	6	0·575	0·531	0·0968	0·0885
Ag 3281	$5\ ^2S_{1/2}$–$5\ ^2P^o_{3/2}$	2370	0·102	6	0·312		0·052	
				10	0·485		0·0485	
Bi 3068	$6\ ^4S^o_{3/2}$–$7\ ^4P_{1/2}$	1910	0·071	3	0·648		0·216*	
Cu 3248	$7\ ^2S_{1/2}$–$4\ ^2P^o_{3/2}$	2600	0·141	4	0·474		0·118	
Fe 2967	$a\ ^5D_4$–$^5F_5$	2900	0·174	4	0·436		0·109	
Pb 2833	$6\ ^3P_0$–$7\ ^3P^o_1$	1840	0·075	5	0·925		0·185	
Sb 2311	$5\ ^4S_{3/2}$–$6\ ^4P_{1/2}$	2130	0·129	3	0·370		0·123	
Sn 2863	$5\ ^3P_0$–$5\ ^3P^o_1$	2700	0·119	6	0·496		0·0827	
Be 2349	$2\ ^1S_0$–$2\ ^1P^o_1$	2900	0·446	3	0·823†			

* Assuming h.f.s. (see Fig. 2.9) $\Delta\nu_L/P = 0\cdot108$ cm^{-1}/atm.
† Result of calculating $\Delta\nu_D$ using equation (6.24).

TABLE 6.11 Results of measuring the Lorentz half-width, calculated per unit of relative absorbance of gas (0°C, 1 atm)

Lines (Å)	Direct method[62–63]			'Growth curves' method[58]			'Hooks' and total absorption method[64]			Combined method		
	Gas	T (°K)	$\Delta\nu_L$ (cm^{-1})	Gas	T (°K)	$\Delta\nu_L$ (cm^{-1})	Gas	T (°K)	$\Delta\nu_L$ (cm^{-1})	Gas	T (°K)	$\Delta\nu_L$ (cm^{-1})
Ag 3281	Ar	1200	0·47	N$_2$	2500	1·2				Ar	2370	0·42
Cu 3248				N$_2$	2500	0·73				Ar	2600	1·12
Fe 3720				N$_2$	2500	0·73						
Fe 2967										Ar	2900	1·16
Ga 4033							Ar	1400	0·87			
Ga 2874										Ar	2400	1·12
In 4101	Ar	1073	1·2				Ar	1300	0·87	Ar	1900	0·93
In 3039										Ar	1840	1·25
Pb 2833	Ar	1220	1·2							Ar	2000	1·14
Tl 3766				N$_2$	2500	2·2						

Note: Where there are blanks the results have not been determined.

perature has also not been published. Theoretical considerations enable us to state that the exponential index describing this relationship must be between 1·5 and 2·0.

In view of this, I proposed that the method of atomizing samples in a cuvette should be used for measuring coefficients of diffusion and investigating the relationship of diffusion to experimental conditions.[21] Actually, as was pointed out on p. 204, under particular experimental conditions diffusion alone causes vapour to be removed from the cuvette.

If the diffusion of vapour through the porous walls of a cuvette is restricted by lining the walls, the use of long, narrow cuvettes and uniform heating make the calculation of the amount of vapour lost from a cuvette an extremely simple diffusion problem, namely that of finding the vapour diffusion in a tube from the centre to open apertures in whose planes the vapour concentration can be assumed to be zero as a result of the condensation of the vapour on the cold surfaces of the cooler, and of convection currents close to the apertures in the heated cuvette.

It was proved earlier (p. 205) that, assuming that there is a linear drop in the density of the vapour from the centre of a cuvette towards its edges, the reduction in the amount of vapour of an element in a cuvette can be described by the equation

$$M_\tau = M_0 e^{-(8D/l^2)\tau} \tag{6.26}$$

where M_0 is the mass of vapour at the initial moment of time, D is the coefficient of diffusion, and l is the length of the cuvette.

The above remarks were developed by Nikolaev.[22] It is interesting to note that the attempt at using the classical problem[23] of the diffusion of a substance from a vapour with a thickness l (the concentration of the substance which is constant at the initial moment of time) to depict the process by which vapour diffuses from a cuvette leads to the equation

$$M_\tau = 0.81 \, M_0 e^{-(\pi^2 D/l^2)\tau} \tag{6.27}$$

which is very nearly the same as equation (6.26). Nevertheless the theory that vapour is uniformly concentrated in a cuvette at the moment when diffusion begins is less justifiable than assuming that there is a linear decrease in the density of the vapour towards the edges. Equation (6.26) was therefore used.

Since, with absorbance A of up to 0·6–0·8, A is proportional to M, equation (6.27) can be rewritten as follows:

$$A_t = A_0 e^{-(8D/l^2)\tau} \tag{6.28}$$

The time constant τ_c characterizing the exponential law for decrease in the amount of vapour in a cuvette is therefore:

$$\tau_c = l^2/8D \tag{6.29}$$

If the angular coefficients of graphs plotted for $\log(A_0/A_t)$ against τ are measured from experimental data, we can find the value of τ_c; using equation (6.29) we can then find the coefficient of diffusion D.

The procedure described above was used for determining the coefficients of diffusion of zinc atoms in argon and helium at different temperatures and pressures. The apparatus used for investigating diffusion differed from the normal type of apparatus intended for analysis in that the signals were recorded with practically no time-lag using a bifilar MPO-2 oscillograph. Examples of the oscillograms obtained were given earlier in Figs. 5.8 and 5.10. Fig. 6.11 shows the variation in the absorbance of the Zn 3076 Å line, calculated from the oscillograms, plotted against time for different foreign gas pressures. The graph shows that this variation actually follows an exponential law. Results indicating that the coefficient of diffusion of zinc atoms is inversely proportional to the pressure of the argon are plotted in Fig. 6.12. These experiments confirm that the departure of samples from cuvettes is of a purely diffusional nature.

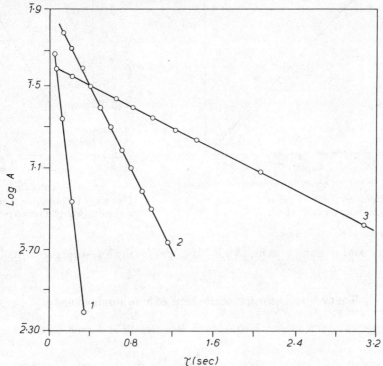

FIG. 6.11. Variation in the absorbance resulting from the diffusion of atoms from a cuvette.[2] 1. 0·22 atm. 2. 1 atm. 3. 4 atm.

Experimental research into the relationship of coefficients of diffusion to temperature (Fig. 6.13) has established that, at temperatures between 1100 and 2600°K, the exponent n in the equation

$$D = D_0 \left(\frac{T}{T_0}\right)^n \tag{6.30}$$

is constant, and is equal to 1·6.

Approximate values of the coefficients of diffusion D_2 are given in Table 6.12, together with experimental values of the coefficients of diffusion D_1; the approximate values of D_2 were calculated using the following equation, given by Arnold:[24]

$$D_2 = \frac{0.00837 \sqrt{(1/M_1 + 1/M_2)}}{(V_1^{1/3} + V_2^{1/3})^2} \cdot \frac{T^{5/2}}{T+C} \qquad (6.31)$$

Here M_1 and M_2 are the molecular weights of the diffusing gases, V_1 and V_2 are the molar volumes of the liquids at boiling point, and C is Sutherland's constant for gas vapour.

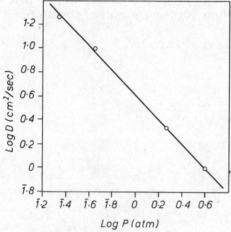

FIG. 6.12. Diffusion coefficient of zinc atoms in argon versus argon pressure.

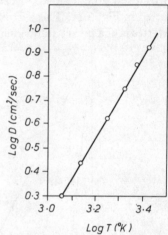

FIG. 6.13. Diffusion coefficient of zinc atoms in argon versus temperature.

For the zinc–argon system, $V_{Zn} = 20.5$ cm³/mol, $V_{Ar} = 29.4$ cm³/mol, and $C = 338$.

TABLE 6.12 Diffusion coefficients of zinc atoms in argon

	1100°K	1350°K	1740°K	2090°K	2330°K	2600°K
D_1 (cm²/s)	2.0	2.8	4.2	5.7	7.1	8.3
D_2 (cm²/s)	1.4	2.1	3.0	4.1	5.0	6.0

The root mean square error representing the reproducibility of the results was about 4 per cent. Allowing for the inaccuracy of the theoretical calculation and the possible systematic experiment errors, associated for instance with the diffusion of vapour through the electrode head (p. 207), the data can be considered to agree satisfactorily.

Measurements of the coefficient of diffusion of atoms of zinc in helium at 1740°K and a gas pressure of 1 atm provided a value of 12.3 cm²/s; this is 2.9 times the corresponding coefficient of diffusion of zinc atoms in argon. Calculations made using equation (6.31) give a 2.8:1 difference between the coefficients of diffusion of zinc atoms in argon and helium, and this agrees well with the experimental results.

CONCLUSION

Finally we shall attempt to estimate the part to be played by atomic absorption spectroscopy in modern analytical chemistry, and to forecast the further development of the method. It is of course extremely risky to discuss any of the prospects for a new and rapidly developing method of analysis, since it is impossible to make sufficient allowance not only for the likelihood of new variations of the method being found, but also for the potential prospects of methods which have for a long time been known and developed. This is confirmed by the unexpected progress, during recent years, in using apparatus assemblies, such as resonance emission and flame sources, which it was thought had been carefully and thoroughly investigated.

On the other hand, in any scientific or technical field forecasts are largely expressions of the subjective opinions of authors and their attitudes to particular problems.

In view of the above statements I believe that my opinions regarding the prospects for developing atomic absorption spectroscopy may appear excessively optimistic or prejudiced to the reader.

When assessing the importance of the atomic absorption method in analytical chemistry, in particular in spectrochemical analysis, it should, in my opinion, be dealt with from three aspects: the development of analysis methods requiring no standards, the simplification of measurements as regards apparatus and methods, and the analysis of ultra-pure substances.

The history of modern spectrochemistry is essentially the history of the continuous search for analysis methods requiring no standards and in which the compositions of samples have no effect on the results of the analysis. In spite of the comparatively long and thorny path which emission spectroscopy has followed in pursuing this line, a general solution to the problem of spectrochemical analysis without using standards has not yet been found. Although attempts are still being made at finding such a method within the framework of the existing technique for emission spectroscopic analysis, there is less and less hope that they will be successful. The necessity for using standards with compositions as close as possible to that of the sample to be determined is therefore accepted by most analysts as being the most obvious and inherent condition for conducting emission analysis.

Against this background the successes achieved in atomic absorption spectrochemical analysis appear most hopeful. It is quite true that, as much experimental research has proved, changing to the absorption method of measuring concentrations is alone insufficient to eliminate entirely the effects of sample composition. It is nevertheless extremely important that most of the effects observed in the presence of third components can be explained, or even interpreted quantitatively.

The use of high temperature reducing flames and graphite cuvettes for atomizing samples, and also of the integral method of measuring absorption, is a considerable advance towards methods of analysis in which standards are not used. As yet we do not possess sufficient factual data confirming the advantages of these methods of atomization, and of measuring signals, for making categorical conclusions. It is quite obvious, however, that the decisive step in developing a method of spectrochemical analysis not involving the use of standards has been taken. Atomic absorption analysis not using standards (employing the analyst's usual meaning for this term) will be introduced in the near future.

Let us now discuss the technical and methodological simplification of analysis provided by the absorption method of measurement. We shall first compare the apparatus particulars for photoelectric methods of emission and absorption analysis.

On the basis of the special features of the atomic absorption method, light source radiation can be stabilized, analysis lines isolated from the source spectrum, and relative changes in intensity recorded, by simpler methods and more reliably than emission photoelectric instruments of the quantum-counter type. By comparison with such instruments, there is no need to use cumbersome spectroscopic devices with large dispersion, no necessity for precision apparatus for determining and controlling the positions of the analysis lines relative to the spectrometer output slits, and no need for high quality sources of light or complicated spectrum excitation generators.

Although, therefore, apparatus for atomic absorption measurements has practically the same particulars, as regards sensitivity, reproducibility and speed of determinations, as photoelectric methods of recording, it is simpler, more compact, and cheaper than apparatus for emission measurements. Equipment mass produced for atomic absorption measurement by practically every optical firm in the world clearly confirms these advantages. The advantages which have been ascribed to the method can to a great extent assist in the automation of the analysis process in whatever fields of engineering and production where it is necessary.

As regards methods, atomic absorption analysis is also more flexible and simpler than emission analysis. We know that, if substances forming a new category are to be analysed by emission spectroscopy, the optimum measurement conditions guaranteeing the required sensitivity, accuracy, precision and economy of analysis must be extremely painstakingly selected. This work includes: the selection of analysis lines on which other lines are not superimposed and which are sufficiently sensitive, the selection of an internal standard and suitable comparison lines, and the selection of spectrum excitation and recording conditions (excitation sources and conditions, methods of introducing samples into the analysis gap, preliminary sparking and exposure times, spectroscopic devices, slit widths, etc.). Often this procedure becomes a prolonged and laborious research project.

The necessity for compulsorily performing this type of research becomes particularly serious when the procedure being developed is intended not for routine analysis of one particular category of specimens, but for the analysis of several samples of random origin or samples with appreciably different compositions. In these cases it becomes either impossible or economically inadvisable to perform accurate quantitative analyses.

Atomic absorption analysis has the main advantage that the number of parameters governing the optimum measurement conditions is much less than with emission measurements. The procedure in the preliminary research stage of developing a method is therefore much simpler, and in some cases it is entirely unnecessary. This advantage is particularly important to the many workers who are not specialists in spectrochemical analysis but require to solve analysis problems of a restricted nature themselves.

For the same reasons atomic absorption permits a high degree of standardization and unification, as regards both the apparatus and the measurement conditions. The importance of this factor can be gauged from all the known difficulties encountered by research workers in reproducing any particular emission procedure worked out by other experts in their own laboratories. Often slight differences in the apparatus, or even in the operator's experience, make it necessary to re-develop the procedure.

Atomic absorption analysis is another matter. In this case the soil is very fruitful for standardizing and unifying the method. The use of standardized apparatus and analysis methods will be conducive to the more effective interchange of experience and research results.

Let us now consider problems associated with the analysis of ultra-pure substances.

Numerous attempts at achieving the sensitivities required for modern techniques (10^{-6}–10^{-8} per cent) by means of the direct emission spectroscopic analysis of an initial material have up to now provided bad results. Although recently methods of improving the sensitivity of emission measurements by using better techniques for recording spectra (spectroscopic apparatus with high resolving power, frequency-selective and scintillation methods of photoelectric recording) have been found, we can hardly hope that a general solution to this problem will be found, in the near future, for the range of concentrations indicated above.

From this point of view the atomic absorption method using a graphite cuvette opens up great prospects. There were two principal factors which made it possible to analyse ultra-pure substances: the high vapour density in the absorbing volume and the low absolute detection limits of elements. Even now, when the peak method of recording, which is far from being the optimum method for making these measurements, is used, relative sensitivities of the order of 10^{-6}–10^{-7} per cent have been achieved. The use of the integration method of recording, and optimization of the measurement conditions, definitely mean that we can progress into the field of even lower concentrations.

It is extremely important that the standardizing procedure should not be complicated (except for the necessity for allowing for unselective optical interferences resulting from molecular absorption by vapour of the parent substance).

The effective utilization of analysis samples in cuvettes has made it possible to achieve sensitivities which are records for spectrochemical analysis (of the order of 10^{-14} g for certain elements); this is very important as regards ultra-microanalysis. It is to be hoped that improvements in the luminosity of light sources, optimization of atomization conditions, and the use of equipment with larger aperture ratios for isolating resonance emission, will result in further progress in this field of analysis.

Here is the point to make the following note. Flames and cuvettes have very different analysis particulars: versatility (the applicability of the method for determining different elements and analysing specimens in different aggregate states), sensitivity and accuracy of measurement, freedom from the effects of sample composition, analysis time, and simplicity of analysis procedure. Since the advantages and limitations of these methods of atomization are largely of a radical nature, we can scarcely hope for any substantial success in overcoming them. Obviously it will be best to expend effort on taking advantage of the possibilities of the two variations of the method, rather than to attempt to solve the wide range of analysis problems on the basis of improvements to any one variation of the method which are not radical, and are therefore comparatively ineffective.

According to the specific advantages of the methods of atomization considered here, we can isolate certain aspects of analysis for which one of the two variants of the method will be preferable.

The flame method will be best for solving the following problems: the routine analysis of samples of the same type, high precision determinations, the direct analysis of organic substances, rapid measurements, the development of the simplified photometers, continuous analysis, and automation of measurements.

The cuvette will be best in the following aspects of analysis: micro-analysis, the analysis of ultra-pure substances, analysis without standards, analysis of toxic substances (for instance radioactive substances), the determination of elements from their resonance lines in the vacuum region of the spectrum, and the determination of low volatility elements (such as aluminium or beryllium) in cases in which for any reason it is difficult to use high-temperature flames.

These fields of analysis do not cover the entire extensive range of analysis problems, but in each case the most important requirements made of the method of analysis can be decided upon, and preference then given to one of the variations.

These are the advantages and some of the possibilities of the atomic absorption method of analysis. Our examination of them will, however, be incomplete if we do not also consider the limitations of the method by comparison with emission analysis. Usually these include the impossibility of determining elements whose resonance lines lie in the vacuum ultra-violet region of the spectrum, the necessity for dissolving solid specimens, and finally the necessity of making successive measurements of concentrations for groups of elements to be determined.

As regards the first two limitations, it seems to me that in this respect we have already made considerable advances, so that we can hope that they will be completely overcome.

The last limitation is still the worst. Quite obviously the successive measurement of concentrations is equally characteristic of any photoelectric device with a single exit slit, either for absorption or for emission measurements. It is therefore more correct to formulate this limitation as the absence, in atomic absorption analysis, of any alternative of the method which is equivalent to the photographic variation of emission analysis. Attempts have been made at devising photographic variations of the atomic absorption method, and it would be useful to continue these attempts. Nevertheless the most radical means of overcoming this limitation is to design multi-channel devices. We have all the prerequisites, both technical and theoretical, for this.

To end this discussion of the part played by atomic absorption spectroscopy in analytical chemistry, it should be emphasized that its uses have already gone far beyond the boundaries of a particular method of spectroscopic analysis. Some of the possible uses of the method in physical research (determination of oscillator strengths, line width and coefficients of diffusion) have been described in this book. The use of the method for determining the vapour pressures of elements, and investigating the thermal stability of compounds and the stability of dilute solutions, is very promising. The possibility of its use for studying processes associated with changes in the concentration of unexcited atoms is interesting; this applies particularly to the use of the method for investigating the processes taking place in sources of light for emission spectrochemical analysis.

It would be possible to list further applications of atomic absorption spectroscopy in different fields of science and engineering. Any forecast, however bold and well thought out, is however unlikely to embrace the full future possibilities of this powerful and elegant method.

Appendix 1

MATHEMATICAL AND PHYSICAL CONSTANTS

π	Pythagorean number	$= 3 \cdot 141593$
$\ln \pi$		$= 1 \cdot 144730$
e	Base of natural logarithms	$= 2 \cdot 718282$
$\log e$		$= 0 \cdot 434294$
N	Avogadro's number	$= 6 \cdot 02252 \cdot 10^{23}\ \text{mol}^{-1}$
c	Speed of light in a vacuum	$= 2 \cdot 997925 \cdot 10^{-8}\ \text{m} \cdot \text{sec}^{-1}$
		$= 2 \cdot 997925 \cdot 10^{10}\ \text{cm} \cdot \text{sec}^{-1}$
e	Charge of electron	$= 1 \cdot 60210 \cdot 10^{-19}\ \text{C}$
		$= 4 \cdot 80298 \cdot 10^{-10}\ \text{CGSE}$
m	Mass of electron	$= 9 \cdot 1091 \cdot 10^{-31}\ \text{kg}$
		$= 9 \cdot 1091 \cdot 10^{-28}\ \text{g}$
h	Planck's constant	$= 6 \cdot 6256 \cdot 10^{-34}\ \text{J} \cdot \text{sec}$
		$= 6 \cdot 6256 \cdot 10^{-27}\ \text{ergs} \cdot \text{sec}$
R	Universal gas constant	$= 8314 \cdot 3\ \text{J} \cdot °\text{C}^{-1}\ \text{kmol.}^{-1}$
		$= 8 \cdot 3143 \cdot 10^{-7}\ \text{ergs} \cdot °\text{C}^{-1}\ \text{mol}^{-1}$
k	Boltzmann's constant	$= 1 \cdot 38054 \cdot 10^{-23}\ \text{J} \cdot °\text{C}^{-1}$
		$= 1 \cdot 38054 \cdot 10^{-16}\ \text{ergs} \cdot °\text{C}^{-1}$
V_0	Molar volume of ideal gases under normal physical conditions (0°C and 760 torr)	$= 22 \cdot 4136\ \text{m}^3\ \text{kmol}^{-1}$
		$= 22 \cdot 4136 \cdot 10^3\ \text{cm}^3\ \text{mol}^{-1}$

These figures were published in *J. Opt. Soc. Amer.* 1964, **54**, 281.

Appendix 2

CORRESPONDENCE BETWEEN SPECTRAL AND ENERGY UNITS AND QUANTITIES

	ν (cm^{-1})	Wavelength (cm)	W (ergs)	W (eV)	W_0 (cal/mol)
1 cm^{-1}		1	$1 \cdot 9857 \cdot 10^{-16}$	$1 \cdot 23968 \cdot 10^{-4}$	$2 \cdot 8578$
1 erg	$0 \cdot 50360 \cdot 10^{16}$	$1 \cdot 9857 \cdot 10^{-16}$		$6 \cdot 2426 \cdot 10^{11}$	$1 \cdot 4391 \cdot 10^{16}$
1 eV	$8066 \cdot 63$	$12396 \cdot 8 \cdot 10^{-8}$	$1 \cdot 6019 \cdot 10^{-12}$		23053
1 cal/mol	$0 \cdot 34992$	$2 \cdot 8578$	$0 \cdot 69488 \cdot 10^{-16}$	$4 \cdot 3378 \cdot 10^{-5}$	

Appendix 3

PHYSICAL AND CHEMICAL CONSTANTS OF ELEMENTS

Symbol	Element	Atomic weight (g/mol)	Abundance[1,2] (per cent)	Melting point[4] (°K)	Boiling point[4] (°K)	Ionization energy[3] (eV)
Ac	Actinium	[227]	(6 . 10⁻¹⁰)	1470	3327	6·89
Ag	Silver	107·87	1 . 10⁻⁵	1234	2483	7·57
Al	Aluminium	26·98	8·80	933	2723	5·98
Am	Americium	[243]		1103	2880	
Ar	Argon	39·95	4 . 10⁻⁶	84	87	15·76
As	Arsenic	74·92	5 . 10⁻⁴	1090 (36 atm) sub.	885	9·81
At	Astatine	[210]		575	607	9·20
Au	Gold	196·97	5 . 10⁻⁷	1336	3243	9·22
B	Boron	10·81	3 . 10⁻⁴	2300	4200	8·30
Ba	Barium	137·34	0·05	987	1913	5·21
Be	Beryllium	9·01	6 . 10⁻⁴	1550	3043	9·32
Bi	Bismuth	208·98	(2 . 10⁻⁵)	545	1833	7·28
Br	Bromine	79·91	1·6 . 10⁻⁴	266	331	11·84
C	Carbon	12·01	(0·1)		5103	11·26
Ca	Calcium	40·08	3·6	1111	1713	6·11
Cd	Cadmium	112·40	(5 . 10⁻⁵)	594	1038	8·99
Ce	Cerium	140·12	4·5 . 10⁻³	1077	3743	6·54
Cl	Chlorine	35·45	0·045	172	239	13·01
Co	Cobalt	58·93	3 . 10⁻³	1768	3173	7·86
Cr	Chromium	52·00	0·02	2148	2938	6·76
Cs	Caesium	132·91	7 . 10⁻⁴	302	963	3·89
Cu	Copper	63·54	0·01	1356	2868	7·72
Dy	Dysprosium	162·50	4·5 . 10⁻⁴	1680	2873	(6·82)
Er	Erbium	167·26	4 . 10⁻⁴	1770	3173	(6·7)
Eu	Europium	151·96	1·2 . 10⁻⁴	1100	1703	5·67
F	Fluorine	19·00	0·027	53	85	17·42
Fe	Iron	55·85	5 . 10	1812	3273	7·90
Fr	Francium	[223]		300	879	3·98
Ga	Gallium	69·72	1·5 . 10⁻³	303	2516	6·00
Gd	Gadolinium	157·25	1 . 10⁻³	1585	3273	6·16
Ge	Germanium	72·59	7 . 10⁻⁴	1210	3103	7·88
H	Hydrogen	1·01	0·15	14	20	13·60
He	Helium	4·00	3 . 10⁻⁷	3	4	24·58
Hf	Hafnium	178·49	3·2 . 10⁻⁴	2495	>5613	5·5
Hg	Mercury	200·59	7 . 10⁻⁶	234	630	10·43

Appendix 3

Symbol	Element	Atomic weight (g/mol)	Abundance[1,2] (per cent)	Melting point[4] (°K)	Boiling point[4] (°K)	Ionization energy[3] (eV)
Ho	Holmium	164.93	$1.3 \cdot 10^{-4}$	1734	2873	(6.9)
In	Indium	114.82	$(1 \cdot 10^{-5})$	429	2273	5.79
Ir	Iridium	192.20	$1 \cdot 10^{-7}$	2727	5573	9.2
I	Iodine	126.90	$(3 \cdot 10^{-5})$	386	456	10.44
K	Potassium	39.10	2.60	337	1033	4.34
Kr	Krypton	83.80	$2 \cdot 10^{-8}$	116	121	14.00
La	Lanthanum	138.91	$1.8 \cdot 10^{-3}$	1193	3743	5.61
Li	Lithium	6.94	$6.5 \cdot 10^{-3}$	382	1603	5.39
Lu	Lutetium	174.97	$1 \cdot 10^{-4}$	1928	3600	6.15
Mg	Magnesium	24.31	2.10	923	1380	7.64
Mn	Manganese	54.94	0.09	1517	2423	7.43
Mo	Molybdenum	95.94	$3 \cdot 10^{-4}$	2890	5833	7.13
N	Nitrogen	14.01	(0.01)	63	77	14.54
Na	Sodium	22.99	2.64	372	1165	5.14
Nb	Niobium	92.91	$1 \cdot 10^{-3}$	2770	4573	6.88
Nd	Neodymium	144.24	$2.5 \cdot 10^{-3}$	1297	3300	(6.31)
Ne	Neon	20.18	$7 \cdot 10^{-9}$	24	27	21.56
Ni	Nickel	58.71	$8 \cdot 10^{-3}$	1728	3003	7.63
Np	Neptunium	[237]		913		
O	Oxygen	16.00	47.2	54	90	13.61
Os	Osmium	190.20	$5 \cdot 10^{-6}$	2973	5773	8.7
P	Phosphorus	30.97	0.08	870 (red) / 317 (white)	553	10.55
Pa	Protactinium	[231]	$(1 \cdot 10^{-10})$	1500		
Pb	Lead	207.19	$1.6 \cdot 10^{-3}$	600	1998	7.42
Pd	Palladium	106.4	$1 \cdot 10^{-6}$	1823	4253	8.33
Pm	Promethium	[147]		1300	3473	(5.9)
Po	Polonium	[210]	$(2 \cdot 10^{-14})$	527	1235	8.2
Pr	Praseodymium	140.91	$7 \cdot 10^{-4}$	1208	3400	(5.76)
Pt	Platinum	195.09	$5 \cdot 10^{-7}$	2043	4803	8.96
Pu	Plutonium	[242]	$1 \cdot 10^{-15}$	913	3508	
Ra	Radium	[226]	$1 \cdot 10^{-10}$	973	1809	5.28
Rb	Rubidium	85.47	0.03	312	961	4.18
Re	Rhenium	186.2	$(1 \cdot 10^{-7})$	3453	6173	7.87
Rh	Rhodium	102.91	$1 \cdot 10^{-7}$	2239	4773	7.46
Rn	Radon	[222]	$(7 \cdot 10^{-16})$	202	211	10.75
Ru	Ruthenium	101.07	$(5 \cdot 10^{-7})$	2773	5173	7.36
S	Sulphur	32.06	0.05	392	718	10.36
Sb	Antimony	121.75	$(4 \cdot 10^{-5})$	903	1653	8.64
Sc	Scandium	44.96	$6 \cdot 10^{-4}$	1811	3003	6.56
Se	Selenium	78.96	$6 \cdot 10^{-5}$	490	958	9.75
Si	Silicon	28.09	27.6	1683	2953	8.15
Sm	Samarium	150.35	$7 \cdot 10^{-4}$	1345	2173	6.6
Sn	Tin	118.69	$4 \cdot 10^{-3}$	505	2543	7.33
Sr	Strontium	87.62	0.04	1043	1653	5.69
Ta	Tantalum	180.95	$2 \cdot 10^{-4}$	3270	5698	7.88

Appendix 3

Symbol	Element	Atomic weight (g/mol)	Abundance[1,2] (per cent)	Melting point[4] (°K)	Boiling point[4] (°K)	Ionization energy[3] (eV)
Tb	Terbium	158·92	$1·5 \cdot 10^{-4}$	1629	3073	(6·74)
Tc	Technetium	[99]		2473	4200	7·23
Te	Tellurium	127·60	$(1 \cdot 10^{-6})$	723	1263	9·01
Th	Thorium	232·04	$8 \cdot 10^{-4}$	2023	4123	6·95
Ti	Titanium	47·90	0·6	1941	3533	6·83
Tl	Thallium	204·37	$(3 \cdot 10^{-4})$	577	1730	6·11
Tu	Thulium	168·93	$8 \cdot 10^{-5}$	1818	1993	(6·6)
U	Uranium	238·03	$3 \cdot 10^{-4}$	1406	4091	(4)
V	Vanadium	50·94	0·02	2173	3723	6·74
W	Tungsten	183·85	$1 \cdot 10^{-4}$	3683	6203	7·98
Xe	Xenon	131·30	$3 \cdot 10^{-9}$	161	165	12·13
Y	Yttrium	88·91	$2·8 \cdot 10^{-3}$	1782	3200	6·38
Yb	Ytterbium	173·04	$3 \cdot 10^{-4}$	1097	1700	6·22
Zn	Zinc	65·37	$5 \cdot 10^{-2}$	693	1180	9·39
Zr	Zirconium	91·22	0·02	2128	3853	6·84

* The figures in round brackets are not very reliable values for the constants; the figures in square brackets are mass numbers for the longest-lived isotopes of radioactive elements.

REFERENCES

1. A. P. VINOGRADOV, *Geokhimiya*, 1956 (1), 52.
2. L. ALLER, *Distribution of Chemical Elements* (IL, 1963).
3. *The Physical and Chemical Properties of Elements* ('Naukova dumka', Kiev, 1965).
4. I. S. KULIKOV, *Thermal Dissociation of Compounds* ('Metallurgia', Moscow, 1969).

Appendix 4

THE MOST SENSITIVE RESONANCE LINES OF ELEMENTS

Elements	Ground state of atom	Wavelength (Å)	Lower line level energy (cm^{-1})	f^*	Sensitivity in flame (μg/ml for 1% abs)
Ag	$^2S_{1/2}$	3280.68	0	0.45	0.1 [4]
		3382.89	0		0.15
Al	$^2P_{1/2}$	3092.71	112	0.23	1.0 [1]
		3961.53	112	0.15	1.3
		3082.16	0	0.22	1.4
		3944.09	0	0.15	2.0
		2373.12	112		3.3
		2367.05	0		4.0
		2575.09	112		8.8
As	$^4S^0_{3/2}$	1890.42	0		1 [4]
		1937.59	0		2
		1972.62	0	(0.07)	3
Au	$^2S_{1/2}$	2427.95	0	0.27	0.23 [11]
		2675.95	0		0.41
B	$^2P_{1/2}$	2497.73	16	(0.65)	50 [1]
		2496.78	0		100
Ba	$^1S_0$	5535.48	0	1.4	8 [4]; 0.4 [1]
		3501.11	0		
		3071.58	0		
Be	$^1S_0$	2348.61	0	0.62	0.024 [1]
Bi	$^4S^0_{3/2}$	2230.61	0		0.68 [11]
		2228.25	0		1.5
		3067.72	0	0.077	2.1
		2061.70	0		5.5
Ca	$^1S_0$	4226.73	0	1.49	0.08 [4]
					0.03 [1]
		2398.56	0		20
Cd	$^1S_0$	2288.02	0	1.2	0.03 [4]
		3261.06	0	0.0018	20

Appendix 4

Elements	Ground state of atom	Wavelength (Å)	Lower line level energy (cm^{-1})	f^*	Sensitivity in flame (μg/ml for 1% abs)
Ce	$^1G_4^0$	5200			30 [12]
		5223·49			30
		5229·35			35
		5567·97			35
Co	$^4F_{9/2}$	2407·25	0		0·16 [5]
		2424·93	0		0·23
		2521·36	0		0·38
		2411·62	816		0·55
		3526·85	0		4·1
		3453·51	3483	0·27	4·2
Cr	$^7S_3$	3578·69	0	0·34	0·11 [11]
		3593·48	0	0·27	0·08
		4254·33	0	0·10	0·20
Cs	$^2S_{1/2}$	8521·10	0	0·80	0·5
		8943·50	0	0·40	
		4555·36	0	0·0027	
		4593·18	0		
Cu	$^2S_{1/2}$	3247·54	0	0·31	0·1 [4]
		3273·96	0	0·16	0·2
		2178·94	0		0·4
		2165·09	0		0·7
		2181·72	0		0·9
Dy	$^5I_8$	4211·72	0		0·70 [2]
		4045·99	0		0·75
		4186·78	0		0·87
		4194·85	0		1·1
Er	$^3H_6$	4007·97	0		0·85 [2]
		4151·10	0		1·3
		3862·82	0		1·3
		3892·69	0		2·4
Eu	$^8S_{7/2}$	4594·03	0	(0·26)	0·75 [2]
		4672·22	0		0·94
		4661·88	0		1·1
Fe	$^5D_4$	2483·27	0		0·1 [4]
		2488·14	416		0·2
		2522·85	0		0·2
		2719·02	0		0·4
		3020·64	0		0·5
		2527·44	416		0·6
		2720·90	416		0·9
		3719·94	0	0·32	1·0
		2966·90	0		1·2
		3859·91	0		2·0
		3440·61	0		2·8
Ga	$^2P_{1/2}$	2874·24	0	0·32	2·3 [4]
		2943·64	826	0·19	2·4
		4172·05	826	0·14	3·7
		4032·98	0	0·13	6·2

Appendix 4

Elements	Ground state of atom	Wavelength (Å)	Lower line level energy (cm^{-1})	f^*	Sensitivity in flame (μg/ml for 1% abs)
Gd	$^9D_2^0$	3684·13	0	(0·46)	17 [2]
		4078·70	533		17
		3783·05	999		17
		4058·22	215		19
Ge	$^3P_0$	2651·58	0	(0·84)	1·5 [1]
		2592·54	557		3·5
		2754·59	1410		4·6
		2709·63	557		5·1
		2691·35	557		9·2
Hf	$^3F_2^0$	3072·88	0	(0·09)	14 [1]
		2866·37	0		17
		2898·26	2357		40
		2964·88	2357		45
		3682·25	0		45
		2940·76	0		56
Hg	$^1S_0$	1849·50	0	1·2	
		2536·52	0	0·03	10 [4]
Ho	$^4I_{15/2}^0$	4103·84	0	(0·39)	1·4 [2]
		4053·93	0	(0·37)	1·9
		4163·03			2·4
In	$^2P_{1/2}^0$	3039·36	0	0·28	0·9 [11]
		3256·09	2213		0·9
		4101·77	0		2·6
		4511·31	2213		2·8
Ir	$^4F_{9/2}$	2088·82	0		7·7 [11]
		2639·71	0		13
		2664·79	0		15
		2849·72	0	(0·056)	18
		2372·77	0		20
		2502·98	0		22
K	$^2S_{1/2}$	7664·91	0	0·69	0·03 [4]
		7698·98	0	0·34	
		4044·14	0	0·0062	5
		4047·20	0		
La	$^2D_{5/2}$	5501·34	0	(0·025)	34 [2]
		4187·32	0		49
		4949·77	0		52
		3574·43	0		120
		3649·53	0		140
Li	$^2S_{1/2}$	6707·84	0	0·71	0·03 [4]
		3232·61	0		15
Lu	$^2D_{3/2}$	3359·58	1994		12 [1]
		3312·11	0	(0·06)	21
		3567·84	0		27

Appendix 4

Elements	Ground state of atom	Wavelength (Å)	Lower line level energy (cm⁻¹)	f*	Sensitivity in flame (μg/ml for 1% abs)
Mg	$^1S_0$	2852·13	0	1·2	0·01 [4]
		2025·82	0		2
Mn	$^6S_{5/2}$	2794·82	0	0·58	0·06 [4]
		2798·27	0	0·42	0·08
		2801·08	0	0·29	0·12
		4030·76	0	0·056	0·8
Mo	$^7S_3$	3132·59	0	(0·2)	0·8 [11]
		3170·35	0		1·1
		3798·25	0		1·3
		3193·97	0		1·4
		3864·11	0		1·7
		3902·96	0		2·4
		3158·16	0		2·8
Na	$^2S_{1/2}$	5889·95	0	0·76	0·03 [4]
		5895·92	0	0·39	
		3302·32	0		5
		3302·99	0		
Nb	$^6D_{1/2}$	3580·28	1050		27 [1]
		3349·07	2154		27
		4079·73	695		32
		4058·93	1050	(0·19)	36
		4123·81	154		40
		4100·92	392		42
Nd	$^5I_4$	4634·24	0	(0·07)	10 [2]
		4896·93	1128		14
		4719·03			21
Ni	$^3F_4$	2320·03	0	(0·1)	0·12 [11]
		2310·96	0		0·18
		2375·54	0		0·46
		2289·98	0		0·54
		2337·49	0		2·1
		3369·57	0		2·6
		2347·52	0		3·1
		3232 96	0		3·9
Os	$^5D_4$	2909·06	0	(0·16)	1·0 [13]
		3058·66	0		1·6
		2637·13	0		1·8
Pb	$^3P_0$	2169·99	0		0·19 [11]
		2833·06	0	0·21	0·49
Pd	$^1S_0$	2447·91	0		0·3 [4]
		2476·42	0	0·051	0·3
		2763·09	0		1·0
		3404·58	6564		1·2
Po	$^3P_2$	2558·01	0		

Appendix 4

Elements	Ground state of atom	Wavelength (Å)	Lower line level energy (cm^{-1})	f^*	Sensitivity in flame (µg/ml for 1% abs)	
Pr	$^4I^0_{3/2}$	4951·36			13 [2]	
		4914·03			19	
		5133·42			23	
		5045·53			42	
Pt	$^3D_3$	2659·45	0		2·2 [11]	
		2175			3·3	
		3064·71	0		4·6	
		2628·03	776		5·3	
		2144·23	0		7·3	
		2830·30	0		7·4	
Ra	$^1S_0$	4825·91	0			
Rb	$^2S_{1/2}$	7800·23	0	0·80	0·1 [4]	
		7947·60	0	0·40		
		4201·85	0	0·00115	10	
		4215·56	0			
Re	$^6S_{5/2}$	3460·47	0	(0·20)	12 [1]	
		3464·73	0		20	
		3451·88	0		33	
Rh	$^4F_{9/2}$	3434·89	0	(0·073)	0·34 [8]	
		3692·36	0		0·57	
		3396·85	0		0·80	
		3502·52	0		1·3	
		3657·99	1530		1·7	
Ru	$^5F_5$	3498·94	0	(0·1)	0·3 [11]	
		3728·03	0			
		2735·72	0			
Sb	$^4S^0_{1/2}$	2175·81	0		0·5 [4]	
		2068·33	0		0·5	
		2311·47	0	0·042	1·2	
Sc	$^2D_{3/2}$	3911·81	168		0·8 [1]; 0·5 [2]	
		3907·49	0	(0·67)	1·1	0·5
		4023·69	168		1·2	0·7
		4020·40	0		1·7	0·9
		3269·91	0		2·8	1·6
Se	$^3P_2$	1960·26	0	(0·12)	1·6 [14]	
		2039·85	1989		11	
		2062·79	2534		43	
Si	$^3P_0$	2516·11	223	(0·26)	1·2 [3]	
		2506·90	77		3·1	
		2528·51	223		3·7	
		2514·32	0		3·8	
		2524·11	77		4·0	
		2216·67	223		4·5	
		2519·21	77		5·5	

Appendix 4

Elements	Ground state of atom	Wavelength (Å)	Lower line level energy (cm^{-1})	f^*	Sensitivity in flame (μg/ml for 1% abs)
Sm	$^7F_0$	4296.74	4021	(0.44)	8.5 [2]
		5200.59	1490		13
		4760.27	812		22
		4728.42	1490		24
Sn	$^3P_0$	2246.05	0		0.7 [9]
		2354.84	1693		1.6
		2863.33	0	0.19	1.3
Sr	$^1S_0$	4607.33	0	1.54	0.2 [4]; 0.06 [1]
		2428.10	0		
		2569.47	0		
Ta	$^4F_{3/2}$	2714.67	0	(0.19)	11 [1]
		2608.63			21
		2775.88			21
		2559.43			24
		2647.47			34
		2661.34			43
		2758.31			43
Tb	$^8H_{17/2}$	4326.47	0	(0.11)	7.2 [2]
		4318.85	0		8.7
		3901.35	2024		12
		4061.59	176		13
		4338.45	0		14
		4105.37	0		26
Tc	$^6S_{5/2}$	4297.06	0		
Te	$^3P_2$	2142.75	0	(0.08)	1 [15]
		2259.04	0	0.0018	7
Th	$^3F_2$	3304.24	0		
		3719.44	0		
		3803.07	0		
		3348.77	0		
Ti	$^3F_2$	3653.50	387	0.14	1.6 [3]
		3642.68	170		1.8
		3371.45	387		2.0
		3199.90	387		2.0
		3752.85	387		2.5
		3191.91	170		2.6
		3741.06	170		2.6
		3354.63	170		2.9
		3186.51	0		3.0
Tl	$^2P^0_{1/2}$	2767.87	0	0.27	0.03 [4]
		3775.72	0	0.13	0.1
		2379.69	0		0.2
		2580.14	0		
Tu	$^2F^0_{7/2}$	3717.92	0	(0.21)	0.67 [16]
		4105.84	0		0.96
		3744.07	0		1.1
		4094.19	0		1.2

Appendix 4

Elements	Ground state of atom	Wavelength (Å)	Lower line level energy (cm^{-1})	f*	Sensitivity in flame (μg/ml for 1% abs)
U	$^5L_6^0$	3584.88	0	(0.16)	80 [2]
		3566.60	620		110
		3514.61	0		200
		3943.82	0		220
		3489.37	0		250
		4153.97	0		250
		3659.16	620		250
V	$^4F_{3/2}$	3185.40	553	(0.4)	1.0 [3]
		3183.41	137		} 1.3
		3183.98	323		
		3060.46	323		2.8
		3066.38	553		2.9
		3056.33	137		3.3
		4379.24	2425		4.4
		3703.58	2425		5.2
		3828.56	137		9.1
W	$^5D_0$	2551.35	0	(0.8)	5.3; 17 [1]
		2681.41	2951		19
		2944.40	2951		12
		2724.35	2951		26
		4008.75	2951		18 28
		2946.98	2951		30
		2831.38	2951		35
Y	$^2D_{3/2}$	4102.38	530		5.0 [1]
		4128.31	530		5.4
		4077.38	0	(0.27)	5.7
		4142.85	0		11
Yb	$^1S_0$	3987.98	0	(0.38)	0.25 [1]
		3464.36	0		0.8
		2464.49	0		1.6
		2671.98	0		10
Zn	$^1S_0$	2138.56	0	1.2	0.03 [4]
		3075.90	0	$1.7 \cdot 10^{-4}$	150
Zr	$^3F_2$	3601.19	1241		15; 18 [1]
		3519.60	0	(0.19)	20
		3011.75	570		30
		3863.87	570		39
		3547.68	570		43
		3623.86	570		45
		3029.52	1241		58
		2985.39	0		65
		3890.32	1241		65
		3509.32	570		78

* The values of f given in parentheses were measured by the emission method.[10] In all other cases the f values given are more reliable, and were measured by the Rozhdestvenskii 'hooks' method or the absorption method. The numbers in square brackets are reference numbers.

REFERENCES

1. M. D. Amos, and J. B. Willis, *Spectrochim. Acta*, 1966, **22**, 1325, 2128.
2. D. C. Manning, *Atomic Absorption Newsletter*, 1966, **5**, 127.
3. J. S. Cartwright, C. Sebens, and D. C. Manning, *Atomic Absorption Newsletter*, 1966, **5**, 91.
4. W. T. Elwell, and J. A. T. Gidley, *Atomic-Absorption Spectrophotometry* (Pergamon, 2nd Edn., 1966).
5. J. S. Cartwright, and D. C. Manning, *Atomic Absorption Newsletter*, 1966, **5**, 114.
6. C. E. Mulford, *Atomic Absorption Newsletter*, 1966, **5**, 63.
7. J. E. Allan, *Spectrochim. Acta*, 1962, **18**, 259.
8. P. Heneage, *Atomic Absorption Newsletter*, 1966, **5**, 64.
9. L. Capacho-Delgado, and D. C. Manning, *Atomic Absorption Newsletter*, 1965, **4**, 317.
10. C. H. Corliss, and W. R. Bozman, *Experimental Transition Probabilities for Spectral Lines of Seventy Elements* (Nat. Bureau of Standards, New York, 1962).
11. W. Slavin, *Atomic Absorption Spectroscopy* (Interscience, New York, 1968).
12. P. E. Thomas, *Resonance Lines*, 1969, **1**, 6.
13. F. J. Fernandez, *Atomic Absorption Newsletter*, 1969, **8**, 90.
14. R. M. Dagnall, K. C. Thompson, and T. S. West, *Atomic Absorption Newsletter*, 1967, **6**, 117.
15. J. Y. L. Wu, H. A. Drall, P. F. Lott, *Atomic Absorption Newsletter*, 1968, **7**, 90.
16. F. J. Fernandez, and D. C. Manning, *Atomic Absorption Newsletter*, 1968, **7**, 57.

REFERENCES
INTRODUCTION AND CHAPTER I

1. T. P. KRAVETS, Absorption of light in solutions of coloured substances. Dissertation. *Izvestiya Moskovskogo inzhenernogo uchilishcha*, 1912, **2**, (6); Works on physics, *Izd. AN SSSR*, 1959.
2. C. FÜCHTBAUER, *Z. Phys.*, 1920, **21**, 322.
3. R. LADENBURG, *Z. Phys.*, 1921, **4**, 451.
4. H. A. LORENTZ, *Proc. Amst. Acad.*, 1915, **18**, 134.
5. W. VOIGT, *Münch. Ber.*, 1912, 603.
6. F. REICHE, *Verhandl Dtsch. Phys. Ges.*, 1913, **15**, 3.
7. R. LADENBURG, and F. REICHE, *Ann. Phys.*, 1913, **42**, 181.
8. C. FÜCHTBAUER, G. JOOS, and O. DINKELACKER, *Ann. Phys.*, 1923, **71**, 204.
9. A. UNSÖLD, *The Physics of Stellar Atmospheres* (translated from the German, IL, Moscow, 1949.
10. A. S. KING, *Astrophys. J.*, 1908, **27**, 353.
11. L. BOVEY, *Spectrochim. Acta*, 1958, **10**, 383; 1959, **11**, 539.
12. S. TOLANSKY, *High Resolution Spectroscopy* (Methuen, London, 1947).
13. T. T. WOODSON, *Rev. Sci. Instrum.*, 1939, **10**, 308; US Patent No. 2227117 (issued in 1940).
14. C. T. J. ALKEMADE, and J. M. W. MILATZ, *J. Opt. Soc. Am.*, 1955, **45**, 583.
15. A. WALSH, *Spectrochim. Acta*, 1955, **7**, 108.
16. B. J. RUSSELL, J. P. SHELTON, and A. WALSH, *Spectrochim. Acta*, 1957, **8**, 317.
17. J. B. ROBINSON, *Analyt. Chem.*, 1960, **32**, 17a.
18. J. A. MAXWELL, *Chem. Canada*, 1963, **15**, 10; E. F. RUNGE, R. W. MINCK, and F. R. BRYAN. *Spectrochim. Acta*, 1964, **20**, 733.
19. S. CH'EN, and M. TAKEO, *UFN*, 1958, **66**, 391.
20. C. F. FÜCHTBAUER, and F. GÖSSLER, *Z. Phys.*, 1934, **87**, 89.
21. H. MARGENAU, and W. W. WATSON, *Phys. Rev.*, 1933, **44**, 92.
22. G. F. HULL, *Phys. Rev.*, 1936, **50**, 1148.
23. S. CH'EN, *Phys. Rev.*, 1940, **58**, 1058.
24. S. CH'EN, and R. B. BENNETT, *Phys. Rev.*, 1960, **119**, 1029.
25. S. CH'EN, A. SMITH, and M. TAKEO, *Phys. Rev.*, 1960, **117**, 1010.
26. E. D. CLAYTON, and S. CH'EN, *Phys. Rev.*, 1952, **85**, 68.
27. W. LENZ, *Z. Phys.*, 1924, **25**, 299; 1933, **83**, 139.
28. V. WEISSKOPF, *Z. Phys.*, 1933, **34**, 1; 1933, **85**, 451; Russian translation: *UFN*, 1933, **13**, 596.
29. E. LINDHOLM, *Arkiv. Mat. Astr. och Fys.*, 1941, **28B**, (3); 1945, **32**, (17).
30. B. V. L'VOV, and G. V. PLYUSHCH, *Proceedings of XV Conference on Spectroscopy* (Minsk, 1963), **2**, p. 159 (Izd. AN SSSR, Moscow, 1964).
31. C. F. FÜCHTBAUER, and H. J. REIMERS, *Z. Phys.*, 1935, **97**, 1.
32. T. Z. NY, and S. Y. CH'EN, *Phys. Rev.*, 1937, **52**, 1158.
33. S. Y. CH'EN, and W. J. PARKER, *J. Opt. Soc. Am.*, 1955, **45**, 22.
34. A. MITCHELL, and M. ZEMANSKY, *Resonance Radiation and Excited Atoms* (Cambridge, 1934).
35. D. L. HARRIS, *Astrophys. J.*, 1948, **108**, 112; Russian translation in a collection of articles 'Present Day Problems in Astrophysics and Solar Physics' (IL, Moscow, 1951).
36. J. C. STRIJLAND, and A. J. NANASSY, *Physica*, 1958, **24**, 935.
37. R. B. KING, and A: S. KING, *Astrophys. J.*, 1935, **82**, 377; R. B. KING, B. R. PARNES, M. N. DAVIS, and K. H. OLSEN, *J. Opt. Soc. Am.*, 1955, **45**, 350.
38. S. E. FRISH, and O. P. BOCHKOVA, *Vestnik LGU*, 1961, (16), 40.
39. B. V. L'VOV, *Optika i spektroskopiya*, 1965, **19**, 507.
40. YU. I. OSTROVSKII, and N. P. PENKIN, *Optika i spektroskopiya*, 1961, **11**, (3).
41. E. HINNOV, and H. KOHN, *J. Opt. Soc. Am.*, 1957, **47**, 151.
42. H. LANDÖLT, and R. BORNSTEIN, *Zahlenwerte und Funktionen*, Vol. I, Table 5 (Berlin, 1952).
43. B. V. L'VOV, *Zavodskaya laboratoriya*, 1962, **28**, 931.
44. B. M. GATEHOUSE, and J. B. WILLIS, *Spectrochim. Acta*, 1961, **17**, 710.
45. J. E. ALLAN, *Spectrochim. Acta*, 1959, **15**, 800.
46. J. E. ALLAN, *Spectrochim. Acta*, 1962, **18**, 259.
47. V. A. FASSEL, and V. G. MOSSOTTI, *Analyt. Chem.*, 1963, **35**, 252.
48. N. P. PENKIN, and S. E. FRISH, *Optika i spektroskopiya*, 1957, **3**, 473.
49. J. B. ROBINSON, *Analyt. Chem.*, 1961, **33**, 1067.

50. O. P. BOCHKOVA, *Izv. AN SSSR, ser. fiz.*, 1954, **18**, 252.
51. R. LADENBURG, and S. LEVY, *Z. Phys.*, 1930, **65**, 189.
52. W. M. ELSASSER, *Harvard Metereological Studies*, Milton, Mass., 1942, (6).
53. A. WALSH, *Advances in Spectroscopy*, Vol II, pp. 1-22 (Interscience, New York, 1961).
54. M. D. AMOS, and P. E. THOMAS, *Analyt. Chim. Acta*, 1965, **32**, 139.
55. A. N. ZAIDEL', N. I. KALITIEVSKII, L. V. LIPIS, and M. P. CHAIKA, *The Emission Spectral Analysis of Atomic Materials* (Fizmatgiz, Leningrad–Moscow, 1960).

CHAPTER II

1. F. PASCHEN, *Ann. Phys.*, 1916, **50**, 901.
2. S. TOLANSKY, *High Resolution Spectroscopy* (Methuen, London, 1947).
3. A. N. ZAIDEL', N. I. KALITIEVSKII, L. V. LIPIS, and M. P. CHAIKA, *The Emission Spectral Analysis of Atomic Materials* (Fizmatgiz, Leningrad-Moscow, 1960).
4. H. N. CROSSWHITE, G. H. DIEKE, and C. S. LEGAGNEUR, *J. Opt. Soc. Am.*, 1955, **45**, 270.
5. B. J. RUSSELL, J. P. SHELTON, and A. WALSH, *Spectrochim. Acta*, 1957, **8**, 317.
6. W. G. JONES, and A. WALSH, *Spectrochim. Acta*, 1960, **16**, 249.
7. A. I. BODRETSOVA, B. V. L'VOV, E. N. PAVLOVSKAYA, and V. K. PROKOF'EV, *Zh. prikl. spektroskopii*, 1965, **2**, 97.
8. G. F. BOX, and A. WALSH, *Spectrochim. Acta*, 1960, **16**, 255.
9. J. E. ALLAN, *Spectrochim. Acta*, 1959, **15**, 800.
10. J. B. ROBINSON *Analyt. Chem.*, 1961, **33**, 1067.
11. O. S. DUFFENDACK, and K. THOMSON, *Phys. Rev*, 1933, **43**, 106
12. Yu. I. TURKIN, *Optika i spektroskopiya*. 1957, **2**, 290; 1959, **7**, 10.
13. D. J. DAVID, *Analyst*, 1960, **85**, 779.
14. B. M. GATEHOUSE, and J. B. WILLIS, *Spectrochim. Acta*, 1961, **17**, 710.
15. O. P. BOCHKOVA, and E. Ya. SHREIDER, *The Spectroscopic Analysis of Mixtures of Gases* (Fizmatgiz, Leningrad–Moscow, 1961).
16. A. I. BODRETSOVA, B. V. L'VOV, and V. I. MOSICHEV, *Zh. Prikl. spektroskopii*, 1966, **4**, 207.
17. S. M. LEVITSKII, *Zh. tekhn. fiziki*, 1957, **27**, 101.
18. J. B. DAWSON, and D. J. ELLIS, *Spectrochim. Acta*, 1967, **23A**, 565.
19. I. M. NAGIBINA, and V. K. PROKOF'EV, *Spectroscopic Devices and Spectroscopy Technique*, p. 209 (Izd.-vo 'Mashinostroenie', Leningrad, 1967).
20. V. VASILISHCHENKO, *Radio*, 1956, (4), 27.
21. K. R. OSBORN, and H. E. GUNNING, *J. Opt. Soc. Am.*, 1955, **45**, 552.
22. W. E. BELL, A. L. BLOOM, and J. LYNCH, *Rev. Sci. Instrum.*, 1961, **32**, 688; Russian translation in *Pribory dlya nauchn. issledovanii*, 1961, (6), 79.
23. R. G. BREWER, *Rev. Sci. Instrum.*, 1961, **32**, 1356; Russian translation in *Pribory dlya nauchn. issledovanii*, 1961, (12), 86.
24. F. A. FRANZ, *Rev. Sci. Instrum.*, 1963, **34** 589; Russian translation in *Pribory dlya nauchn. issledovannii*, 1963, (5), 129.
25. W. F. MEGGERS, and F. O. WESTFALL, *J. Research Natl. Bur. Standards*, 1950, **44**, 447; E. JACOBSEN, and G. R. HARRISON, *J. Opt. Soc. Am.*, 1949, **39**, 1054.
26. E. W. RICHARDS, *Spectrochim. Acta*, 1966, **22**, 159.
27. I. S. MARSHAK, *Pulsed Light Sources* (Gosenergoizdat, Moscow–Leningrad, 1963).
28. *Handbook on lighting engineering*, Ed. by V. S. KULEBAKIN, Vol. I, Izd. AN SSSR, Moscow, 1956.
29. L. S. NELSON, and N. A. KUEBLER, *Spectrochim. Acta*, 1963, **19**, 781.
30. M. N. SMOLKIN, and N. B. BERDINKOV. *Optika i spektrostopiya*, 1963, **14**, 414.
31. J. E. ALLAN, *Spectrochim. Acta*, 1962, **18**, 259.
32. V. A. FASSEL, and V. G. MOSSOTTI, *Analyt. Chem.*, 1963, **35**, 252.
33. H. V. MALMSTADT, and W. E. CHAMBERS, *Analyt. Chem.*, 1960, **32**, 225.
34. N. S. POLUEKTOV, and R. A. VITKUN, *Zh. Analit. Khimii*, 1963, **18**, 37.
35. D. V. ISAEV, Symposium *Automatic gas analysers*, p. 280 (TsINTI, Moscow, 1961).
36. C. B. BELCHER, et al. *Analyt. Chim. Acta*, 1962, **26**, 322; 1963, **29**, 202; 1963, **29**, 340; 1964, **30**, 483.
37. J. E. ALLAN, *Analyst*, 1958, **83**, 466.
38. K. FUWA, and B. L. VALLEE, *Analyt. Chem.*, 1963, **35**, 942.
39. G. G. SLYUSAREV, *The Possible and the Impossible in Optics* (Fitzmatgiz, Moscow, 1960).
40. V. A. SLAVNYI, I. S. ABRAMSON, and YU. V. AFANAS'EV, *Zh. prikl. spektroskopii*, 1964, (1), 26.
41. A. M. BONCH-BRUEVICH, *Radio-electronics in Experimental Physics* (Izd. 'Nauka', 1966).
42. N. O. CHECHIK, S. M. FAINSHTEIN, and T. M. LIFSHITS, *Electronic Multipliers* (2nd edn., Gostekhizdat, Moscow, 1957).
43. V. M. CHULANOVSKII, and N. P. PENKIN, *Izv. AN SSSR, ser. fiz.*, 1945, **15**, 206.
44. N. P. KOMAR', and V. P. SAMOILOV, *Zh. analit. khimii*, 1963, **18**, 1284.
45. A. VAN DER ZIL, *Fluctuations in Radio*

Engineering and Physics (translated from the German, Gosenergoizdat, Moscow–Leningrad, 1958).
46. C. C. YANG, and V. LEGALLAIS, *Rev. Sci. Instrum.*, 1954, **25**, 801.
47. J. U. WHITE, *J. Opt. Soc. Am.*, 1942, **32**, 285.
48. R. W. TABELING, and J. J. DEVANEY, *Developments in Applied Spectroscopy*, Vol. I, p. 175 (Plenum Press, N.Y., 1962).
49. R. C. MILLIKAN, *J. Opt. Soc. Am.*, 1961, **51**, 535.
50. S. E. FRISH, and O. P. BOCHKOVA, *Vestnik LGU*, 1961, **16**, 40.
51. I. V. PODMOSHINSKII, and L. D. KONDRASHEVA, *Proceedings of the X All-Union Conference on Spectroscopy*. Part II, p. 204 (izd. L'vovskogo universiteta, 1958).
52. *Lasers. Bibliography from* **1958** *to* **1963** (Izd. 'Nauka', Moscow, 1964).
53. N. S. POLUEKTOV, and S. E. GRINZAID, *Zavodskaya laboratoriya*, 1963, **29**, 998.
54. B. V. L'VOV, *Inzhenerno-fizicheskii zhurnal*, 1944, **2**, (2), 44; 1959, **2**, (11), 56.
55. K. DOERFFEL, R. GEYER, and G. MÜLLER, *Z. Chem.*, 1963, **3**, 212.
56. B. V. L'VOV, *Zavodskaya laboratoriya*, 1962, **28**, 931.
57. B. V. L'VOV, Atomic absorption spectrochemical analysis using a graphite cuvette for vaporizing substances. Dissertation, Leningrad, 1961.
58. H. SCHWAHN, *Nachrichtentecknik*, 1958, **8**, 158.
59. G. I. NIKOLAEV, A study of some of the prospects of the atomic absorption method of analysis, using a graphite cuvette for vaporising substances. Dissertation. Leningrad, 1963.
60. V. I. DIANOV-KLOKOV, PTE, 1956, (3), 87; 1959, (6), 91.
61. V. M. VOLKOV, *Logarithmic Amplifiers* (GITL Ukr. SSR, Kiev, 1962).
62. J. J. OBERLEY, *Rev. Sci. Instrum.*, 1953, **24**, 125.
63. C. T. J. ALKEMADE, and J. M. W. MILATZ, *J. Opt. Soc. Am.*, 1955, **45**, 583.
64. R. HERRMANN, and W. LANG, *Proceedings of the IX Colloquium Spectroscopium Internationale*, p. 291 (Gaston–Boissier, Paris, 1962).
65. C. A. BAKER, and F. W. GARTON, *J. Atomic Energy Res. Establ.*, 1961, no. R3490.
66. L. R. P. BUTLER, A. STRASHEIM, F. W. E. STRELOW, P. MATHEWS and E. C. FEAST, 'Determination of gold in main solutions by automatic or semi-automatic atomic-absorption spectroscopy'. *XII Colloquium Spectroscopicum Internationale*, Exeter, 1965 (Adam Hilger Ltd., 1966).
67. A. C. MENZIES, *Analyt. Chem.*, 1960, **32**, 898.
68. J. A. GIDLEY, and J. T. JONES, *Analyst*, 1960, **85**, 249.
69. R. L. WARREN, Spectroscopy, p. 27 (Inst. of Petroleum, London, 1962).
70. N. P. IVANOV, L. V. MINERVINA, S. V. BARANOV, L. G. POFRALIDI, and I. I. OLIKOV, *Zh. Analit. Khimii*, 1966, **21**, 1129.
71. N. P. IVANOV, L. V. MINERVINA, and S. V. BARANOV, *Chemical Reagents and Preparations*, Issue 27, Moscow, IREA, 1965, 276.
72. J. V. SULLIVAN, and A. WALSH, *Spectrochim. Acta*, 1965, **21**, 721.
73. H. MASSMANN, *Z. Instrum.*, 1963, **8**, 225.
74. L. R. P. BUTLER, and A. STRASHEIM, *Spectrochim. Acta*, 1965, **21**, 1207.
75. A. WALSH, Australian Patent Specification 163586 (23041/53).
76. H. MASSMANN, 'Spurenanalyse mittels atomabsorption in der Graphitküvetten nach L'vov mit einem Mehrkanalspectrometer'. II Internationales Simposium 'Reinststoffe in Wissenschaft und Technik', Dresden, 1965.
77. D. S. DRABKIN, and N. P. IVANOV, Priority Certificate No. 197775, dated 9 June, 1967.
78. N. P. IVANOV, D. S. DRABKIN, and B. M. TALALAEV, *Chemical Reagents and Preparations*, Issue 30, Moscow, IREA, 1967, 517.
79. V. L. GINZBURG, and G. I. SATARINA. *Zavodskaya laboratoriya*, 1965, **31**, 249.
80. J. H. GIBSON, W. E. GROSSMAN, and W. A. COOK, *Analytical Chemistry*, p. 288 (Elsevier Publ. Comp., 1963).
81. F. A. KOROLEV, *Vestnik MGU*, 1958, (3), 97.
82. A. WALSH, *Spectroscopy*, p. 13 (Inst. of Petroleum, London, 1962).
83. R. RIKMENSPOEL, *Rev. Sci. Instrum.*, 1965, **36**, 497; Russian translation in *Pribory dlya nauchn. issledovanii*, 1965, (4), 85.
84. *Light Pipes for Transmitting Images*, edited by K. I. BLOCK and V. B. VEINBERG (Moscow 1961).
85. *The Continuous Glass Fibres*, edited by M. G. CHERNYAK (Izd. 'Khimiya' Moscow, 1965).
86. J. V. SULLIVAN, and A. WALSH, *Spectrochim. Acta*, 1965, **21**, 727.
87. J. A. BOWMAN, J. V. SULLIVAN, and A. WALSH, *Spectrochim. Acta*, 1966, **22**, 205.

CHAPTER III

1. C. T. J. ALKEMADE, and J. M. W. MILATZ, *J. Opt. Soc. Am.*, 1955, **45**, 583.
2. B. J. RUSSELL, J. P. SHELTON, and A. WALSH, *Spectrochim. Acta*, 1957, **8**, 317.
3. B. M. GATEHOUSE, and A. WALSH, *Spectrochim. Acta*, 1960, **16**, 602.
4. A. S. KING, *Astrophys. J.*, 1908, **27**, 353.
5. B. V. L'VOV, *Inzhenerno-fizicheskii zhurnal*,

1959, **2**, (2), 44; *Spectrochim Acta*, 1961, **17**, 761.
6. L. S. NELSON, and N. A. KUEBLER. *Spectrochim. Acta*, 1963, **19**, 781.
7. *Thermodynamic Properties of Individual Substances*, vols. I–II, Handbook edited by V. P. Glushko (Izd. AN SSSR, Moscow, 1962).
8. A. GAYDON, *Dissociation Energies and Spectra of Diatomic Molecules* (Chapman & Hall, London, 1953).
9. *Chemical Bond Breakdown Energies. Ionization Potentials and Electron Affinity.* Handbook edited by V. N. KONDRAT'EV (Izd. AN SSSR, Moscow, 1962).
10. K. E. SHULER, and J. WEBER, *J. Chem. Phys.*, 1954, **22**, 491.
11. W. D. HAGENAH, K. LAQUA, and V. MOSSOTTI, 'Atomic-absorption analysis with volatilisation by laser'. *XII Colloquium Spectroscopicum Internationale*, Exeter, 1965 (Adam Hilger Ltd., London, 1966).
12. A. V. KARYAKIN, V. A. KAIGORODOV, and M. V. AKHMANOVA, *Zh. analit. khimii*, 1965. **20**, 145; *Zh. prikl. spektroskopii*, 1965, **2**, 364.
13. B. V. L'VOV, and G. G. LEBEDEV, *Zh. prikl. spektroskopii*, 1967, **7** 264.
14. S. L. MANDEL'SHTAM, *Introduction to Spectroscopic Analysis* (Gostekhizdat, Moscow, 1946).
15. YA. D. RAIKHBAUM, and V. D. MALYKH, *Optika i spektroskopiya*, 1960, **9**, 425.

16. M. SLAVIN, *Ind. Eng. Chem., Anal. Ed.*, 1938, **10**, 407.
17. V. D. MALYKH, and M. A. SERD, *Optika i spektroskopiya*, 1964, **16**, 368.
18. J. A. RAMSAY, *J. Exptl. Biol.*, 1950, **27**, 407; 1953, **30**, 1.
19. J. A. RAMSAY, S. FALLOON, and K. MACHIN, *J. Sci. Instrum.*, 1951, **28**, 75.
20. H. L. KAHN, G. E. PETERSON, and J. E. SCHALLIS, *Atomic Absorption Newsletter*, 1968, **7**, 35.
21. R. H. WENDT, and V. A. FASSEL, *Analyt. Chem.* 1966, **38**, 337.
22. YU. I. KOROVIN, *Zh. analit. khimii*, 1961, **16**, 494.
23. B. V. L'VOV, *Zav. laboratoriya*, 1962, **28**, 931.
24. YU. I. BELYAEV, L. M. IVANTSOV, A. V. KARYAKIN, FAM-HUNG-FEE, and V. V. SHEMET, *Zh. analit. khimii*, 1968, **23**, (2).
25. P. W. J. M. BOUMANS, Theory of Spectrochemical Excitation (Adam Hilger, London, 1966).
26. W. ATWILL, *International Electronics*, 1964, **7** 18.
27. V. G. MOSSOTTI, K. LAQUA, and W. D. HAGENAH, *Spectrochim. Acta*, 1967, **23B** 197.
28. B. V. L'VOV, *Atomic Absorption Spectrochemical Analysis* (Izd-vo 'Nauka', Moscow, 1966).

CHAPTER IV

1. R. MAVRODINEANU, *Appl. Spectroscopy*, 1959, **13** 139; 1960, **14** 17.
2. M. MARGOSHES, *Analyt. Chem.*, 1962, **34** (5), 221R.
3. J. A. DEAN, *Analyst*, 1960, **85** 621.
4. J. A. DEAN, *Flame Photometry* (McGraw-Hill, New York, 1960).
5. R. HERRMANN, and C. T. J. ALKEMADE, *Flammenphotometrie* (2nd edn., Springer-Verlag, Berlin, 1960).
6. W. SCHUHKNECHT, *Die Flammenspektralanalyse* (F. Enke, Stuttgart, 1961).
7. M. FILCEK, *Z. Pflanzenernähr., Düng., Bodenkunde*, 1959, **85**, 112.
8. R. HERRMANN, and W. LANG, *Proceedings of the IX Colloquium Spectroscopicum Internationale*, p. 291 (Gaston-Boissier, Paris, 1962).
9. S. NUKIYAMA, and Y. TANASAWA, *Soc. Mech. Engrs. Japan*, 1938–1940, **4**, **5**, **6**; translated from Japanese into English: E. HOPE, *Experiments on the Atomization of Liquids in an Air Stream* (Dept. of National Defense, Ottawa).
10. O. E. CLINTON, *Spectrochim. Acta*, 1960, **16**, 985.

11. V. C. O. SCHÜLER, and A. V. JANSEN, *J. S. Afric. Inst. Mining and Metallurgy*, 1962, **62**, 790.
12. L. R. P. BUTLER, *J. S. Afric. Inst. Mining and Metallurgy*, 1962, **62**, 786.
13. J. W. ROBINSON, and R. J. HARRIS, *Analyt. Chim. Acta*, 1962, **26**, 439.
14. R. N. KNISELEY, A. P. D'SILVA, and V. A. FASSEL, *Analyt. Chem.*, 1963, **35**, 910.
15. J. E. ALLAN, *Spectrochim. Acta*, 1961, **17**, 467; 1962,**18**, 605.
16. R. LOCKYER, J. E. SCOTT, and S. SLADE, *Nature*, 1961, **189**, 830.
17. J. W. ROBINSON, *Analyt. Chem.*, 1961, **33**. 1067.
18. R. L. WARREN, *Spectroscopy*, p. 27 (Inst. of Petroleum, London, 1962).
19. E. PUNGOR, and M. MAHR, *Talanta*, 1963, **10**, 537.
20. E. HINNOV, and H. KOHN, *J. Opt. Soc. Am.*, 1957, **47**, 151.
21. A. G. GAYDON, and H. D. WOLFHARD, *The Flame, their Structure, Radiation and Temperature.* (2nd edn., London, 1960).
22. R. MAVRODINEANU, *Spectrochim Acta*, 1961. **17**, 1016.

23. M. R. BAKER, and B. L. VALLEE, *Analyt. Chem.*, 1959, **31**, 2036.
24. W. H. FOSTER and D. N. HUME, *Analyt. Chem.*, 1959, **31**, 2028.
25. K. FUWA, R. E. THIERS, B. L. VALLEE, and M. R. BAKER, *Analyt. Chem.*, 1959, **31**, 2039.
26. H. C. ESHELMAN, J. A. DEAN, O. MENIS, and T. C. RAINS, *Analyt. Chem.*, 1959, **31**, 183.
27. V. A. FASSEL, R. H. CURRY, and R. N. KNISELEY, *Spectrochim. Acta*, 1962, **18**, 1127.
28. V. A. FASSEL, R. N. MYERS, and R. N. KNISELEY, *Spectrochim. Acta*, 1963, **19**, 1187.
29. J. E. ALLAN, *Analyst.*, 1958, **83**, 466.
30. D. J. DAVID, *Nature*, 1960, **187**, 1109.
31. B. M. GATEHOUSE, and J. B. WILLIS, *Spectrochim. Acta*, 1961, **17**, 710.
32. J. E. ALLAN, *Spectrochim. Acta*, 1962, **18**, 259.
33. V. A. FASSEL, and V. G. MOSSOTTI, *Analyt. Chem.*, 1963, **35**, 252.
34. W. SLAVIN, and D. C. MANNING, *Analyt. Chem.*, 1963, **35**, 253.
35. F. B. DOWLING, C. L. CHAKRABARTI, and G. R. LYLES, *Analyt. Chim. Acta*, 1963, **28**, 392; 1963, **29**, 489.
36. J. W. ROBINSON, *Analyt. Chim. Acta*, 1962, **27**, 465.
37. N. S. POLUEKTOV, and R. A. VITKUN, *Zh. analit. khimii*, 1963, **18**, 37.
38. YU. V. ZELYUKOVA, and N. S. POLJEKTOV, *Zh. analit, khimii*, 1963, **18**, 435.
39. K. FUWA, and B. L. VALLEE, *Analyt. Chem.*, 1963, **35**, 942.
40. J. B. WILLIS, *Nature*, 1965, **207**, 715.
41. D. C. MANNING, *Atomic Absorption Newsletter*, 1966, **5**, 127.
42. C. T. J. ALKEMADE, *Analyt. Chem.*, 1966, **38**, 1252.
43. M. L. PARSONS, W. J. MCCARTHY, and J. D. WINEFORDNER, *J. Chem. Educ.*, 1967, **44**, 217.
44. B. V. L'VOV, and G. V. PLYUSHCH, *Zh. prikl. spectroscopii*, 1969, **11**, 217.
45. R. HERRMANN, and W. LANG, *Arch. Eisenhüttenwesen*, 1962, **33**, 643.
46. J. D. WINEFORDNER, and T. J. VICKERS, *Analyt. Chem.*, 1964, **36**, 1947.
47. *Chemist's Handbook*, vol. 1 (2nd edn, vol. I Goskhimizdat, Leningrad–Moscow, 1962).
48. M. V. LYKOV, *Spray drying* (Pishchepromizdat, Moscow, 1955).
49. J. DROWART, G. DE MARIA, R. P. BURNS, and M. G. INGRAM, *J. Chem. Phys.*, 1959, **30**, 318; 1960, **32**, 1366.
50. N. B. VARGAFTIK, *Handbook on the Thermal Physical Properties of Gases and Liquids* (Fizmatgiz, Moscow, 1963).
51. V. D. MALYKH, and M. A. SEDYKH, *Optika i spektroskopiya*, 1964, **16**, 368.
52. I. V. VEITS, and L. V. GURVICH, *DAN SSSR*, 1956, **108**, 659.
53. *The Thermodynamic Properties of Individual Substances*, p. 775. Handbook edited by V. P. GLUSHKO (*Izd. AN SSSR*, 1962).
54. I. RUBEŠKA, B. MOLDAN, and I. VALNY, *Analyt. Chim. Acta*, 1963, **29**, 206.
55. W. T. ELWELL, and J. A. GIDLEY. Atomic Absorption Spectrophotometry (Pergamon Press, London, 1961).
56. C. A. BAKER, and F. W. J. GARTON, Atomic Energy Res. Establ., No. R 3490, 1961.
57. T. R. ANDREW, and P. N. R. NICHOLS, *Analyst*, 1962, **87**, 25.
58. W. LEITHE, and A. HOFER, *Mikrochim. Acta*, 1961, No. 2, 268.
59. A. C. MENZIES, *Analyt. Chem.*, 1960, **32**, 898.
60. F. J. WALLACE, *Analyst*, 1963, **88**, 259.
61. J. B. WILLIS, *Spectrochim. Acta*, 1960, **16**, 259.
62. D. J. DAVID, *Analyst*, 1959, **84**, 536.
63. M. D. AMOS, and J. B. WILLIS, *Spectrochim. Acta*, 1966, **22**. 1325, 2128.
64. A. P. D'SILVA, R. N. KNISELEY, and V. A. FASSEL, *Analyt. Chem.*, 1964, **36**, 1287.
65. J. D. WINEFORDNER, and T. J. VICKERS, *Analyt. Chem.*, 1964, **36**, 161.
66. J. D. WINEFORDNER, and R. A. STAAB, *Analyt. Chem.*, 1964, **36**, 1367.
67. J. M. MANSFIELD, and J. D. WINEFORDNER, *Analyt Chem.*, 1965, **37**, 1049.
68. C. T. J. ALKEMADE, International Conference on Spectroscopy, College Park, Md., 1962.
69 J. V. ROBINSON, 'Mixtures of solvents used in atomic absorption spectroscopy'. XX International Congress on Theoretical and Applied Chemistry. Abstract of papers. Sections E and F, p. 9 (Moscow, 1965).
70. D. A. DAVIS, R. VENN, and J. B. WILLIS, *J. Sci. Instrum.*, 1965, **42**, 816.
71. K. E. KNUTSON, *Analyst*, 1957, **82**, 241.
72. S. R. KOIRTYOHANN, and C. FELDMAN, *Developments in Appl. Spectroscopy*, Vol. 3 pp. 180–188 (Plenum Press, New York, 1964).
73. H. L. KAHN, *J. Chem. Educ.*, 1966, **43**, A7.
74. E. A. BOLING, *Spectrochim. Acta*, 1966, **22**, 425.
75. R. H. WENDT, and V. A. FASSEL, *Analyt. Chemistry*, 1965, **37**, 920; 1966, **38**, 337.
76. J. A. RAMSAY, *J. Exptl. Biol.*, 1950, **27**, 407; 1953, **30**, 1.
77. J. A. RAMSAY, S. FALLOON, and K. MACHIN, *J. Sci Instum.*, 1951, **28**, 75.

78. H. KAISER, *Spectrochim. Acta*, 1951, **4**, 351.
79. H. KAISER, and H. SPECKER, *Z. Anal. Chem.*, 1955, **149**, 46.
80. A. N. ZAIDEL', *Fundamentals of Spectral Analysis* (Izd-vo 'Nauka', Moscow, 1965).
81. H. KAISER, *Trace Characterization, Chemical and Physical* (Natl. Bur. Stand., Washington, 1967).
82. W. SLAVIN, S. SPRAGUE, and D. C. MANNING, *Atomic Absorption Newsletter*, 1964, **3**, 11.
83. H. L. KAHN, *Atomic Absorption Newsletter*, 1964, **3**, 89.
84. E. E. PICKETT, and S. R. KOIRTYOHANN, *Spectrochim. Acta*, 1968, **23B**, 235.
85. V. A. FASSEL, and D. W. GOLIGHTLY, *Analyt. Chem.*, 1967, **39**, 466.
86. Perkin–Elmer prospectus for the British Exhibition of Scientific and Industrial Instruments, Moscow, 1968.
87. M. D. AMOS, and J. B. WILLIS, *Spectrochim. Acta*, 1966, **22**, 1325.

CHAPTER V

1. B. V. L'VOV, M. A. KABANOVA, D. A. KATSKOV, G. G. LEBEDEV, and M. A. SOKOLOV, *Zh. prikl. spektroskopii*, 1968, **8**, 200.
2. B. V. L'VOV, *Inzhenerno-fizicheskii zhurnal*, 1959, **2**, (2), 44.
3. B. V. L'VOV, *Inzhenerno-fizicheskii zhurnal*, 1959, **2**, (11), 56.
4. B. V. L'VOV, *Zavodskaya laboratoriya*, 1962, **28**, 931.
5. G. I. NIKOLAEV, and V. B. ALESKOVSKII, *Zh. analit. khimii*, 1963, **18**, 816.
6. G. I. NIKOLAEV, *Zh. analit. khimii*, 1964, **19**, 63.
7. G. I. NIKOLAEV, and V. B. ALESKOVSKII, *Zh. tekhn. fiziki*, 1964, **34**, 753.
8. G. R. FONDA, *Phys. Rev.*, 1928, **31**. 260.
9. E. S. GOLOVINA, *DAN SSSR*, 1952, **85**, 141.
10. *Carbon-Graphite Structural Materials*, Collection of articles, No. 1 edited by S. E. VYATKIN (Izd. 'Metallurgiya', Moscow, 1964). V. N. KRYLOV, and YU. N. VIL'K. *Carbon-Graphite Substances and their Use in the Chemical Industry* (Izd. 'Khimiya', Moscow, 1965).
11. *The Thermodynamic Properties of Individual Substances*, vols. I–II. Handbook edited by V. P. GLUSHKO (Izd. AN SSSR, Moscow, 1962).
12. E. HINNOV, and H. KOHN, *J. Opt. Soc. Am.*, 1957, **47**, 151.
13. I. V. VEITS, and L. V. GURVICH, *DAN SSSR*, 1956, **108**, 659.
14. N. P. PENKIN, *ZhETF*, 1947, **17**, 355; *Izv. AN SSSR, Ser. fiz.*, 1947, **2**, 217.
15. J. E. ALLAN, *Spectrochim. Acta*, 1962, **18**, 259.
16. A. N. ZAIDEL', N. I. KALITIEVSKII, L. V. LIPIS, and M. P. CHAIKA, *The Emission Spectral Analysis of Atomic Materials* (Fizmatgiz, Leningrad–Moscow, 1960).
17. YU. I. KOROVIN, *Zh. analit. khimii*, 1961, **16**, 494.
18. *Trace Analysis.* Papers presented at a Symposium on Trace Analysis, Nov. 1965. Edited by J. H. YOE, and H. J. KOCH.
19. S. L. MANDEL'SHTAM, and V. V. NEDLER, *Zh. prikl. spektroskopii*, 1961, **10**, 390.
20. S. L. MANDEL'SHTAM, *Zh. prikl. spektroskopii*, 1964, **1** (1), 5.
21. H. MAECKER, and T. PETERS, *Z. Physik*, 1954, **139**, 448.
22. N. K. RUBTSOVA, *ZhETF*, 1947, **17**, 1005.
23. E. I. NIKONOVA, and V. K. PROKOF'EV, *Optika i spektroskopiya*, 1956, **1**, 298.
24. V. V. NALIMOV, *The Use of Mathematical Statistics in Analysing Substances* (Fizmatgiz, Moscow, 1960).
25. B. V. L'VOV, *Zh. prikl. spektroskopii*, 1968, **8**, 517.
26. H. MASSMANN, Spurenanalyse mittels Atomabsorption in der Graphitküvetten nach L'vov mit einem Mehrkanalspectrometer. II Internationales Simposium 'Reinststoffe in Wissenschaft und Technik', Dresden, 1965.
27. V. A. SLAVNYI, I. S. ABRAMSON, and YU. V. AFANAS'EV, *Zh. prikl. spektroskopii*, 1964, **1**, 310.
28. YU. I. BELYAEV, and L. M. IVANTSOV, *Modern Methods of Analysis* (a collection of papers), p. 20, Edited by D. I. RYABCHIKOV (Izd. 'Nauka', Moscow, 1965).
29. G. I. NIKOLAEV, *Zh. analit. khimii*, 1965, **20**. 445.
30. B. V. L'VOV, *Spectrochim. Acta*, 1961, **17**, 761.
31. A. N. ZAIDEL', *Fundamentals of Spectroscopic Analysis* (Izd. 'Nauka', Moscow, 1965).
32. S. R. KOIRTYOHANN, and E. E. PICKETT, *Analyt. Chem.*, 1965, **37**, 601.
33. A. C. MENZIES, *Analyt. Chem.*, 1960, **32**, 898.
34. B. V. L'VOV, *Atomic Absorption Spectrochemical Analysis* (Izd-vo 'Nauka', Moscow, 1966).
35. C. S. RANN, and A. N. HAMBLY, *Analyt. Chem.*, 1965, **37**, 879.
36. B. V. L'VOV, *Spectrochim. Acta*, 1969, **24B**, 53.
37. R. DITCHBURN, *Physical Optics* (Izd-vo 'Nauka', Moscow, 1965).

38. G. MIE, *Ann. d. Phys.* 1908, **25**, 377.
39. S. R. KOIRTYOHANN, and E. E. PICKETT, *Analyt. Chem.*, 1966, **38**, 585.
40. W. FINKELNBURG. *Kontinuierliche Spektren* (Springer, Berlin, 1938).
41. H. MASSMANN, *Zeitschrift für analyt. Chemie*, 1967, **225**, 203.
42. H. HERTSBERG, *Spectra and Structure of Diatomic Molecules* (IL, Moscow, 1949).
43. J. E. ALLAN, *Analyst.*, 1958, **83**, 466.

CHAPTER VI

1. S. E. FRISH, *The Optical Spectra of Atoms* (Fizmatgiz, Moscow–Leningrad, 1963).
2. H. LANDOLT, and R. BÖRNSTEIN, *Zahlenwerte und Funktionen*, Vol. I, Part I (Berlin, 1950).
3. O. P. BOCHKOVA, and E. YA. SHREIDER, *The Spectroscopic Analysis of Mixtures of Gases* (Fizmatgiz, Moscow, 1963).
4. K. R. OSBORN, and H. E. GUNNING, *J. Opt. Soc. Am.*, 1955, **45**, 552.
5. A. N. ZAIDEL', and E. P. KORENNOI, *Optika i spektroskopiya*, 1961, **10**, 570.
6. A. N. ZAIDEL', and E. P. KORENNOI, *Zavodskaya laboratoriya*, 1963, **29**, 1449.
7. A. WALSH, *Spectrochim. Acta*, 1955, **7**, 108.
8. J. A. GOLEB, and Y. YOKOYAMA, *Analyt. Chim. Acta.*, 1964, **30**, 213.
9. J. A. GOLEB, *Analyt. Chem.*, 1963, **35**, 1978.
10. B. V. L'VOV, and V. I. MOSICHEV, *Zh. prikl. spektrosktroskopii*, 1966, **4**, 491.
11. L. A. SPEKTOROV, *Trudy fiziko-mat. f-ta Kirgızskogo un-ta*, 1955, **3**, 168.
12. R. K. COLE, D. F. HALL, R. F. KEMP, and J. M. SELLEN, *Aerospace Sciences Meeting* (AIAA 8603-638-KU-000, New York, 1964).
13. *Ionic, plasma and arc rocket engines*, Collection of articles (Gosatomizdat, Moscow, 1961.
14. YU. I. OSTROVSKII, G. F. PARCHEVSKII, and N. P. PENKIN, *Optika i spektroskopiya*, 1956, **1**, 822.
15. A. N. NESMEYANOV, *The Vapour Pressures of Chemical Elements* (Izd. AN SSSR, Moscow, 1961).
16. B. V. L'VOV, and G. V. PLYUSHCH, *Proceedings of the XV Conference on Spectroscopy (Minsk, 1963)*, vol. 2, p. 159 (Izd. AN SSSR, Moscow, 1964).
17. B. V. L'VOV, *Optika i spektroskopiya*, 1965, **19**, 507.
18. N. P. PENKIN, and N. YU. YU. SLAVENAS, *Optika i spektroskopiya*, 1963, **15**, 154.
19. YA. D. RAIKHBAUM, and V. D. MALYKH, *Optika i spektroskopiya*, 1960, **9**, 425.
20. V. M. GOL'DFARB, and E. V. IL'INA, *Optika i spektroskopiya*, 1961, **11**, 445.
21. B. V. L'VOV, *Inzhenerno-fizicheskii zhurnal*, 1959, **2**, (2), 44.
22. G. I. NIKOLAEV, and V. B. ALESKOVSKII, *Zh. tekhn. fiziki*, 1964, **34**, 753.
23. R. BERRER, *Diffusion in Solid Bodies* (IL, Moscow, 1948).
24. J. H. ARNOLD, *Industr. and Engng Chem.*, 1930, **22**, 1091.
25. G. D. BELL, M. H. DAVIS, R. B. KING, and P. M. ROUTLY, *Astrophys. J.*, 1958, **127**, 775.
26. A. S. VALTERS, E. I. NIKONOVA, and G. P. STARTSEV, *Optika i spektroskopiya*, 1964, **16**, 717.
27. YU. N. KUZNETSOV, *Zavodskaya laboratoriya*, 1964, **30**, 1521.
28. N. P. KOMAR', *Zavodskaya laboratoriya*, 1963, **29**, 1052.
29. J. A. GOLEB, *Analyt. Chim. Acta.*, 1966, **34**, 135.
30. C. FÜCHTBAUER, and H. BARTELS, *Z. Phys.*, 1921, **4**, 337.
31. G. L. VIDALE, Technical Information Series. Nos. R 60SD330, R 60SD331, R 60SD333, R 60SD390, General Electric Company, Philadelphia, USA.
32. A. N. ZAIDEL', and E. YA. SHREIDER, *Vacuum Ultra-violet Spectroscopy* (Izd. 'Nauka', Moscow, 1967).
33. J. E. ALLAN, Fourth Australian Spectroscopy Conference, Aug. 1963.
34. S. SPRAGUE, D. C. MANNING, W. SLAVIN, *Atomic Absorption Newsletter*, 1964, **3**, 27.
35. W. SLAVIN, C. SEBENS, and S. SPRAGUE, *Atomic Absorption Newsletter*, 1965, **4**, 341.
36. A. WALSH, *Proc. of the X Colloquium Spectroscopicum Internationale*, pp. 13–26 (Washington, 1963).
37. H. MASSMANN, *Z. für analit. Chemie*, 1967, **225**, 203.
38. H. MASSMANN, *2. Internat. Symp. Reinstoffe in Wiss. u Techn.*, Dresden, 1965.
39. W. R. S. GARTON, M. S. W. WEBB, and P. C. WILDY, *J. Sci. Instrum.*, 1957, **34**, 496.
40. T. A. CHUBB, and H. FRIEDMAN, *Spectrochim. Acta*, 1956, **8**, 121.
41. H. PRUGGER, and W. ULMER, *Z. agnew. Phys.*, 1959, **11**, 467.
42. T. A. JACOBS, B. H. CARSON, and K. R. GIENDT, *Appl. Optics*, 1965, **4**, 754.
43. K. W. MEISSNER, *Ann. Phys.*, 1925, **76**, 124.
44. H. B. DORGELO, *Z. Phys.*, 1925, **34**, 766.
45. O. P. BOCHKOVA, *Izv. AN SSSR, ser. fiz.*, 1954, **18**, 252.
46. J. A. GOLEB, *Analyt. Chem.*, 1966, **38**, 1059.
47. D. C. MANNING, and W. SLAVIN, *Atomic Absorption Newsletter*, 1962, **1**, 39.

48. J. A. GOLEB, *Analyt. Chim. Acta.*, 1966, **36**, 130.
49. P. DAVIDOVITS, and N. KNABLE, *Res. of Sci. Instrum.*, 1964, (7), 857.
50. K. G. KESSLER, and W. G. SCHWEITZER, *J. Opt. Soc Amer.*, 1959, **49**, 199.
51. M. BOTTEMA, *Nederl. tijdschr., natuur.-kunde*, 1960, **26**, 188.
52. C. H. CORLISS and W. R. BOZMAN, *Experimental Transition Probabilities for Spectral Lines of Seventy Elements* (Natl. Bureau of Standards, New York, 1962).
53. P. F. GRUZDEV, and V. K. PROKOF'EV, *Optika i spektroskopiya*, 1966, **21**, 255.
54. G. M. LAWRENCE, J. K. LINK, and R. B. KING, *Astrophys. J.*, 1965, **141**, 293.
55. N. P. PENKIN, *J. Quant. Spectrosc. Radiat. Transfer*, 1964, **4**, 41.
56. N. G. MOROZOVA, and G. P. STARTSEV, *Optika i spektroskopiya*, 1964, **17**, 327.
57. E. HINNOV, *J. Opt. Soc. Am.*, 1957, **47**, 151.
58. E. HINNOV, and H. KOHN, *J. Opt. Soc. Am.*, 1957, **47**, 156.
59. W. BEHMENBURG, and H. KOHN, *J. Quant. Spectrosc. Radiat. Transfer*, 1964, **4**, 163.
60. B. V. L'VOV, *Zh. prikl. spektroskopii*, 1968, **8**, 517.
61. B. V. L'VOV, and G. G. LEBEDEV, *Proceedings of the Symposium on Theoretical Spectroscopy* (Erevan, 1966).
62. E. D. CLAYTON, and S. Y. CH'EN, *Phys. Rev.* 1952, **85**, 68.
63. S. Y. CH'EN, A. SMITH, and M. TAKEO, *Phys. Rev.*, 1960, **117**, 1010.
64. N. P. PENKIN, and L. P. SHABANOVA, *Optika i spektroskopiya*, 1967, **23**, 22.

INDEX

Absorbance
  definition, 6
  direct measurement, 23, 111–2
  effect of repeated reflection, 105–6
  error in measurement, 95–6, 102
  function $\phi(A_s)$, 21–2, 30, 279
  length of absorbing volume, 170, 212
  unit, 164
Absorbing volume, *see* Analysis volume
Absorption
  discovery, 1
  measurement, 16–22, *see* also under the various methods—Equilibrium method, Integration method, and Peak method
Absorption coefficient, 6
Absorption line, 1, 6–7
  half-width, 7
  number of, 33–4
  profile, 1, 7–14
  see also Hyperfine structure and Broadening of lines
Accuracy of analysis, 245
Acetylene flames, *see* Fuel gases
Adaptors, 158–61
Aleskovskii, V. B., 223, 238
Alkemade, C. T. J., 2, 94, 112, 122, 135, 171, 185N
Allan, J. E., 32, 77, 83, 107, 143, 146, 154, 155, 170, 171, 184, 185, 254
Aluminium
  cathodes, 39, 42, 43, 54, 56
  determination
    by cuvette, 222–4, 242–3
    by flame, 153, 155–6, 177–8
    in pure metals, 238
    in water and acids, 245
  dissociation of compounds, 129
  effect on magnesium, 183–5, 187, 190–1
  hyperfine splitting, 56
  resonance lines, 32
Amos, M. D., 139, 155, 167, 185
Amplifiers
  logarithmic, 109–12, 189
  narrow-band-pass, 109
Analysis lines, 31–3, *see* also Resonance lines
Analysis signal, 120
  and sensitivity, 162–4
Analysis volume
  length, 157–8, 170, 212
  of arc, 125
  of cuvette, 124
  of flame, 122
  of hollow cathode, 124
  of pulsed lamp, 126

Andrew, T. R., 183
Arc
  atomization by, 125
  emission analysis, 239–40
Arsenic determination, 255
Arnold, J. H., 288
Astrophysics, 1, 2, 19
Atomic absorption method, limitations of, 292–3
Atomic fluorescence method, 171–4
Atomization methods, 122–8
  arc, 125–6
  cuvette, 124–5, 201–4, 222–5, 242
  flame, 122–3, 175–89
  fractional vaporization, 242
  hollow-cathode discharge, 124
  pulsed lamp, 126
  pyrotechnic mixture, 128
Atomization rate
  in cuvette, 201–4, 222–5
  in flame, 175–89
Attenuation of light in cuvette, 230–7
Atwill, W., 127

Background concentration of electrons, 134
Baker, C. A., 113, 114, 148
Ballast resistors, 52
  thermal noise, 99
Bartels, H., 278
Beckman burner, 135, 140–2, 154, 158, 159
  mixing tubes, 141–2
Behmenburg, W., 282
Bell, W. E., 65
Belyaev, Yu. I., 125
Beryllium 2349 A line
  Lorentz effect, 284
  oscillator strength, 280
Bismuth determination
  in granite, 248
  optical interference, 235–6
  with adaptor, 161
Bloom, A. L., 65
Bochkova, O.P., 19, 79 N, 105, 257
Bodretsova, A. I., 37, 43, 54, 70, 101
Boling, E. A., 140
Boltzmann's law, 7, 31, 34
Boosted output lamps, 69–71
  multi-element, 74
Boron
  determination, 169
  isotopic analysis, 273–5
Boumans, P. W. J. M., 126
Bowman, J. A., 94

Box, G. F., 40, 52, 79, 100
Brewer, R. G. 65
Broadening of lines, 7–15
  Doppler broadening, 9, 10, 14, 16, 21, 24, 25, 282–4
    cuvette, 246
    flame, 220
    hollow-cathode lamps, 36, 48, 54, 57
    isotopic analysis, 260
    oscillator strengths, 279–80
    radio-frequency lamps, 64
  Lorentz broadening, 10–15, 16, 21, 282–4
    collision theory, 11–13, 16, 17, 282–4
    determination
      of Lorentz widths, 282–4, 285
      of oscillator strengths, 279–80
    growth curves, 282–4
    statistical theory, 11
  natural broadening, 8–9
  self-absorption, 105, 216, 218
    hollow cathodes, 31, 46–50, 54, 270–2
    metal vapour lamps, 69
    spherical lamps, 67
Bromine determination, 256
Burners, 138–43
  Beckman, 135, 140–2, 154, 158, 159
  Boling, 140
  elongated, 139–40, 155
  forced feed, 140–1
  Méker, 138
  slot, 138–9, 140, 155, 157–8
Butler, L. R. P., 73, 83, 113, 139

Cadmium
  analysis lines, 35–66
  determination
    in zirconium dioxide, 244, 248
    limits of detection, 229
    with adaptor, 159–61
    with air-acetylene flame, 170–1
    with cuvette, 216
  diffusion in cuvette, 211
  self-reversal, 49, 69
Calcium determination,
  effect of phosphorus, 136
  heated cell, 85
  hollow cathode, 39
Caesium
  determination in plasma beam, 275–6
  determination of vapour concentration, 278
  spherical lamps, 65, 68
Calibration graphs, 23, 28, 162
  continuous emitters, 30–1
  cuvettes, 215–8
  isotopic analysis, 263–73
  nitrogen in gas mixtures, 259
  resonance monochromators, 85
  self-absorption, 270, 272–3
  slot burners, 140

Carbon
  arc
    absorption analysis, 125
    emission analysis, 239–40
  dissociation of CO, 214
  particles in flames, 86, 134, 151
  vapour pressure, 214
Chambers, W. E., 109
Chechik, N. O., 99N
Chlorine determination, 256
Chromium determination
  application to hollow cathode, 39
  pulsed power supply, 63
Clinton, O. E., 138
Colbalt determination
  optical modulation, 94
  separation of lines, 80
  with adaptor, 160
Cole, R. K., 275
Combined method
  line widths, 282
  oscillator strengths, 279
Competing ions, addition of, 186–7
Complex-forming reagents, 186–7
Contamination of samples, 250–1
Confidence levels, 161–2
Continuous spectrum sources
  calibration curves, 30–1
  continuous power, 76–8
  pulsed power, 75–6
Copper determination
  adaptor, 159–61
  hollow cathode, 49, 61
  resonance monochromation, 84–5
Crosswhite, H. N., 37
Cuvette, see Graphite cuvette
Cyanogen flame, 138, 147, 148, 150, 153, 155

David, D. J., 50, 154
Davies, D. A., 136
Dawson, J. B., 58–9, 61, 62
Dean, J. A., 135, 145, 153
Detector noise and multi-pass systems, 103, 107
Detector synchronized, 112
Detection limits, see Limit of detection
Deuterium lamps, 77
Devaney, J. J., 103
Dianov-Klokov, V. I., 110
Dieke, G. H., 37
Diffusion, coefficients of, 209, 284–93
Dissociation, 128–32, 213–5
  aluminium compounds, 129
  CO in cuvettes, 214
  degree of dissociation, 130–1
  lithium compounds, 129
  molecular gases, 256
  oxides and hydroxides, 130–1
  in cuvettes, 214–5
Ditchburn, R., 231
Doerffel, K., 109

Doppler effect, *see* Broadening of lines
Double-beam spectrophotometer, 112-3
Dowling, F. B., 155
Drabkin, D. S., 74
Droplet
   size, 175, 178-80
   temperature, 176-7
   vaporization time, 177-180

Einstein, A., 1, 5
Electrolytic application of metals to hollow cathodes, 39
Ellis, D. J., 58-9, 61, 62
Elongated burners, 139-40, 155
Elwell, W. T., 181
Emission analysis
   atomic fluorescence, 171-3
   carbon arc, 239-40
   comparison with absorption methods, 2, 33-5, 290-1
      atomization, 115, 126
      dimensions of emitting cloud, 35
      limits of detection, 165-6, 171-3
      sensitivity in ultra-violet, 35
      with carbon arc, 239-40
      superimposition of lines, 33-4
      systematic errors, 35
      temperature variations, 34-5
   carbon arc method, 239-40
   flame photometry, 135, 165-6
   integration method, 120-1, 124, 126
   intensity, 32-3, 120
Emission lines
   effect of solvents, 149
   intensity, 6, 32-3
   maxima, 23-4
   non-monochromicity, 26-9
   self-reversal, 69
   superimposition, 33-4
Emissive power of source, 69
Emissive stability of source, 69, 78, 95-8, 100-1
Equilibrium method, 117-8, 120, 121
   arc, 126
   flame, 122-3, 164-6, 219, 238-9
   hollow-cathode discharge, 124
   laser, 127
Error
   law of accumulation, 246
   mean square, 102
   spectrophotometric measurements, 95-102
Eshelman, H. C., 153
Ethanol as solvent, 153-4
Excited states
   lifetimes, 2
   populations, 7

Fabry-Perot
   étalon principle, 79
   interferometer, 17, 47
Fassel, V. A., 32, 77, 104, 141-2, 153, 154, 155 156, 157, 162, 165

Feast, E. C., 113
Feldman, 161
Fibre optics, 84
Filcek, M., 136, 185
Filters, 78-9
   multiplex, 79
Flame, number of atoms in, 169
Flame emission photometry, 32, 135, 165-6
Flame method (absorption), 122-3, 135-92, 292
   combustion speeds, 138
   diffusion and premixed flames, 138,140,142,158
   effect of foreign elements, 174-89
   effect of solvents, 143, 149-53
   electron concentration, 134
   integration method, 189-92
   limits of detection, 161-74
   noise, 140, 141
   pathlength, 157-61
   temperature and composition, 146-53
   vacuum ultra-violet region, 254
   see also Burners and Fuel gases
Fluorine determination, 256
Fluorspar, 253-4
Fonda, G. R., 201
Foreign elements, effect on measurements
   in cuvettes, 221-5
   in flames, 134, 174-89
Foreign gas,
   effect on absorption lines, 10-15, 24-5, *see also* Broadening of lines
   in cuvettes
      distribution system, 198
      effect on atomization rate, 202-3
      effect on calibration curve, 216, 218-9
      effect on diffusion losses, 209, 210-1
   in hollow-cathode lamps, 42-4
      interference, 50-1
      line-broadening, 47-50
   in spherical lamps, 65-6, 68
Foster, W. H., 149
Fractional distillation, 241, 224-5
Fractional vaporization, 241-4
Franz, F. A., 65
Fraunhofer lines, 1
Frish, S. E., 19, 105
Füchtbauer, C., 1, 17, 278
Fuel gases, 138
   acetylene-air flame
      burners, 139-40, 141-3
      droplet vaporization, 177-8
      foreign elements, 181-2, 185
      refractory oxides, 154-6, 214
   acetylene+nitrogen+oxygen flames, 139, 141, 143
   acetylene-nitrous oxide flame, 139, 151, 185
   acetylene-oxygen flame, 147, 161
      droplet vaporization, 158-9, 178-9
      refractory oxides, 153-4, 155-6
      temperature and composition, 152
   butane flame, 138, 139

Fuel gases—*continued*
  cyanogen flame, 138, 147, 148, 150, 153, 155
  hydrogen flame, 138, 148–51, 161
    effect of solvents, 150–1
  propane flame, 138, 139
  town gas flame, 138, 139, 181–2, 183
Fuwa, K., 83, 159, 160–1, 170–1

Gallium
  form of cathode for determination, 38–9
  resonance lines, 32
Garton, F. W., 113
Gases, analysis of
  inert, 256
  molecular absorption, 256–7
  non-resonance lines, 258
Gatehouse, B. M., 52, 53, 124, 155, 171
Gaydon, A., 135
Getter, 37–8
Geyer, R., 109
Gibson, J. H., 78
Gidley, J. A., 114, 181
Glushko, V. P., 178N
Gold
  analysis lines, 35
  determination, 113, 159
  foil with hollow-cathode lamps, 39
Goleb, J. A., 265, 266, 273
Golightly, D. W., 165
Graphite, detection of impurities, 243, 247
Graphite cuvette, 124–5, 193–252
  apparatus, 124–5, 193–8
    caps, 209–10
    chamber, 195–6
    diameter, 210, 237
    electrodes, 196
    gas distribution system, 198–9
    length, 209, 220, 237
    lining, 205–6
    optical system, 193–4
    power supply, 196–7
    recording system, 194–5
    slits, 231–4
  applications
    diffusion coefficients, 286–8
    line widths, 282–4
    oscillator strengths, 124, 279–82
    ultra-pure substances, 291
    ultra-violet region, 255
  measurements
    absolute sensitivity, 225–30
    accuracy, 245
    adsorption on walls, 251–2
    analysis time, 252
    atomization rate, 201–4
    calibration, 215–25
    diffusion losses, 204–12
    dissociation of compounds, 213–5
    distribution of atoms, 212–3, 216
    internal standards, 247

Massmann's method, 213
optical interference, 230–7
procedure, 198–201
pulsed method, 125
processing results, 200–1
purification, 198–9
reproducibility, 245
sources of error, 246–51
Grinzaid, S. E., 109
Ground level
  multiplet, 31
  population, 7, 31, 32
Growth curves, 19, 282–4
Gunning, H. E., 79, 260
Gurvich, L. V., 178–9

Hagenah, W. D., 127
Half-width, 7–8, *see* also Broadening of lines
Hall, D. F., 275
Halogens, determination of, 253–6
Harris, R. J., 140
Helium, isotopic analysis, 260, 268–73
Herrmann, R., 112, 135–8, 171
Hinnov, E., 25, 146, 214, 284
Hofer, A., 184, 185
Hollow-cathode lamps, 36–63
  applications
    isotopic analysis, 263, 265–6, 273
    Walsh's method, 22–3, 26, 34
  atomic vapour, 44–5
  ballast resistances, 52
  cathode forms, 38–9, 42, 43, 72–3
  cleaning systems, 36–8
  cooled lamps, 36, 57, 124
  design and manufacture, 36–40, 56–7
  d.c., 40–54
  emission intensity, 42–6, 66–7
  getter, 37–8
  gold foil, 39
  high-amplitude pulse current, 58–63
  history of development, 36–7
  lamp life, 36–7, 40
  lamp noise, 53
  multi-element lamps, 72–3
  optical modulation, 94
  power supply, 39–40
  pulsed current, 54–63
  recording methods, 124
  sealed lamps, 37
  second-order collisions, 45
  self-absorption, 26, 46–7
  semi-enclosed cathodes, 42, 43
  signal-to-background ratio, 50–1
  spectral characteristics, 40
  sputtering, 57–7, 124
  stability, 51, 100
  uncooled lamps, 37, 45
Hume, D. N., 149
Hydrogen lamps, 77
Hyperfine splitting, 26, 28

Hyperfine structure, 2, 15, 36, 54-5
　and broadening, 15-6
　and isotopic analysis, 274, 277
　and nuclear moments, 15
　Hg line, 262
　isotope shift, 15

Illuminating (optical) systems, 85-91
　collimated beam, 86, 87-8
　design of, 89-91
Indium, use with hollow cathode, 38
Inert gases
　absorption lines, 259
　determination, 256, 260
　impurities, 257-8
　in cuvettes, 209
　in pulsed lamps, 76, 126
　in spherical lamps, 65, 68
Infra-red spectrum, use in helium isotopic analysis, 268
Ingram, M. G., 178
Integrating circuit, 191
Integration method, 189-92, 291
　and fractional vaporization, flames, 123
　for magnesium, 190-1
　for microscopic amounts, 238
　measuring area under curve, 191, 200-1
　of recording concentrations, 119, 120, 121, 128
　with arc, 125-6
　with cuvettes, 125, 200-1, 219, 220, 224-5
　with hollow cathodes, 124
　with laser, 127
Internal standards, 187
　in cuvettes, 247
　palladium, 113
Inverse sensitivity, 162-4
Inverse spectrograph, 73-4
Iodine determination in the ultra-violet, 255
Ionic lines, 45
　Mg lines, 58
Ionic rocket engines, 275-6
Ionization
　degree of, 132-3
　effect of foreign elements, 188
　in atomic vapour, 132-4
　suppression of, in alkali earths, 188-9
Ionization chambers, 256, 257
Ionized components, excess of, 222
Iron
　intensity of lines in hollow cathode, 35
　optical modulation of iron line, 94
　resonance lines, 80
Isotopes, hyperfine line structure, 277
Isotopic analysis, 260-75
　determination
　　boron, 273-5
　　from ion resonance lines, 275
　　helium, 268-73

　　hydrogen, 267-8
　　lithium, 263-6
　　mercury, 260-3, 275
　　uranium, 266-7
Isotopic filtration, 274-5
　for helium, 268, 274
　for lithium, 265, 274
　for uranium, 266-7, 274
Isotopic shift, 15, 260, 261
　use of infra-red spectrum in, 268
Ivanov, N. P., 68
Ivantsov, L. M., 125

JACO monochromator, 165
Jansen, A. V., 139
Jones, W. G., 37

Kahn, H. L., 123
Kaiser, H., 161-2, 163
Karyakin, A. V., 127
Katshov, D. A., 59, 247, 248
Kemp, R. F., 275
Khartszov, A. D., 206, 229, 255, 257
King, A. S., 2, 19, 124
　furnace, 2, 221, 278
King, R., 19
Knutson, K. E., 153
Kohn, H., 25, 146, 214, 282, 284
Koirtyohann, S. R., 161, 165, 233, 235
Kondrasheva, L. D., 105
Korennoi, E. P., 263, 264
Korolev, F. A., 79
Kravets, T. P., 1
Kuebler, N. A., 76, 127

Ladenburg, R., 1, 21, 262, 275
Lagrange-Helmholtz law, 81, 87, 88, 89
Lambert's law, 6
Lamp life in resonance monochromators, 85
Lamps
　high-frequency discharge, 34
　Lyman, 76, 126
　hollow cathode, see Hollow cathode lamps
　metal-vapour discharge, 68-9
　noise, 69
　with separate atomization and excitation, 69-70
Lang, W., 112, 136, 137, 138, 171
Laqua, K. 127
Laser, 5, 65
　for atomization, 127-8
　mirror systems, 107
Law of mass action, 129
Lead
　adsorption in pipette, 251-2
　analysis lines, 35
　determination, 159, 161
Lebedev, G. G., 22, 59, 224, 280, 283
Legagneur, C. S., 37
Legallais, V., 82, 100

Leithe, W., 184, 185
Lenz, W., 12
Levilokii, S. M., 58
Levy, S., 21
Light sources
    continuous, 36, 75-8
    hollow-cathode, *see* Hollow-cathode lamps
    line, 36
    modulation, 91
    multi-element, 71
        boosted output, 74-5
        hollow cathodes with alloys, 72, 73
        optical combination of light from separate discharges, 72, 73-4
        radio-frequency lamps, 72
        ring cathodes with separate power supplies, 74
        separate discharges, 72, 73-4
        with rings, 72, 73
Limit of detection, 161-173
    absolute limit, 163
        relative, 163, 164
    calibration of, 169-72
    confidence levels, 162
    definition, 161
    for elements with stable oxides, 155-6
    for various elements, 164-7
    in atomic flourescence, 171-2
    inverse sensitivity, 162-3
Lindholm, E., 12, 29
'Line' absorption, 16
Lithium
    dissociation of compounds, 130
    isotopic analysis, 263-6
Lithium iodide, molecular absorption, 234
Lockyer, R., 143, 146
Lorentz effect, *see* Broadening of lines
Lündegardh nebulizer, 135, 145
L'vov, B. V., 13, 20, 22, 24, 37, 42, 43, 54, 59, 64, 65, 69, 70, 74, 78, 109-10, 123, 125, 160-1, 184, 189, 193, 195, 206, 224, 229, 231, 234, 241, 248, 250, 251, 255, 257, 268, 279-80, 286
Lykov, M. V., 176
Lyman lamp, 76, 126
Lynch, J., 65

Macleod, H. A., 79N
Magnesium
    analysis lines, 35
    determination, 159, 160, 161, 181-2, 187
        by integration method, 190-1
        by resonance monochromation, 84, 85
        effect of aluminium, 184-5
        in air-acetylene flame, 153, 154
    lines in hollow-cathode lamps, 44
    self-absorption, 54, 55, 58
    self-reversal, 49
Mahr, E., 145
Malmstadt, H. V., 109

Malmstadt photometer, 79
Malykh, V. D., 178
Mandel'shtam, S. L., 116, 239-40
Manganese
    determination, 167
    electrolysis onto hollow cathodes, 39
Manning, D. C., 155, 156, 191, 265
Margoshes, M., 135
Marshak, I. S., 76
Massmann, H., 73, 74, 213, 226, 234, 247, 255
Matthews, P., 113
Mavrodineanu, R., 135
Méker burner, 138
Menzies, A. C., 113, 114, 161N, 184, 216
Mercury
    analysis lines, 35
    continuous monitoring of vapour, 79
    determination, 109
        by excitation of phosphor, 79
        in ultra-violet, 256
        of neutral atoms in plasma beams, 275-6
        of vapour in air, 2
        isotopic, 79-80
        use of adaptors, 159
    diffusion, 284
        of vapour out of cuvette, 205, 209, 210
    in spherical lamps, 67
    isotopes, 15, 20, 64
    isotopic analysis, 260-3, 275
        hyperfine line structure, 262
        oscillator strength of Hg 2537 line, 17
Metal-vapour discharge lamps, 68-9
Microanalysis, 238
    in cuvettes, 229
Mie, G., 231
Milatz, J. M. W., 2, 94, 112, 122
Millikan, R. C., 104
Mist, increasing dispersion of, 186
Mist droplets
    and solvents, 144-6
    dispersion of, 137
Modulation of light source emission, 91-4
    electrical method, 93
    mechanical method, 93
    optical, 93-4
Molecular absorption of light in cuvettes, 230-1, 233-4
Moldan, B., 181
Molybdenum
    cathodes, 61
    determination in air-acetylene flames, 154-5
Monochromators, 36, 78-85
    absorption filters, 79
    angular dispersion, 81
    aperture ratio, 81, 82
    diffraction grating, 81, 82
    interference filters, 79
    linear dispersion, 80
    multiplex filters, 79

Monochromators — *continued*
  power ratio, 80
  prism instruments, 81, 83
  resonance, 84
  slits, 80, 81
Mosicher, V. I., 54, 70, 268
Mossotti, V. G., 77, 104, 127, 155, 156
Müller, G., 109
Multi-channel devices, 72, 292–3
  medium quartz spectrographs, 83–4
Multi-pass systems, 103–7
  with concentric flames, 103–5
  with restricted angular aperture sources, 105–7

Nagibina, I. M., 62
Nalimov, V. V., 245, 246
Nanassy, A. J., 17
Natural broadening, 8–9
Nebulization, 122, 123
  effect of foreign elements, 174–89
  effect of solvents, 143–6
Nebulizers, 135
  Beckman burners, 135
  compressed air, 135
    angled, 135–6
    concentric, 135–6
  forced-feed, 175
  Lündegradh, 135, 145
  pneumatic, 136
Nedler, V. V., 239–40
Nelson, L. S., 76, 127
Nesmeyanov, A. N., 278
Neutral atoms
  in beams of ions, 20
  in gas phase, 275–8
Nickel determination, 160
Nikolaev, G. I., 110, 203, 209, 222, 223, 238, 249, 286
Nitrogen determination in gases, 259
Noise
  amplifier, 95
  photodetector (shot), 99, 100
  thermal in photomultipliers, 99

Oscillator strengths, 31, 32
  absolute measurement, 278–82
  determination, 19
  interpretation, 1
  pressure of foreign gases, 17
Osborn, K. R., 79, 260
Ostrovskii, Yu. I., 20, 278
Oxides of metals, vapour pressure, 180
Oxy-acetylene flame, *see* Fuel gases
Oxygen determination in gas mixtures, 257
  molecular absorption, 253
Ozone determination, 112

Palladium as internal standard, 113–4
Paschen, F., 36

Pavlovskaya, E. N., 37, 43
Peak method of absorption
  cuvette, 200–1, 216–7, 220, 222–3, 225–30, 238–9, 291
  flame, 123
  fractional vaporization, 242–244
  graphite furnace, 125
  laser, 127
  pulsed lamps, 126–7
  recording, 118–9
Penkin, N. P., 20, 221, 278
Phosphorus
  effect on determination of alkali earths, 183
  in ultra-violet, 255
Photoelectric measurements, 76, 77
Photo-emulsion, *see* Schwartzchild effect
Photographic methods for atomic absorption, 292
Photomultipliers
  dark current, 99–100, 101
  errors due to thermal noise, 99
  shot noise, 100—1
Pickett, E. E., 165, 235
Pipette, adsorption in, 251–2
Plank's constant, 5
Plasma
  concentration of atoms in, 239–42
  determination of neutral atoms, 275–8
  flame
    ionization in, 157
    noise, 157
  jet
    atomization, 128
    d.c., 123
    induction-coupled, 123
    plasmatron, 123
  tube, induction-coupled, 157
Platinum foil on hollow cathodes, 39
Plutonium spectrum, 2
Plyushch, G. V., 123, 131, 189
Podmoghinskii, I. V., 105
Poiseuille's law, 145
Poluekstov, N. S., 79, 109, 135, 159
Polythylene vessels, adsorption, 252
Polychromators, 85
Potassium
  application to hollow cathodes, 39
  determination by photometer, 79
Potassium iodide, molecular absorption, 233
Power supply
  for hollow-cathode lamps, 39–40
  for graphite cuvetter, 196–7
  stabilized, 40
Pressure broadening, *see* Broadening of lines
Prokof'ev, V. K., 37, 43, 62
Pulsed lamps, 126–7
  spherical, 75
  tubular, 75–6
Pungor, E., 145

Pyrolytic graphite, 205-6
Quality factor of amplifier, 92
Quartz
  absorption of, 252, 253
  optical, 254
Quantum radiation theory, 1, 5
'Quenching', 172

Radio-frequency lamps, 63-9
  glow discharge, 63
    halide-filled, 64
  ring discharge, 63, 65
  spectral microtube, 69
  spherical, 65-8, 173
Radioactive substances, analysis of, 229
Raikhbaum, Ya. D., 116
Rains, T. C., 153
Ramsey, J. A., 189, 191
Rare-earth elements, determination of, 153
Ratiometric methods of measuring absorbance, 97-8
Rayleigh scattering region, 231
Recording devices, 36
  time constant, 164
Recording systems,
  and sample composition, 116-7
  automatic instrument for cuvettes, 217
  band-pass, 91-3
  design, 107-9
  for use with graphite cuvette, 194-5
  methods of recording, 118-22
    equilibrium, 118, 120, 121, 122
    integration, 119, 120, 121, 122
    peak, 118-9, 121, 122
Refractory compounds, determination of, 157
Refractory oxides, 153-5
  dissociation of, 214
Reiche, F., 1
Reproducibility, definition of, 245
Resonance lines
  absolute oscillator strengths measured, 279-82
  broadening, 8
  definition, 7
  gases, 256, 258
  helium, 268, 270
  hydrogen, 267
  ions, 275
  listed, 299-305
  natural width, 9
  non-metals, 255
  populations, 7
Resonance monochromator, 84-5
Resonance radiation, 7
Richards, E. W., 64
Rikmenspoel, R., 82
Robinson, J. B., 32, 42, 50, 140, 143, 155, 159, 185
Rozhdestvenshii 'hooks' method, 280, 284
Rubeška, I., 181

Rubidium
  isotopic analysis, 274-5
  Lorentz effect, 11
Russell, B. J., 112

Saha's equation, 133, 167
Sample heterogeneity and analysis error, 247-50
Sapphire windows, 254
Scale expansion, 97
Scanning spectrometer, 114
Scattering of light, 230-1
  in cuvette, 230-3
  Rayleigh scattering region, 231
Schreider, E. Ya., 79N
Schuhknecht, W., 135
Schüler, V. C. D., 139
Schwartzchild effect, 59, 62, 126
Second order collisions, 45, 58
Sector disk, 109, 112
  use for chopping, 59-60
Sedykh, M. A., 178
Selenium determination in the ultra-violet region, 254
Self-absorption, 19, 105, 216, 218
  hollow cathodes, 31, 46-50, 54, 270-2
  metal vapour lamps, 69
  spherical lamps, 67
Self-reversal, 49
Sellen, J. M., 275
Semi-enclosed cathode, 42, 43, 54
Sensitivity, 161-73
  absolute, 121
    definition, 225
    in cuvettes, 225-30
  analogy with optics, 162
  and concentration of foreign elements, 180-5
  and cuvette cross-sections, 212
  and nebulizer design, 136, 138
  and solvents, 143-4, 146
  calculation of, 169-73
  in plasma, 157
  relative, in cuvettes, 237-45
  with adaptors, 159, 160, 161
  with oxy-acetylene flame, 156
  with slot burners, 140, 158
Shelton, J. P., 103, 112
Shift of line maximum, 10-12, 24-5
  isotopic, 14
Shot effect, 96, 99
  and modulated emission, 91
  in photomultipliers, 100
Shot noise
  in detector, 69
  in photomultipliers, 53
  in recorder with laser atomization, 127
Silicon
  determination of vapour pressure, 278
  effect on determination of alkaline earths, 183
Silver
  determination, 159

Silver — *continued*
  foil on hollow cathodes, 39
  in hollow cathodes, 61
  internal standard, 247
Skin effect, 67
Slavin, M., 120–1, 155, 156, 162, 165, 191, 265
Slit width
  and signal-to-background ratio, 101
  and shot noise, 101
Slot burners, clogging 158
Sodium
  application to hollow cathode, 39
  determination, 159, 181–2
    by photometer, 79
    of vapour pressure, 278
  impurity in air, 133–4
Solar spectrum, 1
Solvents
  C/O ratio, 151
  effect on flames, 147–53
  effect on nebulization, 143–5
  for increasing dispersion, 186
Spectrographs
  Hilger medium quartz, 83, 107, 113
  use as monochromators, 82
Spectrometer
  multichannel, 234
  scanning, 114
Spectrophotometers, 36–
  Alan's, 107–8
  design, 107–
    double-beam, 112–3, 113–4
    integrating, 114
    logarithmic amplifiers, 109–111
    two-channel, 113–4
    use of narrow band-pass amplifier, 109
  Malmstadt and Chambers', 109
  Poluektov and Grinzaid's, 109
  Poluektov and Viklin's, 109
  use with graphite cuvetter, 193–4
  see also Light sources, Monochromators,
Spektrov, L. A., 270
Spent (exhaust) gases
  adaptors, 158–61
  molecular absorption, 233–4, 235
  use of, 170–1
Spherical radio-frequency lamps, 65–8, 173
Spherical rubidium lamps, 80
Sprague, S., 191
Spray chambers, 136, 137
  Beckman, 138
  vortex air circulation, 137
Sputtering, 36, 40, 42, 44, 49, 118
Staab, R. A., 173
Standards, analysis without, 280–1, 289–90, 292
Strasheim, A., 73, 83, 113
Strelow, F. W. E., 113
Strijland, J. C., 17

Strontium
  competing ion in calcium determination, 186
  determination, 161
Sullivan, J. V., 70, 84, 85, 94, 225
Surface tension of solutions, 175
Sutherland's constant, 288
Sulphur determination in ultra-violet, 255
Synchronous detector, 93
  with pulsed supplies to hollow cathode, 63

Tabeling, R. W., 103
Tanasawa, Y., 137
Tantalum foil in cuvettes, 205–6
Tellurium determination in ultra-violet region, 254
Thallium determination, 159 161
  optical interference, 235–6
Thermal evaporation from cathode, 36
Thomas, P. E., 166N
Tin, use with hollow cathode, 38
Total energy method, 121
Toxic substances, analysis of, 229, 292
Transit time, 120
  and droplet vaporization time, 178–80
  definition, 116
  in arc, 123
  in flames, 122, 219
  in induction-coupled plasma, 123
Transitions, 5
  absorption, 5
  and molecular absorption, 236
  number of possible, 33–4
  spontaneous emission, 5
  stimulated emission, 5, 6
Tungsten
  atomization of, 201
  radio-frequency discharge tube, 64
Tungsten strip filament lamp, 76, 77
Turkin, Yu. I., 48

Ultra-pure substances, analysis of, 229–30, 291–2
Ultra-violet, 76, 77, 157
  and atomic flourescence, 173–4
  spectra in hollow-cathode lamps, 50–1
  vacuum, 253–6, 292
Uranium
  in cleaning traps, 37
  isotopic analysis, 266–7
  spectrum, 2

Vallee, B. L., 83, 148, 159, 160–1, 170–1
Valny, I., 181
Vaporization, fractional, 241–4
  selective, 186–8
Vapour pressure, saturation determination, 278
Veits, I. V., 178–9
Venn, R., 136
Vidale, G. L., 278
Viscosity of solutions, 175
Vitkun, R. A., 79, 109
Voigt, W., 1

Voigt equation, 14
Volkov, V. M., 111
Wallace, F. J., 187
Walsh, A., 2, 33, 40, 52, 70, 73, 79, 84, 85, 94, 100, 103, 112, 122, 124, 255, 264
Walsh's method, 2, 22–31, *see* also Hollow cathode lamps
Warren, R. L., 114, 143
Water vapour, determination in air, 256–7
Weisskopf, V., 12
Wendt, R. H., 157
Willis, J. B., 52, 53, 136, 139, 140, 155, 156, 167, 171, 185, 188
White, J. U., 103
Winefordner, J. D., 162, 169, 171, 173

Xenon
  anomalous line broadening in lamp filled with, 68
  determination, 256
  line broadening, 274
Xenon lamps, high-pressure, 77

Yang, C. C., 82, 100
Yokoyama, Y., 265, 266

Zaidel', A. N., 162, 263, 264
Zeeman effect, 2
Zeeman filter, 275
Zelyukova, Yu. V., 159
Zinc
  analysis lines, 35
  as internal standard, 247
  atomization rate, 203
  determination, 160, 161
    in aluminium, 249
    in cuvette, 222–4
    in other metals, 243
  diffusion in gases, 287–8
  diffusion rates, 209–10
Zirconium
  as getter, 37
  in radio-frequency discharge tubes, 64
Zirconium dioxide, determination of elements in, 243–4, 247–8, 250